ANTHROPOLOGISTS IN THE STOCK EXCHANGE

# ANTHROPOLOGISTS IN THE STOCK EXCHANGE

A Financial History of Victorian Science

MARC FLANDREAU

THE UNIVERSITY OF CHICAGO PRESS

CHICAGO AND LONDON

MARC FLANDREAU is professor at the Graduate Institute for International Studies and Development in Geneva under joint appointments in the departments of history and economics. He is the author of *The Making of Global Finance 1880–1913* and *The Glitter of Gold* and is the editor of *Money Doctors: The Experience of International Financial Advising 1850–2000*.

The University of Chicago Press, Chicago 60637
The University of Chicago Press, Ltd., London
© 2016 by The University of Chicago
All rights reserved. Published 2016.
Printed in the United States of America

25 24 23 22 21 20 19 18 17 16     1 2 3 4 5

ISBN-13: 978-0-226-36030-0 (cloth)
ISBN-13: 978-0-226-36044-7 (paper)
ISBN-13: 978-0-226-36058-4 (e-book)
DOI: 10.7208/chicago/9780226360584.001.0001

Library of Congress Cataloging-in-Publication Data

Names: Flandreau, Marc, author.
Title: Anthropologists in the stock exchange : a financial history of Victorian science / Marc Flandreau.
Description: Chicago : The University of Chicago Press, 2016. | Includes bibliographical references and index.
Identifiers: LCCN 2015048927| ISBN 9780226360300 (cloth : alk. paper) | ISBN 9780226360447 (pbk. : alk. paper) | ISBN 9780226360584 (e-book)
Subjects: LCSH: Anthropology—England—History—19th century. | Anthropological Society of London—History—19th century. | Stock Exchange (London, England)—History—19th century. | Learned institutions and societies—Corrupt practices—England—London—History—19th century. | Stock exchanges—Corrupt practices—England—London—History—19th century.
Classification: LCC GN17.3.G7 F53 2016 | DDC 301.0942/09034—dc23 LC record available at http://lccn.loc.gov/2015048927

♾ This paper meets the requirements of ANSI/NISO Z39.48-1992 (Permanence of Paper).

Pour Julie, avec qui nous inventâmes
naguère un Capitaine Pim Pam Poum:
ces *Nouvelles Aventures du Capitaine Pim*,
une histoire de science et d'injustice.

I believe that the real history of cannibalism has not yet been told.
—Dr. Berthold Seemann, member of the Anthropological Society
of London, March 14, 1865

# CONTENTS

In writing this book, I have tried to abstain as much as possible from value judgments, a difficult thing to do in a story devoted to corrupt science, white-collar crime, finance, politics, colonies, and cannibals. While my story speaks to recent controversies about the relationships between money, politics, and the social sciences, I am emphatically not giving any lesson. I am not advising on the appropriate incentives to better insulate social science researchers from conflicts of interest. Doing so could perhaps be helpful, but it runs the severe risk of diluting my main message, which is that the contours of the social sciences themselves are born from the precise contours of the conflicts of interest that surround them. The social sciences are fundamentally interested in, and can in no way abstract themselves from, the web of material interests where they find their resources. This is a conclusion that emerges clearly from the work of a myriad of previous writers, among whom are found Michel Foucault, Bernard S. Cohn, Pierre Bourdieu, and Timothy Mitchell, and it is taken here as a solid starting point.

I hope to make a contribution by adding a perspective that comes from my own background, training, and writing in finance and economics. The tools of financial economics, since they have been developed to deal with value, should be able to tackle a problem which I claim is at the heart of the social sciences—the problem of value. This seems a simple and natural notion, and yet it is one nobody has followed so far. The economics of science (let alone the economic history of science) thus far belongs to the ineffable. A financial history of science has never been tried before.

There is a certain idealist way of construing the scientist where the material, the interest, and the dollar are taboo. In response, there exists another way, that of the pragmatic realist, where the material, the interest, and the dollar are necessary. When it comes to the dollar's underpinning of things,

I have noted some of my colleagues and students increasingly saying that they "have no problem with that." This lack of a problem is not very different from the taboo—another form of lack. Thus, in arguing as it does for a history of anthropology based in material interests, this book runs against this general current. I suggest considering seriously the interests of science as actor and science as predator.

I can see why and how this could be upsetting. For instance, while some sympathetic readers have encouraged me to downplay the white-collar criminality part of my tale, others equally sympathetic but in a different way have wondered what is left of science at the end of my foray—of science as you and I like to think of it, the pure, disinterested (and largely imaginary?) sociability that would have been practiced by aristocratic or bourgeois scientists of the eighteenth century and reinvented as a professional ideal in the nineteenth and twentieth centuries. Following a presentation I made of a paper that was to grow into the present book, one colleague asked whether, beyond the various traffickers and white-collar criminals I had described in my talk, there were also "real" scientists—"des vrais" was the way he nicely put it.

But what is a real anthropologist? As we shall see, this subject summoned flows of ink and saliva in the 1860s and still does. This book, in addressing that question, has reversed it: I ask why the more shady characters in the following tale have been so conspicuously omitted from earlier narratives even when it is quite obvious that the story of anthropology cannot be understood without them. At least I hope I'll have made that clear to the reader by the end of the book. It is precisely the attempt at circumscribing anthropology to those characters who have "left a real mark in the history of anthropology," as one leading historian has put it, that has produced the current history of the origins of anthropology.[1] But this history suffers from a "survivor bias." It is a chronicle of scientific landmarks. The creation of landmarks or totems erected to great men and great ideas may provide profitable tourist attractions, topics for first-year university courses, and fundraising opportunities, but as a guide for historical explorations, the approach is invariably bound to discover that the starting point is not the end point.

Writing history should not be about commemoration, because commemoration feeds oblivion. Instead, by "de-provincializing" the peripheral anthropologist and the white-collar criminal, one is bound to resuscitate the lost world in which "real" anthropology was born, explaining why my cast has so many antiheroes (and a few supposed heroes, who the reader will discover were not heroes after all). Perhaps the most intriguing was Hyde Clarke, a man whose dissimulative capacities, as his name suggested, were outstanding. Journalist, engineer, traveler, statistician, accountant, and an anthropologist himself, typical of those crossover characters that the period was so apt at producing, Clarke turned out to be the great slayer of a significant tribe of Victorian anthropologists. More than many others, he deserves attention for having made Victorian anthropology by the determined way he sought to unmake the Anthropological Society of London. As a result, he left a deep imprint on the organizational foundation of British anthropology. Without him, the Royal Anthropological Institute might never have existed the way it has, a fact that has been downplayed but recognized before, if only in footnotes.

And yet, even though this same Hyde Clarke has been all but omitted from the main body of previous narratives, his own language and expressions were used (without quotation marks) to characterize the Anthropological Society of London in subsequent academic research. When it came to the Anthropological Society and its record, finance, credit, and above all scientific honesty, historians have relied on Hyde Clarke. The curse he threw on the Anthropological Society in the summer of 1868, in a column for a literary newspaper where he likened the society's leaders to a "clique," lingers today in the works of modern historians who commonly use the term. Was Hyde Clarke such an authority that his words were to be taken without a careful review or even a grain of salt? And if he was an unquestionable authority, why has he not received more attention? It is curious, given Clarke's importance in this respect, that he does not even have an entry in the British *Dictionary of Modern Biography* (now an Oxford dictionary).

Something, evidently, is being repressed, and as repression generally goes, it is both unconscious and revealing of the unconscious. I claim that in this repression, a lot of knowledge is hidden, including a better understanding of what anthropologists truly were and social scientists truly are. I will try and explore the staging of anthropology in the 1860s and show the strings and the string pullers. The making of Victorian anthropology was a process of displaying individuals and issues, and the displaying itself had the effect of leading observers to overlook the stage. The string pullers worked either in the shadows or in such a blinding light that they could not be observed easily. Using special glasses, one suddenly discovers either new individuals or the same individuals doing things different than initially thought. Thus my story's broadened cast. The result is not only about palatable anthropologists but about all anthropologists, the hidden and the displayed, the material and the immaterial. To paraphrase J. S. Mill's expression about one of the characters in this story, it is about, too, those who were "enthusiasts rather than scholars"—although of course any scholarly achievement

requires enormous amounts of enthusiasm.[2] I have therefore produced a
history that is neither optimistic nor pessimistic about the problems raised
by the entanglement of science and capital. It is not the story of the irre-
mediable corruption that some, having modern times in their eyes, fear to
be the destiny of the social sciences. Nor is it the story of the triumph of
untouchable scientists over conflicts of interest, not at all. In optimistic accounts, good triumphs over evil while in pessimistic accounts, the opposite
happens. The following tries to be as true a history as is feasible, and in true
accounts, evil takes care of evil, and good is a rose offered and received.

This lack of desire to take the moral high ground explains why, in order
to guide me in such a perilous trip, I accepted being led by exceptionally interesting natives—my dear Cannibals. Previous scholars vilipended them,
but for me, what mattered most was that they came from an exceptionally
interesting part of the world and an exceptionally interesting time of history. I was blessed to come across this group of scientists of terrible repute
haunting a hidden recess of Trafalgar Square within a stone's throw of the
centers of Victorian power. I am clearly not endorsing them. And if the reader
is upset that I deal with them when they have been shunned and looked on
with contempt by my predecessors, I would respond that without these detestable sherpas to carry my luggage, I would not have gone anywhere.

Many things can and will be said about the method adopted in this
book—or lack thereof. I confess to writing it almost in one sitting and in a
very unintentional way (but then was assigned to revise it). I confess to not
speaking with an editor until I had a manuscript. I confess to transcribing,
often with excessive passion and in a language that is not mine, a story that
imposed itself on me in the very process of writing and was organized, if
not naturally, at least in a stream of consciousness. I confess to not having
the kind of reliable knowledge about the pedigree of Victorians that only,
I am sure, an Oxford or Cambridge education can provide. The long reading list I was provided with in order to revise my manuscript cannot have
fixed my fundamental ignorance. I confess my ignorance of anthropology,
an ignorance that has actually heightened my surprise at finding so much
confusion among the cognoscenti and might itself be taken as a good reason
for writing this book.

Yet the ignorant are blessed, for ignorance is a remediable form of blind-
ness. Thus let me make a blanket apology to all those whom this book will
rattle. For this book, by a student of history and of economics, is, I hope,
profoundly about things bigger than the Victorian, the anthropological, or
even the British. It is a critique of the social sciences in general, and of the

way they inspire policy and reflect deep social and political structures. Anthropology is a metaphor, perhaps the best metaphor I have come across in my intellectual life, to express ideas that have long been with me, long before I heard of Hyde Clarke. Thus the subject matter of this book is much broader than an arcane dispute between anthropologists and ethnologists. And if the cast will feature—beyond the white-collar criminals and the scientists—a Miskito king, Abyssinian hostages, American secret agents, Victorian clubs, Benjamin Disraeli, and the stock exchange, it is because this book is in fact about modernity and its birth.

This is a lame but sincere excuse for the fact that I could not find a linear way to tell my story. In fact, I find myself very much in the same position as children's writer Pierre Gripari, author of *Tales of the Rue Broca*.[3] He is an apt authority to summon at this stage since Rue Broca in the Fifth Arrondissement of Paris got its name from Paul Broca, founder of the Société d'Anthropologie de Paris in 1859, creator of a human skin color chart, and a source of inspiration and reference for the Anthropological Society of London. In the tale of the White-Haired Devil, Gripari states that he is at a loss because, to tell the story, he needs first to tell that of the Contrarian Wife; but before he can tell the story of the Contrarian Wife, he must tell the story of How the Wife Became Contrarian. Likewise for me: to tell the story of the White-Haired Armchair Anthropologist, I need to tell first the story of How the Anthropologist Was Put in an Armchair. But to tell the story of How the Anthropologist Was Put in an Armchair, I need first to tell the story of How Financial Criminals Became Anthropologists, as well as several ancillary stories from which the reader, just like myself, is in peril of losing her marbles.

One should make a virtue out of a necessity. Reality is sinuous. I suggest that stories, even when what they tell is of the most convoluted nature, can convey a simple form of truth. Something must be said about the simplicity of the narrative and the truth it can deliver. I am not sure history is an academic discipline in the same way as anthropology, political science, sociology, and economics. Perhaps it is more like philosophy, a philosophical way to think of things that pass through the world. And thus this book really argues against the way history has been deconstructed and reconstructed, most recently, as an academic science. It should reclaim its dimension as a genuine form of writing, which does not mean that it is devoid of concepts and conceptualizations. Beyond the scholarly interest I hope I will provoke, and the many errors that I keep finding in the manuscript (and those that others will certainly spot), many tales in this book can be taken as attempts to show the nefarious effects of splitting stories into anthropological, political, sociological, or economic components.

This splitting argues that reality would exist in the same way as stones in the Zen garden: one different, incomplete perspective for each observation point. This book is driven and inspired by the need to provide ways to reclaim the garden in the way a cubist painter would have represented it. Rather than arguing that different perspectives coexist, I am really arguing—narrating—how they were articulated; specifically, how the stock exchange *led* to anthropology. To paraphrase the character in Eugène Ionesco's *La Leçon* who claims that "arithmetic leads to philology and philology leads to crime," this history, in a nutshell, is a way to demonstrate how the stock exchange led to white-collar criminality and how white-collar criminality led to anthropological science—and vice versa.[4] I only hope that my straight narrative of sinuous characters will be as enjoyable to read as it was to write and that it will not be just one of How the Author Lost His Dear Reader.

# ACKNOWLEDGMENTS

Unbeknownst to me, this book was for a long time in the making. It came about when an angel visited me and, touching me lightly, whispered about the meaning of letting go, and that we hold things we love with an open hand.

A few people helped me along the quest. As my deep debt to them is often a private one, they will find my gratitude properly handwritten on the cover page. But some careful and dedicated readers of the first draft provided special insight, suggestions for additional readings, amicable contradiction, and above all inspiration in relation to their own work and research. This kind of debt is public and must be acknowledged publicly. I thus warmly thank my two most careful initial readers, Carolyn Biltoft and Rui Pedro Esteves. I also thank my colleagues at the Graduate Institute—Grégoire Mallard, Amalia Ribi Forclaz, and Davide Rodogno—for their many shared thoughts and feedback.

Scholars already working on some of the characters in this tale have provided me with most valuable insights. I should in particular acknowledge the feedback of Michael Hammerson and Charles Priestley (on Bedford Pim), of Mark Patton (on John Lubbock), and of Mr. Umit Sakmar (on Hyde Clarke's links with Freemasonry). They should be absolved from my own additions and ruthless extrapolations.

I am tremendously grateful to the editors of the University of Chicago Press, T. David Brent, Joe Jackson, and Ellen Kladky, for their interest and support for a book that crosses so many lines in so many ways. I also owe them for discovering hard-to-find referees, one of whom (still anonymous) helped me with especially insightful comments and criticism—and a tutorial in anthropology.

My copyeditor Jeffery Johnson deserves more than a special mention for the way he improved the manuscript and with unending patience and good humor kept making wonderful proposals for transposing into the English language the numerous Gallicisms of the original manuscript. He also helped me uncover many inconsistencies of substance or form. Through his dedicated work, I have learned more than I could possibly summarize here.

The research has drawn on many different sources. Warm thanks are due to those of my students who assisted me with the investigation and invariably put their intelligence and good humor to the task. I especially thank Sam Segura-Cobos as well as Joanna Kinga-Sławatyniec, Riad Rezzik, Gabriel Geisler Mesevage, and Maylis Avaro.

The welcome of the Royal Anthropological Institute is also gratefully acknowledged. I especially thank Sarah Walpole, archivist, and David Shankland, director. Sarah was much obliging in sharing with me her insights on a lot of scattered material and until the last minute feeding me with her discoveries. They helped to remove a number of errors in the manuscript. I bear full responsibility for the remaining ones.

Other archivists to whom I owe substantial debts include Julie Carrington (Royal Geographical Society); Jennie De Protani (Athenaeum Club, London); Timothy Engels and Patricia Figueora (Church Collection, Hay Library, Brown University); Janet Foster (Royal Statistical Society, London); David Hollander (legal resources, Princeton University Library); Dan Mitchell (University College London); Rachel E. Robinson (Tulane University); and Helen Wiltong (Bodleian, Oxford). I finally thank the whole team of archivists of the British National Archive at Kew Gardens for the support they provided to Sam Segura-Cobos and myself in the treasure hunt in Bedford Pim's files in the Admiralty and Foreign Office records.

A special mention is due to Monsieur René Derupt from 4DigitalBooks for the kindness with which he provided high quality, digitized images of the etching of the king of Miskito and of the map of Bedford Pim's Central America.

My gratitude goes as well to the always efficient team in the library of the Graduate Institute of International Studies and Development at Geneva. Martine Basset, Yves Corpataux, and Marc Le Hénanf showed their unending treasures of kindness and patience, not least when they helped me with retrieving many diverse works and by renewing multiple interlibrary loans as I digested the material far too slowly.

The Graduate Institute of International Studies and Development finally deserves special mention for its financial support and interdisciplinary atmosphere; the departments of international history, economics, and anthropology and sociology are just a few steps away from one another. I warmly thank, in particular, the Graduate Institute's director, Professor Philippe Burrin, to whom I owe a considerable debt. Thanks to him, I benefited from the comfort and tranquility of the shores of Lac Léman, where some secrets about the stock exchange are carefully kept. Perhaps for this reason, Geneva has been for centuries a propitious haven to those seeking to think critically about capitalism.

# The Stock Exchange Modality

In 2013, a few weeks after the first lines of this book were laid down, the academic world was stirred by news that famed anthropologist Marshall Sahlins of the University of Chicago had resigned from the U.S. National Academy of Sciences, citing his objections to the academy's military partnerships and to its electing as member Napoleon Chagnon, a controversial anthropologist who wrote *Noble Savages: My Life among Two Dangerous Tribes—The Yanomamo and the Anthropologists*. In public statements about his resignation, Sahlins referred to work by himself and other critics of Chagnon such as Patrick Tierney, author of *Darkness in El Dorado: How Scientists and Journalists Devastated the Amazon*. In that book, Tierney charged that Chagnon had contributed to a bloody and destructive war on the Yanomamo people by mingling with a corrupt political insider and an entrepreneur with a reputation for illegal gold mining who also was a known opponent of Indian land rights. The episode provided a new twist in an increasingly strident and increasingly public controversy about the conflicts of interests of anthropologists and their tendency to succumb to commercial, imperial, and military uses of anthropology.[1]

But the fact that anthropology has been misused by imperial powers is no recent discovery. An entire branch of modern anthropological history is devoted to exploring the unsavory relations between the science of man and colonial rule, military or civilian. Indeed, there is nothing new in the realization that colonization and colonizers drew inspiration and advice from anthropology. The links between anthropology and empire are old and infamous. Occupying armies have long relished expert assessments from dedicated anthropologists. A much-quoted example is E. E. Evans-Pritchard, whose research of the Nuer in southern Sudan had been commissioned by the government of Anglo-Egyptian Sudan and was embedded in that government's repressive politics, "which included bombing and machine-gunning of camps," as Evans-Pritchard described it later. As historical counterpart to the anthropologists at the elbow of General David Petraeus, former commander of the United States Central Command of operations in Iraq and Afghanistan, who himself holds a Ph.D. in international relations from Princeton's Woodrow Wilson School of Public and International Affairs, the first fieldwork manual ever to include a chapter on ethnological exploration was published in 1849 under the auspices of the British Navy—the *Manual of Scientific Inquiry Prepared for the Use of Her Majesty's Navy and Adapted for Travellers in General*.[2]

Another well-known example is anthropologist Bronislaw Malinowski, who in 1929 wrote an essay advocating something he called "practical anthropology" as a significant and valuable output of his discipline—an output oriented toward public policy. A lot has been made of the fact that Malinowski drew funding from the private Rockefeller Foundation and of how this would have enabled him to distance himself from the British colonial state. But as his 1929 essay makes clear, Malinowski's newly founded practical anthropology, by asserting the superiority of the academic expert over the colonial administrator as the legitimate owner of anthropological knowledge, was intended to benefit anthropologists by educating colonizing authorities, in effect creating an ecology between colonialism and science. As for the Rockefeller Foundation, historians now think that it was a scout for an emerging U.S. imperialism.[3]

Another example, given by Adam Kuper at the opening of a chapter on anthropology and colonialism in his *Anthropology and Anthropologists*, is the painting that decorated independence leader Kwame Nkrumah's office when he became president of Ghana in 1960. The painting showed Nkrumah himself wrestling with the last chains of colonialism as three figures flee the scene—a capitalist, a priest, and an anthropologist, who carry respectively a briefcase, a Bible, and *African Political Systems*, which Evans-Pritchard had coauthored.[4]

On the other hand, against this narrative of the anthropologist inadvertently or deliberately succumbing to the conflicts of the trade, a tradition of research and reflections has instead narrated the (why not, triumphant) efforts of anthropologists to preserve their integrity. A classic illustration of this view is provided in *Anthropology and the Colonial Encounter*, where critical anthropologist Talal Asad felt justified in declaring, "I believe it is a mistake to view social anthropology in the colonial era as primarily an aid to colonial administration, or as the simple reflection of colonial ideology." According to Asad, while the colonial encounter did provide the occasion for doing anthropology and must be kept in mind to understand the conditions of production of anthropology in history, the "final reckoning" would not show the contributions of anthropology to have been so "crucial for the vast empire which received knowledge and provided patronage." This would be true, according to Asad, because "bourgeois consciousness, of which social anthropology is merely one fragment, has always contained within itself profound contradictions and ambiguities—and therefore the potential for transcending itself."[5]

In another variant of the optimistic narrative, anthropology in the service of the state is rescued by professionalism. As the institutions of anthropology consolidated after 1870, a distinction gradually emerged between scientific experts following an academic path and those experts embedded in the colonial bureaucracy—a separation epitomized by Malinowski's practical anthropology. Like the experts within the colonial bureaucracy, academic anthropologists had practical preoccupations. But unlike the former, they were pure in their motives and kept so by academic institutions and their pledge to professional standards (the American Anthropological Association, or AAA, for one thing, has a code of ethics). That the rich or powerful have an interest in the outputs of the science of man should not mean that anthropology is doomed. Despite desires, purpose may remain pure.[6]

We have a natural urge for the story of conflicts of interests in science to end well. One cannot avoid having some sympathy for the anthropologist's quandary because it is emblematic of a universal problem with the social sciences (as the subtitle of David H. Price's book *Weaponizing Anthropology: Social Science in the Service of the Militarized State* underscores). Reasons for the universal prevalence of this problem are simple. To the extent that truth has social significance, its production and ownership have tremendous value. A patented truth sells better (the important word here is *patented*) and as a result commandeering truth provides many benefits.

The example of Columbus may illustrate the argument, especially since he was so conspicuously referred to in the writings of many of the characters in my tale. The first European to navigate along the Mosquito Coast, with which I will deal in later chapters, hypothesized that the earth was round rather than flat. We may ask why did he undertake a voyage and put himself at risk instead of publishing a paper in *Nature*? The superficial answer is that in 1492, there was no such thing as *Nature*. An alternative was emphasized by historian Pierre Vilar. We know from Columbus's travel log that he was obsessed with the finding of gold, pearls, and gems. In other words, the economics of truth—the relation between truth and value—is a critical part of the process of knowledge and discovery and should be an essential object to study. Can one understand the history of truth if we do not understand how its fruits are produced and distributed? The Columbus experience reminds us there is more than one way to own a hypothesis, and his was not peer reviewing but a fleet of galleons. In other words, peer reviewing has something in common with the chartering of the *Pinta*, *Niña*, and *Santa Maria*.[7] And if this conclusion is admitted, then one cannot quite see from which balance sheet Asad could derive his "final reckoning" showing the "contributions" of anthropology as not so crucial to imperialism.

The simple point is that there cannot be such a thing as a disinterested science. There might be institutions that somehow insulate scientists from the surrounding interests and conflicts and help tie them to their purpose as Ulysses to his mast, but it is obvious that these institutions must contend with incredibly powerful external pressures and forces, forces that often provide the rope itself. Whether social scientists use it to tie or to hang themselves is an everyday challenge. The record of the economic profession before and after the subprime crisis may be taken as indication that the dilemmas identified and complained about by Price, Sahlins, and others are indeed universal, with anthropology providing a good metaphor and case study but certainly not a scapegoat. But for the historian, it remains that the relations between forms of knowing and forms of owning are an important and valid research agenda. Moreover, a number of more recent works have strengthened the notion of a convergence of interests between science and power, resulting in the identification of a ubiquitous coupling of knowledge and empire suggestively described by historian Londa Schiebinger as a "colonial science complex."[8]

As the existence of a lively contemporary and past discussion on the subject of colonialism and anthropology suggests, exploration of the relationships between forms of knowing on the one hand and colonialism and empire on the other is hardly a new subject. Nonetheless, this book does not intend to be simply a study of this fascinating nexus in a given historical context—that of the capital market. Rather, by exploring simultaneously a period and a set of modalities of knowledge in a new way, it seeks to derive a novel theoretical perspective on the making of anthropology through a renewed understanding of the making of its institutions. I call this perspective the "stock exchange modality," the true subject of this book and a topic that I argue has significance beyond the period and context studied here. And because of the importance of the stock exchange modality to the story I am going to tell and to the way I shall tell it, I need to devote the rest of this introduction to explaining what I mean by this.

The bulk of previous research on the history of anthropological knowledge has focused (and continues to focus) on the links between bureaucratic authority and knowledge and whether the two are seen as cooperating or conflicting. As a result, existing work has borne to various extents the imprint of French historian Michel Foucault's perspective on knowledge as part of an apparatus of control. Perhaps the most famous study of early colonial knowledge that fits this bill is Bernard S. Cohn's set of classic articles later published as *Colonialism and Its Forms of Knowledge*, which was actually developed in parallel with Foucault's work rather than derived from it. A pioneer of the history of the early development of anthropology and its relation with the British colonizing of India, Cohn focused on the East India Company and showed how knowledge of India was constructed as a product of the concerns of the East India Company's bureaucracy. The pervasive concern to cut out the intermediary (the "cunning Banians" or any other) and end dependence upon external brokers was of critical importance in the primitive phase of knowledge accumulation by the East India Company.

As Cohn demonstrated, the first great scholars of Indology, such as William Jones and Henry Thomas Colebrook, were senior employees of the company and developed their expertise as a response to its taxing and law-making concerns. Knowledge may have eventually been stored in repositories outside of the company. Colebrook, for one thing, played a key role in the launching of the Asiatic Society of London in 1823, but this knowledge still originated in the company and was eventually valued there. It was a knowledge that developed in a monopolistic context, to be used by a commercial bureaucracy that both governed and exploited. Early ethnologists were typically powerful bureaucrats. John Crawfurd, for instance, had been colonial governor of Singapore and Java during the British interregnum. The relation was between an exploitive bureaucracy (or a set of exploitive bureaucracies, for the Navy was not the same thing as the East India Company) and its forms of knowledge accumulation. This goes a long way toward explaining why the 1849 *Manual of Scientific Inquiry* with its chapter on ethnological data gathering had been—and had to be—published under the authority of the commissioners of the Admiralty.

This framework of the bureaucratic modality has come in handy in studying later periods of the British Empire, long after the East India Company dissolved and its assets and liabilities were taken over by the British crown. In the twentieth century, a bureaucracy with monopoly power again provided for governing the empire, as historians of British anthropology such as James Urry have argued.[9] Reflecting this, recent case studies of colonial knowledge have typically focused on the interwar years when this

knowledge-follows-the-sword pattern enabled observers such as Asad to argue that anthropologists found the world already colonized and could thus
plead not guilty. Indeed, after 1900 the institutions of empire had routine
behind them, and the British Anthropological Institute had become Royal.
There was now such a thing as the colonial administrator with a well-
defined job—the administering of the colony—and there was such a thing as
the anthropologist with a well-defined job—the production of anthropology.

By then, the metaphor of empire-as-fieldwork, which Helen Tilley has
put in the title of her beautiful book, *Africa as a Living Laboratory*, seemed
sui generis: colonization was fieldwork. One could train as an administrator
or as an anthropologist and be led to believe that such categories existed
naturally and would exist forever. To display the marriage, there was an
Imperial Bureau of Anthropology. As with any marriage, it was natural that
there would be disputes. By then, the origins of colonial administration and
anthropology were long forgotten, and anthropologists could picture themselves walking a tightrope between providing information to colonial authorities and risking their academic integrity. They could mindlessly enjoy
their seemingly immanent dilemma and think that there was nothing else
beyond trying not to succumb to it. The reasons for this tightrope-walking
were lost.[10]

The mid-nineteenth century, with which this book deals (say the period
1840 to 1880), provides an economic and political context that does not fit
well into such categories, however. It was characterized by a more amorphous system, when empire started expanding substantially beyond the reach
of state-chartered companies with state-like monopolies and before the
formalization of the British Empire in the 1870s. The work of historian
Edward Beasley, focusing on the creation of the Colonial Institute in 1868
and the subsequent growth of an empire of information, has identified the
importance of the organization of knowledge that took place at this point
and how this permitted the emergence of an "imperialism of information
management."[11] We'll see that this book shares with Beasley the idea that
1868 was a turning point in the evolution of an imperial knowledge system.
But as this book will suggest, Beasley's attempt to separate this transformation from the underlying economic and financial motives is more open to
question.

The relevance of the financial context to the shaping of imperial knowledge can be seen vividly by focusing on the specific case of anthropology. The
1860s were an age of rapid commercial expansion, which created and mul-
tiplied opportunities for getting rich through long-distance projects. For that

reason, what anthropologists Peter Pels and Oscar Salemink have called the "ethnographic occasion" (meaning the unintended fieldwork that provided the primitive anthropologist with material to write upon) was often the product of competing, multi-pronged mercantile motives that shattered traditional locations of anthropology. The enormous implications of such evolutions for the existing system of anthropological knowledge are not always properly understood. Unlike in the preceding period, substantial anthropological knowledge now prospered outside of the remit of colonial bureaucracies. Unlike in the subsequent period, neither empire nor anthropology were fully constituted from an institutional point of view.

In terms of political economy, this was an era later defined by historians such as John Gallagher, Ronald Robinson, D. C. M. Platt, and, more recently, P. J. Cain and A. G. Hopkins as one of informal empire, understood to have had its heyday around 1840–1870 at the time immediately before the development of institutionalized imperial rule in the late nineteenth century. Informal empire conjured a series of more or less soft-power technologies on which British domination rested. Gallagher and Robinson's classic paper "The Imperialism of Free Trade" argued persuasively that British commerce was an inherent part of the apparatus, and the later work of Platt as well as the generalizations of Cain and Hopkins have suggested the same for finance. Indeed, this period was one that saw the rise and centrality of the London Stock Exchange in British foreign domination.[12] But neither the "free" commerce that expanded globally from the 1840s at an unprecedented pace nor the finance that did pretty much the same were bureaucracies in the conventional sense, even if they did include bureaucratic elements. They were not hierarchies but competitive, inchoate markets, suffused by their own forms of violence, corruption, and exploitive technologies, different from those that had characterized the big chartered companies and bound to produce a knowledge of their own.

Thus while commerce and the London Stock Exchange had concerns and goals that were relevant to overseas territories, those just could not be the same as the ones that had been relevant to the East India Company's administration, which wanted to use knowledge of the colonized to, say, increase the efficiency of revenue collection or provide natives with legal services that would tie them to British rule.[13] And these concerns and goals came to be rather encompassing. English novelist Anthony Trollope made the London Stock Exchange the central character of his famous 1875 novel, *The Way We Live Now*. But already in 1857 the French anarchist Pierre-Joseph Proudhon had devoted a well-informed practical treatise to the subject matter of the capital market, arguing that finance had pervaded everything: "Only finance has the power to summon passions, to excite enthusiasm or hatred, to make the hearts beat, to reveal life. It is for finance that the army is on duty, the police on watch, the University in teaching, the Church in prayers, the people at work and in sweat, it is for finance that the sun shines, crops mature, and everything grows and fructifies."[14]

As a result, the social sciences that came to be used and valued at this juncture were influenced and commandeered by the stock exchange and its concerns. This imparted original features to forms of knowledge. I argue that the stock exchange left a deep imprint on the social sciences that emerged at that time, not only in the substance that was being studied but perhaps more fundamentally in the design of the institutions of science. Because science was valuable, it invited capture and manipulation according to the rules and dynamics of the capital market. One consequence of this was the involvement of colorful characters and white-collar criminals. This book shows, then, that within and beyond the various forms of knowledge identified in previous research—the historiographic, the travel, the survey, the enumerative, the museological, and the surveillance modalities of knowledge—room should be made for the neglected and ill-understood stock exchange modality.[15] As we shall see, this modality has been and is primarily concerned with the problem of valuation. Exploring the stock exchange modality enables us to introduce important questions about the value of the social sciences to the capital market.[16]

In this book, the stock exchange modality is described in the traits of its Victorian avatar—the "art of puff" or the promotion of bubbles. These were mercantile years when everything from the quality of a novel to that of a learned society—to say nothing of the joint stock company—was ballooned and leveraged so as to be sold and distributed with a profit for the promoter. The art of puff defined the set of techniques used by a number of prodigals and projectors to promote arts, science, and companies.[17] Puffing had its specialists, its rules, and its remedies. There were ways to create bubbles and ways to puncture them, and these were more important than the precise objects to which they were applied. There was thus a profound continuity between the puffing of seemingly heteroclite objects, a continuity that created further opportunities for the encounter of anthropology and the stock exchange. This explains why a combined history of bubbles puffed and punctured, whether scientific or financial, might hold important insights. It enables us to see the technological, political, and sociological underpinnings of the promotion of railways, steamboats, telegraphs, and learned societies.

Doing so delivers a new explanation or perspective on the history of British anthropology in the mid-nineteenth century, one that departs from existing narratives such as that proposed by the late American historian George W. Stocking, a leading scholar of British anthropology. Stocking explained the birth of the Anthropological Institute of Great Britain and Ireland (known today as the Royal Anthropological Institute), created in 1871, as a piece of cultural history—the result of a controversy between an Ethnological Society and an Anthropological Society. And this controversy, Stocking argued, essentially revolved around an ideological battle. It consisted in taking sides in the American Civil War, for or against slavery.[18] Neither Stocking's narratives nor those of his followers deal in any way with the stock market, yet I shall show that some of the key characters of the anthropological dispute that Stocking has emphasized were first and foremost tied to the stock exchange. If it means anything to follow Stocking's own exhortations to take history and the context seriously, then the capital market should feature in the narrative of how the Anthropological Institute was born.[19]

I will argue that the pressures on anthropology during those critical years from 1863 to 1871, which have attracted previous attention, belonged squarely to the remit of the stock exchange modality. The same individuals who puffed explorations of Africa or the promotion of mines or railways in Central or Latin America also puffed anthropology. And the rampant puffing in turn triggered fierce battles over the ownership of the territories of science—the ownership of learned societies as ownership of truth, and the ownership of truth as an instrument for financial deal-making. Moreover, I shall argue that these processes had a tremendous impact in shaping empire as it emerged in the later part of the nineteenth century, designing it in ways that other scholars working on later periods have taken as a given. In fact, both institutions (late nineteenth-century empire *and* late nineteenth-century anthropology) were shaped at the same time and cannot be understood separately. Empire and British anthropology were jointly structured in the early 1870s, and their joint structuring responded to circumstances that developed in the capital market during the mercantile and financial expansion between 1840 and 1870.

As suggested, one result of the expansion of the mercantile economy was the multiplication of ethnological occasions, which in turn resulted in the appearance of original forms of white-collar criminality. The individuals who shuffled between the trading rooms of the London Stock Exchange and the debriefing sessions in London-based learned societies had an interest in certain truths, which affected the price and marketability of the assets they sold. As anthropology became an instrument of legitimacy, its traction expanded enormously. This became the origin of an anthropological bubble, which began not only to escape political control but also to openly threaten and challenge existing bureaucracies, knowledge infrastructures, and scientific sociability. The final stage came when anthropology was transformed into an instrument against the incumbent Liberal government in 1864–65. In this conflation of forces can be found the reasons for the chronological coincidence of deep transformations in anthropology and in the organizational machinery that channeled capital exports.

Such conclusions are important for the history of British imperialism writ large. According to high school historiography, the high point of Victorian imperialism began with Disraeli's speech at the Crystal Palace in 1872, which narrated empire as something valuable and therefore worth spending resources on. It was not just a cost, burden, and danger for the British state as the old Liberal tradition had maintained. But the event had a context. Before this occurred, the London Stock Exchange had been the theater of a mania for foreign government debt, eventually crashing in 1872–73 and revealing scandals that would lead to a parliamentary investigation in 1875. To monitor foreign debtors, the British Board of Trade sanctioned the Corporation of Foreign Bondholders in 1873, a bipartisan institution featuring prominent Liberal and Conservative members of Parliament as well as bankers and brokers. Such was the context in which the Anthropological Institute was itself established in 1871. This book will argue that the forces that brought the Anthropological Institute into being were tightly interwoven with those that produced Disraeli's Crystal Palace speech, the preceding foreign debt boom, and the creation of the Corporation of Foreign Bondholders.

One important theme of this book is that the specific historical relations between the stock exchange and anthropology provides an instance of what Carl Jung has referred to as archetypes, universal images or patterns with an archaic quality. The archetype my book will reveal is that of "brokerage," and it is intimately related to the stock exchange modality. Anthropologists are indeed fond of using the brokerage concept. They like to study brokers belonging to the groups they observe and rely upon them for information, as the title of a recent book edited by Simon Schaffer, Lissa Roberts, Kapil Raj, and James Delbourgo, *The Brokered World, Go-Betweens and Global Intelligence, 1770–1820*, illustrates. Moreover, anthropologists like to think of themselves as brokers, sometimes refining their role as in the expression "culture brokers." The concept is almost always used somewhat offhandedly, as a quasi-natural notion. This is suggested by the frequent parallel with similar concepts such as intermediaries, translators, mediators, middlemen, and so on. But unlike its many equivalents, "brokers" is a metaphor that conjures up a site par excellence where the broker reigns supreme—the stock market. No wonder that it's also the preferred metaphor.[20]

Thus the stock exchange modality has been hiding in plain sight. Indeed, the trade brokerage role of the anthropologist—both word and substance—was involved in an overwhelming number of early ethnological occasions. Think for instance of Willem Bosman, a trader and member of the Dutch East India Company operating in Africa who articulated the first anthropological theory of the fetish. According to William Pietz, Bosman, like other Western traders and explorers before, beginning with Columbus, was intrigued by striking discrepancies in the valuation of given commodities across different cultures (in effect the source of the profits which traders were making). Bosman thus developed an explanation of the worship of fetishes that relied on the African's "capricious fancy" as opposed to the true price system set by the rational merchant in Europe. This would have provided reason why Africans chose "trifles" rather than "valuables." Of course, as Marx would emphasize later, the Western merchant had also his own fetishes, his own trifles.[21]

Likewise, historian of anthropology Bernard S. Cohn's classic chapter "The Command of Language and the Language of Command" narrates the processes whereby the East India Company gradually took control of Indian languages as essential carriers of knowledge. As he describes, seventeenth- and eighteenth-century ledgers of the company contain the "names and functions of Indians who were employed by the Company or with whom it was associated, on whom the British were dependent for the information and knowledge to carry out their commercial ventures." This reliance on external informants such as brokers and translators occasionally led to the use of long chains of intermediaries. Sir Thomas Roe, during his mission to the Moghul's court at Agra to obtain protection for the East India Company, once employed an Italian to whom he spoke in Spanish. The Italian spoke Turkish, and this was then translated by an officer of Mughal Emperor Jahangir's court, who knew both Turkish and Persian. This created anxiety and frustration regarding the extent to which the intermediary really obeyed the instruction of the British principal. Cohn emphasizes Roe's "suffering" in the hands of what he called his *Broker*, who, Roe complained, "will not speak but what shall please; yea they would alter the Kings letter because his name was before the Mughals, which I would not allow." Roe was longing for what he described as a proper interpreter who would enable him to cut out the intermediary. Evidently in Roe's mind, brokers and interpreters

were not quite the same thing, the latter being solely concerned with transcribing the instructions of his principal in a foreign idiom (an agent) and
the former owning part of the instructions through his independent knowledge (an intermediary).[22]

This brief political economy of anthropological translation is important
because it has the potential to explain both the geography of anthropological production and its shifts (two major themes of this book) without the
anachronistic use of later categories such as professionalism. In fact, as we
shall soon see, focusing on the brokerage role of anthropologists enables us
to understand such categories as professionalism in a broader framework.
My starting point is to think of anthropologists as businessmen, so to speak,
with concerns that are not far from those of the commercial or financial
intermediary. Let us consider them literally as brokers. This is legitimate
because they derive at least part of their revenue from their role as intermediaries, as all good brokers should. According to historians Roseanne
and Ludo Rocher, Henry Thomas Colebrooke, one of the founders of British
Indology, was the son of a failed East India Company director who had received a token of support for his progeny in the form of employment of his
sons in India. For the son, production of science *inside* the East India Company became an instrument for personal financial restoration. The knowledge he acquired by trying to build a corpus of Sanskrit law was valued by
the colonizing bureaucracy in which he worked.[23] By promoting such individual efforts as Colebrooke's, the East India Company was intent on cutting
out the intermediary as Roe had wished. The objective was the elimination
of the manipulative local broker and his replacement by an agent embedded
in, and thus more dependent upon, the East India Company. More generally, the history of the changing geography of anthropological production
resulted from subsequent adjustments in the knowledge needs of global exchange and in the trading structures supporting this exchange.

Thus with the end of trading monopolies and big chartered companies, with the Sepoy Mutiny of 1857, and with the enormous expansion of
global trade after 1840, existing forms of anthropological knowledge were
transformed. The myriad agents who followed on the back of the quantitative progress of "globalization" experienced new ethnological encounters. Reflecting this, the frequency of any country or state in appearing
in British newspapers—be it Honduras, Tasmania, Japan, or Abyssinia—
peaked during the 1860s. Delving further into the reasons for this, one invariably discovers the contribution provided by association with financial
undertakings. Because of this, any relevant narrative of the evolution in the

THE STOCK EXCHANGE MODALITY          13

geography of ethnographical encounters derives to a large extent from the spatial history of capitalism.

Use of the brokerage metaphor to describe the function and work of anthropologists is also, literally, a guide and source for understanding the political economy that provided a foundational basis to this form of knowledge. As intermediaries, brokers cater to what is known in the jargon of modern economists as a "double-sided" platform. In less pedantic terms, they are situated at the center of a kind of coupling system, with the sell side on one end and the buy side on the other. Evidently, because of the adverse interests of buyers and sellers (one wants to sell high, the other wants to buy low), brokers perform some kind of arbitration between these conflicting interests. Depending on tastes, circumstances, and opportunities, they can decide to ally themselves more closely with one end than with the other.

This logic can explain a paradox identified by Asad in his *Anthropology and the Colonial Encounter* quoted above. According to him, anthropologists always have had to contend with irreconcilable judgments—now a "Red" siding with the insurgency, now the handmaid of imperialism feeding counter-insurgency with intelligence—and he suggests that somehow this reveals the absurdity of any attempt to characterize the anthropologist as either of the two. But from the vantage point of the argument I am developing here, there is no reason for choosing between these narratives, which are in fact two sides of the same coin. Brokers are brokers, and their honesties and dishonesties toward either end of the bargain are in the nature of the trade. In the end, the fact that they are part of a bargain demonstrates the difficulty they may have in thinking beyond the very exchange relation in which they are involved, the same relation they are bound to support and endorse even when they chastise the buyer or the seller (or precisely for that reason). This provides yet another definition of what I mean by the stock exchange modality.

From this follows a last point of great analytical importance. Focusing on anthropologists as brokers enables us to provide an alternative to an anachronistic approach that would attempt to gauge Victorian anthropologists with the yardstick of modern science. It is well known to the historians of finance that, while the concept and reality of brokerage is easily recognizable in all mercantile and capitalist polities, the variety of actual designs is baffling. At the broadest level, recent studies of historical stock exchanges have revealed the concern of market authorities regarding the behavior of brokers. Because of the need to ensure a certain quality of transactions, stock exchange authorities have sought to subject market operators—the

brokers—to norms of behavior that would protect the interests of the trading parties and in turn that of the exchange. An exchange notorious for siding with one end of the market, whether buy or sell, would have a hard time
retaining the clientele it undermines. Privileging any part heralds the end
of the broker.

Now, previous authors have emphasized the subsequent professionalization of anthropology in the late nineteenth century. This reference to professionalization often served to replace hard-to-read amateurs with academics
with whom modern academics can see eye to eye. But this was in no way an
escape from the spell of the stock exchange modality because the very setting
of ethical and professional standards can be analyzed as just a response to it.[24]
Just like the authorities of the stock exchange are concerned with retaining
both buyers and sellers, the managers of anthropology and the scientific societies they ran worried that the behavior of some of their members would
eventually undermine the ability of the discipline to play its brokerage role
effectively. Like the authorities of the stock exchange, they thus attempted to
subject members of the profession to a set of rules that were continually ne-
gotiated and renegotiated in a constant effort to protect both the source of anthropological knowledge (the indigenous people among whom the fieldwork
is conducted) and the consumers (readers, students, academic colleagues, or
the U.S. Navy). From there it follows that much can be learned, not by starting with anthropology imagined or reimagined but by locating ourselves as
close as possible to the trading relation at the heart of the anthropological
exchange. From there it follows that the only yardstick that is usable is the
market context at the time when the exchange took place.

The previous digression on brokerage may explain why analysis of the stock
exchange modality has implications far beyond the history of empire or the
history of anthropology narrated here. Understanding the forces that produced simultaneous political and institutional upheavals in the science of
man and in the politics of empire in the 1860s and early 1870s will help us
understand more perennial relations between truth, value, and power. This
is my main task here. The project considered here is not to deconstruct the
social sciences to show that they have the quality of an illusion. It is to reconstruct them from the surrounding web of material and real interests and,
as a result, show the stuff science is made of. The object, really, is to produce a better understanding of the political economy that presided over the
birth of a discipline that is described today as a social science—or, equiva-
lently, a historical perspective on the way it became part of the body politic.

Consequently, the following history of the production of anthropological truths is primarily concerned with understanding how they were commoditized. It departs from conventional historical practice and its dichotomies (history of ideas vs. material culture). It departs from most critical theory and its disdain of the material (everything is construction). My goal here is to weave together the meaning and the quantitative. I believe that analyzing the very material forces at work in the production of knowledge structures can shed light on the meanings we attach to knowledge. My hypothesis is that the pursuit of knowledge and riches are related to each other in a more curious way than previously recognized. The property rights that can be laid over knowledge and its forms have a peculiar nature of their own, and understanding how knowledge is owned is perhaps more important than understanding how it is produced. On this point, one should perhaps ponder the fact that it is the same word—speculation—that is used for the art of conjecturing in science and finance both.

And thus, I speculate that thinking in terms of the stock exchange modality might illuminate some aspects of modern social sciences that anthropology once again epitomizes. Viewed from this vantage point, the Chagnon affair would not be solely about professional indiscretion. There is something deeper at work, and I suggest reading in it a resurgence of the stock exchange modality. Today, it is not difficult to come across City people who pride themselves on having graduated from an Ivy League institution or from Oxford with a degree in a subject that has at first glance little to do with a career in finance—something exotic like anthropology, for instance. The voracity with which Wall Street and the City have been snatching up such graduates suggests that the alleged exoticism is puffed up. In fact, *Financial Times* star journalist Gillian Tett, herself a Ph.D. in anthropology, makes no secret of the connection she sees between anthropology and finance.[25] And if language again is any guide and a personal recollection is permitted, I remember a phrase I heard when I was working in Paris for the now-defunct American investment bank Lehman Brothers. In an early talk, I was warned by the head of Global Economics in London, to whom I reported, that, being based in France, there was the risk that I would go native. The head of Global Economics apologized for the expression.

In the same vein, we may mention that, according to Karen Ho, an anthropologist and the author of *Liquidated: An Ethnography of Wall Street*, modern visions of capitalism are shaped by anthropological and social discourse just as much as by the global talk of big corporations.[26] This is not very far from the impression under which Proudhon had found himself some 150 years ago, when he emphasized that everything, material or not, was growing in response to the call of the stock exchange. In what follows, I want to go back to a foundational moment of finance and anthropology and unpack the forces whereby the stock exchange came to make everything grow for it, including anthropology. If anything, this book suggests that the phenomenon whereby social sciences find their inspiration in the money market has deep roots. Although hidden from sight—as all modes of knowing must be—the stock-exchange modality, whose origin my book explores in the context of the Victorian era, is alive and well.

# Writing about the Margin

In early 1863, a group of self-styled anthropologists split from the older Ethnological Society to set up their Anthropological Society of London. They had two main leaders. One was James Hunt, a thirty-year-old speech therapist who had been former secretary of the Ethnological Society of London. The other was the already famous British explorer Captain Richard Francis Burton, a former employee of the East India Company, a passionate translator of foreign languages, notably erotic folklore, and a founder of the modern anthropology of the body.[1] Hunt would hold the fort while Burton would travel. He would organize discussion and take care of the society's publications while Burton concentrated on his books.

As reward, Hunt was made president of the society with responsibility for running it. Burton, as vice-president, would provide social and scientific backup. The society then endowed itself with vice-presidents, honorary secretaries, and a list of Council members as would any learned society worthy of its name. Facilities were found at 4 St. Martin's Place near Trafalgar Square in a building owned by the Royal Society of Literature, which also hosted the meetings of the Ethnological Society of London. In the summer of 1864, the rental was expanded to secure space for a small museum and library.[2] With the collaboration of Nicholas Trübner, a publisher contemporaries described as "an intermediary between Europe and the East," the "Anthropologicals" launched several periodicals: the *Journal of the Anthropological Society of London*, which kept members informed of meetings and discussions; the *Anthropological Review*, which contained articles, surveys, and speculation on current topics (see figure 1); and the semi-periodic *Memoirs Read before the Anthropological Society of London*, where the longer and more significant papers were published. There was also in 1866

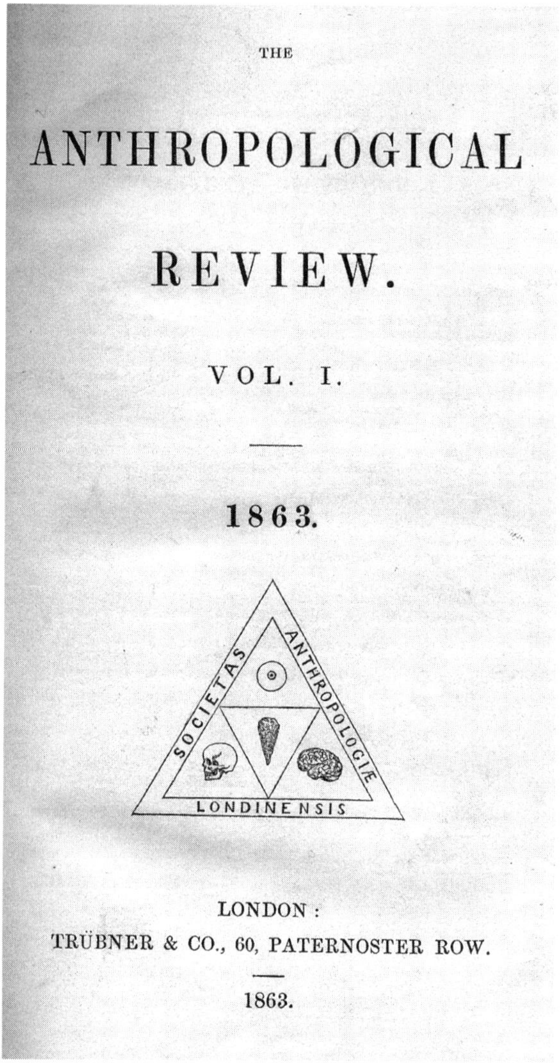

THE

ANTHROPOLOGICAL

REVIEW.

VOL. I.

1863.

SOCIETAS ANTHROPOLOGIÆ LONDINENSIS

LONDON:
TRÜBNER & CO., 60, PATERNOSTER ROW.

1863.

FIGURE 1. Cover page of first issue of *Anthropological Review* (vol. 1, 1863).

the short-lived *Popular Magazine of Anthropology* and a number of translations of influential Continental anthropologists.[3]

A fascination for symbols—their production and interpretation—would remain a distinctive trait of the Anthropologicals. On the society's logo, the Latin phrase *Societas Anthropologiae Londinensis* was displayed along the sides of a Masonic triangle, and within the triangle, the three Masonic dots were respectively represented by a skull, a brain, and an eyeball, all surrounding a second triangle in the center of which was a flint implement.[4] Among other ghoulish fetishes, the Anthropologicals also kept in their headquarters an articulated skeleton, visible through the window of the building much to the displeasure of the Christian Union Institute across the road, which asked the society to remove it from their sight. The society responded it was willing to put up a blind to shut out the view if the Christian Union would pay for it. From this, it seems, came the name of the group of men who led the society: the "Cannibal Club."[5] Among the ostensible goals of the group, perceptible in the name they had given themselves, was shock and provocation. Anthropologists sought to delineate a new, raw form of knowledge, set apart from ethnology (according to some Cannibals—certainly according to Burton) and its good intentions, conformism, and bigotry. Anthropology became a synonym for a sort of scientific crudity. In a preface written anonymously for her husband Richard Burton's *Highlands of Brazil* (1869), Isabel Arundell dwelt on his penchant for the raw (half in apology, half in praise), and warned readers of the need "to steer through these anthropological sand-banks . . . as best as he or she may."[6]

During the eight following years, the two organizations—the Anthropological Society and the Ethnological Society—indeed lived a feuding existence. For all practical purposes, both societies focused on similar subjects, and members of both could be considered students of the science of man, a definition on which they almost unanimously agreed. The involvement of Burton in the Anthropological Society, with his reputation as a rabid fieldworker who had once disguised himself as an Indian merchant to penetrate forbidden Mecca, also provided the Cannibals with some of their branding. They proved themselves willing to discuss and publish on such matters as hermaphrodites or practices of clitoral elongation. The rival institutions fought for audience and membership, output and legitimacy, and above all leadership in the science of man. The early annual reports of the Anthropological Society kept score, reporting the number of papers presented respectively by anthropologists and ethnologists at the annual meetings of the British Association for the Advancement of Science, where the latest scientific ideas were floated and enjoyed the light of publicity. But even when public relations were acrimonious, there was also an almost permanent discussion of merger. Repeated attempts were made, the first in 1864. Another significant effort collapsed in the late spring and early summer of 1868, when the two societies failed to create a "Society for the Promotion of the Science of Man."

This episode in fact caused much ink to flow, for the failed amalgamation triggered a controversy in which these students of man started throwing insults at one another much sharper than the stone implements or Indian arrows they were fond of studying. The dispute was started by a trenchant letter to Hunt, written by one Hyde Clarke, a member of both societies, in which he denounced the Anthropological Society's poor governance standards. The letter was published on August 15, 1868, in the *Athenaeum*, a literary criticism journal that boasted of being the "first literary journal to have made honesty its aim." Hyde Clarke's letter started an extended controversy, for he had questioned Hunt's character, painting him and his Cannibals as individuals prepared to derive personal benefits from exploiting the readers of the Anthropological Society's publications. They had exaggerated membership, doctored the books, and effectively put the society's Fellows in debt.[7] This led to a protracted and much advertised dispute, with the Anthropological Council trying but failing to exclude Hyde Clarke from the society.

In August 1869, a few months after the dust of the *Athenaeum* controversy settled, James Hunt passed away. He did so at the untimely age of thirty-six, an age more befitting the life expectancy of a Jamaican slave than that of the member of a learned society whose tendency to try and be immortal is well known.[8] His death was followed by new openings for reunification from the Ethnological Society's leader, Thomas Huxley, and this time the overtures met with success. Thus were ethnology and anthropology happily reconciled in early 1871. Together, they gave rise to an Anthropological Institute under the presidency of John Lubbock, a prominent member of the Ethnological Society who ran the so-called X-Club, the Ethnological Society's counterpart to the Cannibal Club.[9] Lubbock was a former research assistant to Charles Darwin, an entomologist who carried ants with him in his travels, a protector of ruins, a banker, a member of Parliament, the creator of the Bank Holiday once known as "St. Lubbock" day, an archaeologist, and the coiner of the terms *Neolithic* and *Paleolithic*.[10] To secure the merger, compromises had been necessary, and one was the use of the term *anthropology* in the new organization's name. This was no small concession. *Anthropology* had been a rude word for ethnologists in Lubbock's group, and they did not conceal their distaste. "Anthro . . . —(I never can spell the horrid word)" was how his wife, Ellen Lubbock, famously put it.[11] In fact it is still unclear to students of the subject why the word triumphed while Hunt was out of the way.

## CHANCE, COINCIDENCE, AND ENCOUNTERS

Seemingly unbeknown to the men grappling with the vicissitudes of anthropology, another drama played out on another Victorian stage. Starting in the 1850s, the London Stock Exchange, still burdened with bad memories of foreign investment since the crisis of 1825, was the scene of a new investment boom that focused on infrastructure projects such as railways, telegraphs, and navigation companies. As the European economy recovered from the economic and political turmoil of 1848, the continent discovered that its world leadership had further increased. In the 1850s the Ottoman Empire's disintegration accelerated, providing Europeans with the possibility of a shorter and more secure route for their Asian voyages through a planned canal at Suez. Europe retained an enormous edge over the other expansionary power of the time, the United States, which brandished the Monroe Doctrine, annexed Texas and California, and looked toward South America. During the 1860s, the United States became preoccupied by a bloody Civil War that was to consume most of its energies, interrupt its domination of cotton exports, and enable the troops of the French emperor to land in Mexico in an attempt to introduce an Austrian ruler.

The versatility of European capitalism was enormous. If cotton could not be obtained from the United States following destruction of the plantation system, other possibilities existed from creating or expanding existing outlets. Cotton might just as well be obtained from Egypt, India, or Abyssinia. Europe's stock markets, centered in London, and Europe's formidable annual contingent of emigrants to the Americas still held the capacity to design the world. Foreign issues reached the floor of the London Stock Exchange in accelerating succession despite the crisis of 1866, which unlike previous crises such as the one in 1825 did not cause a foreign debt crash. This was because the Bank of England generously poured in liquidity and kept the ball rolling for another a seven-year period apt to inspire biblical metaphors. Speculation resumed, and the foreign investment boom continued until 1873, when Germany's adoption of the gold standard caused a major drain on world gold resources and, some argued, by forcing the Bank of England to raise its interest rate to defend its gold reserves, eventually put the financial market in reverse. In early May 1875, Bernard Cracroft's *Weekly Stock and Share List* would contain a much-discussed chart showing the vicissitudes in the price of government-guaranteed railway bonds of Honduras, San Domingo, Costa Rica, Paraguay, and Bolivia. "Investors in these ambitious securities," one journal commented humorously and with evident geographical gusto, could have the "pleasure of tracing in this chart

their capricious flight down to the time when, after uncertain zigzags in
the empyrean of the latitude of 80, or even in the case of the high vaulting
Honduras and the yet more aspiring Paraguayan of 90, they all with one
accord swoop down to 20 and 10, except Honduras, which seems unable to
rest even there."[12]

When the foreign debt bull market of the 1860s and early 1870s was put
in bear mode, problems with the way securities had been previously originated and distributed became apparent. In the myriad projects that had been
hatched at the London Stock Exchange and sold to investors worldwide, it
became apparent that there were many fraudulent (or as we would say today, toxic) products. The way railways, telegraph, and navigation companies
had been structured revealed subpar financial ethics. This was just a few
years after Herbert Spencer coined the expression "railway morals." And for
evident reasons: railways had none. The problem was so exemplary of the
tendencies of Victorian society that foreign railway promotion became the
background for Trollope's famous novel *The Way We Live Now*, written between May and December 1873 and published in 1875 precisely at the time
when those aspiring foreign securities had all with one accord swooped
down to 20 and 10 in Cracroft's chart.[13]

As Trollope described with immense sociological and economic acumen
in his novel, the collection of railway scams had been inflated or "puffed"
by the use of fabulous narratives and poetic figures, and the "puffers" who
manipulated the project or "jobbed" them in the stock exchange inspired
Trollope's imaginary railway from Salt Lake City to Vera Cruz and his hero,
Melmotte. Trollope's narrative rides the bubble. Melmotte's fortune rises
in the wake of the railway to Vera Cruz puff, which lifts him from the
stock exchange to Parliament. With the bubble eventually exposed, the railway stock crashes, punctured, and Melmotte commits suicide after a good
dinner—though this last part may have been melodramatic compared to
real-life models, who instead took a boat to France, where they enjoyed the
good dinner safe from complications. The depredations of foreign finance
fixated the minds of contemporaries and led to many discussions in the press
and in public debates. A parliamentary commission, the famous Select
Committee on Loans to Foreign States, was set up to inquire into the matter. It issued its much-discussed report in 1875, the same year Trollope's
book was released.

Beyond literary style, there were a few differences between Trollope and
the select committee. In Trollope's novel, Melmotte schemes a private joint
stock company. There were indeed many such projects, but the Select Committee on Loans to Foreign States decided to focus on one particularly preva-

lent type of machination that backed a railway line with a foreign government loan. There were several reasons for this financial engineering strategy, one being that foreign government debt issues, unlike companies, were subjected to laxer disclosure requirements. Joint-stock companies had to apply to the Board of Trade (which played a kind of regulatory role), name at least seven promoters, and disclose the list of their shareholders. Nothing of the like was required for issuers of government debt.

The committee discussed the machinations at length, and provided extremely detailed commentary and data on the techniques employed by the abusers. It received criticism however, for failing to take decisive action beyond the usual recommendation for enhanced transparency.[14] In place of this professed failure a scheme evolved ostensibly to protect investors and deal with the wrongdoings that were being committed in this market. The arrangement, initially known as the Council of Foreign Bondholders, was first advertised in the immediate aftermath of the crisis of 1866, then floated in a rudimentary form in the fall of 1868 by Isidor Gerstenberg, a London stockbroker and veteran at bondholder activism. The Council operated informally for a few years and was formalized in 1873 as the Corporation of Foreign Bondholders. The incorporation coincided with growing public rumors of pervasive fraudulent behavior and calls for moralization of foreign finance. Thus was the corporation born with remarkable timing, ready to redress the evil that was about to be revealed by the Select Committee on Loans to Foreign States and by Trollope.

Like any respectable Victorian institution, the corporation had an executive board known as the Council, which featured in a most senior position the very same man who had just been appointed to the helm of the Anthropological Institute: Sir John Lubbock. He would act as vice-chairman of the Council until 1889, then chairman between 1890 and 1898. This was no mere honorific title, since the corporation's archives show that Lubbock was an influential man in the organization from the first years. Indeed, Lubbock's name appears among the group of eight promoters whose signatures are found on its application for incorporation in the archives of the Board of Trade.[15]

I am not aware of any work that has emphasized this fact. In Patton's careful biography, the activities of Lubbock at the Corporation of Foreign Bondholders feature briefly in one separate and final chapter and are obviously not his subject. The chapter is called "The Sins of Saint Lubbock" because an accusation of gerrymandering and nepotism from a member of the Council created difficulties for Lubbock. In his classic *Ethnology in the Museum*, William Ryan Chapman finds Lubbock in many places and writes that this testified to his "organizational talents" that would have been "recognized from the first."[16] He was indeed into everything and, as far as science is concerned, never too deeply into anything. Patton's biography gives the impression of a scientific tourist who mixed up archaeology and Schliemann-spotting and whose division of prehistory into Paleolithic and Neolithic he drew from secondhand literature. But it remains that Lubbock was a man of multiple facets, which modern commemorations state as a fact to admire rather than as a riddle to solve.

It is not clear why previous research has ignored or brushed aside the coincidental role of John Lubbock in both finance and anthropology. If we conceive him as an anthropologist or ethnologist or paleographer, then his involvement in the Corporation of Foreign Bondholders may have arisen because he was also a banker and thus may have had a natural interest in the corporation as a financial institution. Or if we construe him as a banker—and thus likely involved in humanitarian activities—why would he not be involved in science too, another disinterested activity? Historians are so used to finding financiers in the various humanitarian and well-meaning groups active at the time that they barely pay attention anymore to such coincidences, accepted as facts of life. The role of the money interest in the Aborigines' Protection Society is a case in point, and the well-known link between the Aborigines' Protection Society and the Ethnological Society of London suggests that Lubbock's varied interests were hardly exotic. But then, one may ask, why gerrymander and be nepotistic?

Lubbock's involvement as chairman of many societies might seem all too natural given that he has been described as the ultimate "armchair anthropologist."[17] If so, he might just have been perfect for sitting as chair of the Anthropological Institute, the Corporation of Foreign Bondholders, or of any of the many other learned societies and charities to whose councils he belonged. And why not? That would be it if intersection between the parallel lines of bondholding and anthropology were limited simply to Lubbock. But it is not. The day-to-day management of the Corporation of Foreign Bondholders, which prominently included relations with the media, was in the hands of its secretary, who until the 1880s happened to be the energetic Hyde Clarke, the same man who had started the controversy about the Anthropological Society in August 1868 and was to join the council of the newly amalgamated Anthropological Institute in 1871.

When Hyde Clarke's letter accusing the Cannibals of managerial wrongdoing was released by the *Athenaeum*, he was within weeks of becoming involved by Isidor Gerstenberg as secretary of the Council of Foreign Bondholders, which as indicated above started operating in 1868, and their rela-

tion went back to January 1867 at least, when Hyde Clarke had advised
Gerstenberg on how to organize the practical side of a foreign investors
protection body. In fact, Hyde Clarke preceded Lubbock by a few years, for
it was only the formal incorporation of 1873 that brought Lubbock into the
Bondholders' Council. Yet in 1868 a tight link between Lubbock and Hyde
Clarke already existed, in science rather than finance. In the spring of 1868,
Hyde Clarke was among the cadre of members who dominated the Ethnological Society of London and was one of the three Ethnological delegates,
along with Thomas Huxley and George Balfour, involved in the abortive
amalgamation talks between Anthropologicals and Ethnologicals. In 1869
his name appeared on the cover of the society's journal, *Transactions of
the Ethnological Society of London,* as one of a select team of editors that
included the grandees Huxley, Busk, Lubbock, Lane Fox (later Pitt Rivers),
and Wright. Yet when coming within sight of such coincidences, previous
anthropological history has been looking the other way. When George W.
Stocking, the foremost pioneer of the cultural history of Victorian anthropology, discussed the *Athenaeum* controversy, he simply mentioned Hyde
Clarke as "one anthropologist" and did not even care to reveal the identity
of the individual in the main body of the text.[18] Why dig further if the intention is to focus only on those "significant" anthropologists?

  And this is by no means the last instance of anthropo-financial overlap.
As the finger goes down the list of men who sat on the Council of the Corporation of Foreign Bondholders in 1873, additional names come up. For instance, George Balfour, another of the three delegates for the Ethnological
Society in the 1868 merger talks, would also sit on the Council. Thus, two
out of the three Ethnological Society delegates who discussed the amalgamation ended up with senior positions in the Corporation of Foreign Bondholders. And along with Lubbock, we readily have three important protagonists of the ethno-anthropological controversy who were also key players in
the bondholding and investing drama.

  Pointing to overlapping council memberships is not the only way to underscore the entanglements between anthropology and foreign finance. As
mentioned, contemporaries had witnessed in 1875 the resounding collapse
of a group of foreign securities. Two of the aspiring securities that were
to collapse abjectly in 1872–73 had been originated or distributed with the
help of a prominent anthropologist or ethnologist. The Bolivian security of
1872 had been originated by one George Earl Church, who aimed at constructing a railway that would connect Bolivia and Brazil in the upper Amazon region. Susanna B. Hecht, who has devoted insightful paragraphs to Church in
her wonderful *The Scramble for the Amazon,* emphasizes Church's "tastes in

history and ethnography . . . Today Church's name survives only as a footnote in Amazonian studies," but he was in fact, she claims, a "ubiquitous imperial explorer."[19] And indeed, his library, which I have used for writing this book, is a prominent source of information on the output of nineteenth-century explorations of Latin America, ranging from the construction of railways to the writing of anthropology.

Another loan inspired by anthropology whose collapse was documented in Cracroft's *Weekly Stock and Share List* was a railway project through Honduras originally conceived by a major figure in American anthropology, Ephraim George Squier. By the time it came under the focus of the Select Committee on Loans to Foreign States, the Honduras railway had been reinvented by a group of men that featured prominently (and as we shall see infamously) one Bedford Clapperton Pim, a member of the Anthropological Society's Council who had subsequently joined the merged Council of the Anthropological Institute. When one explores the minutes of the Anthropological Society's Council discussion that led them to reject the suggested amalgamation terms in June 1868, one discovers several key resolutions that were moved by Bedford Pim, evidently an important member of the Cannibal Club.

Yet another protagonist of the amalgamation controversy was William Bollaert, who was, like Hyde Clarke, associated with both the Ethnological (of which he was a corresponding member) and the Anthropological (in whose Council he sat for several years) Societies. Bollaert, a member of the first Council of the Anthropological Society in 1863, was a specialist of Latin America Indians and a firm supporter of the view that European immigrants had brought syphilis to natives and not the other way round as some had suggested. He was also a longtime associate of Isidor Gerstenberg, the architect of the Corporation of Foreign Bondholders. They partnered in the Ecuador Land Company, a colonization company created in 1859 that Gerstenberg chaired and ran with another leader of the bondholding industry, the London broker John Field.[20]

## WHERE THE WILD THINGS ARE

The accumulation of little findings like these (rendered even more intriguing by the editor of the *Economist* Walter Bagehot's insistence that the typical foreign debt projector had "probably never heard of Darwin") prompted my rewriting the tale of anthropology's early years.[21] The Lubbock connection alone could be coincidence, but three connections is a pattern, and I have found many more than three. This book is the foster parent of those many orphan facts. Despite their unusual and unfriendly looks, they seemed to me as if they readily formed a family, or tribe, or race, as a Victorian would have said. On the one hand, not being an anthropologist at all, I should have no pretense to rewrite the history of early anthropology, which others with more legitimate titles have written well. On the other hand, why did previous historians of anthropology neglect these findings?

There has indeed been resistance to acknowledging the relevance of the facts with which this book deals. It is revealing, for instance, that after all my exploration, I could only come across but one instance of a scholarly work that has explicitly recognized and emphasized the relation between anthropology and financial promotion. It is found in an interesting, long-forgotten paper on the American anthropologist Ephraim George Squier published in 1966 by a leading scholar of the history of banana republics, Charles L. Stansifer. Both a fellow and honorary fellow of the Anthropological Society of London in the 1860s, Squier, who had launched in 1871 his own short-lived Anthropological Institute of New York, had previously devoted several years of his life during the 1850s to the development of the Honduras Inter-Oceanic Railroad project.[22] Stansifer's article might have been considered pioneering had it been followed by additional research. Indeed, Stansifer noted rightly that before him "there [had] been a good deal of mystery concerning Squier's private and promotional interest in Central America," and he added that previous students of Squier had been able "only to hint darkly as to his true interests and motives."[23] Yet although the material on which Stansifer drew is still available at Bancroft Library in Berkeley and at Tulane University, it has been ignored since Stansifer's paper. A recent biography of Squier presents his interests in Honduras as more of a hobby, although as Stansifer had emphasized, the hobby became Squier's main obsession during at least ten years of his life.

Beyond psychological resistance, there is also a technological explanation for the earlier neglect of the facts on which my narrative is based. Traditional history is constrained by the repository, the archive. But more often than not, should the relevant characters have left material, it is unlikely that this will include the sort of facts that make the stuff of financial trafficking. For instance, Lubbock's biographer Mark Patton reported privately his pessimism regarding the possibility of investigating "the sins of Saint Lubbock" from Lubbock's own diary: "Bear in mind," he told me, "that the [Lubbock] archive has been 'redacted,' I suspect twice (once by Lubbock himself and once by his second wife, Alice, after his death)— anything that might be thought to show Lubbock in a less than favourable light . . . has been removed."[24]

The advice, which proved premonitory, holds true for so many papers of the time. What I suggest calling *social editing*—the process and technologies through which society stores, displays, or destroy communication artifacts—stands in the way of a proper understanding of the relationships between objects that have been construed as disconnected from one another. The relationship between science and finance offers a case in point. In fact, as this research progressed and came to focus on a number of examples of financial projectors and prodigals who were *also* anthropologists, a chasm started widening between, on the one hand, the accumulating evidence of involvement in wrongdoing (the volume of litigation bearing witness to controversial record) and on the other hand the process of social editing as illustrated by the ways entries in the *Dictionary of National Biography* were later written, invariably omitting references to trials. In fact, for many of my heroes, checking the conventional places such as the *Dictionary* provides no trace of their more perplexing achievements. Sometimes, even checking the Admiralty or Foreign Office archival records has proven a challenge, since the material motivating sanctions was eventually removed, with a subsequent return to grace.

To make sense of the existence of so many intersections between allegedly parallel lines, it is therefore necessary to rethink the geometry of historical deposits. The task at hand looks more perhaps like that of the archaeologist than that of the historian, who can sit comfortably in front of Lubbock's papers in the hope of writing the history of the scientist-banker. Traces are scattered and mixed with one another, and figments of past traffics projected in curious ways and places, to be read as forensic science does blood splatters. Indeed, many of the gentleman-swindlers discussed in this book have not been recognized as such thus far, even when their rogue character is quite evident for whomever takes the time to observe, read, and above all *connect* sources.

The methodological point is to focus more on signifiers rather than on the signified. The important element is not whether such or such individual was a swindler or an honest man but rather who said he was so, where he is spotted, and which language surrounds his appearance. The reasons why such a manner of writing history has taken time to come are understandable. What is at stake is the possibility of writing a history of the marginal, a history that would de-provincialize truth and criminality from the outer boundaries of capitalism and would put them right at its center. Yet writing about the marginal or the interstitial is theoretically possible but very costly. One can take notes of the interstitial, and go back to the notes, and try to organize them, but if we take such notes, we shall end up noting everything, for everything is potentially both a remnant and a figment, which defeats the purpose of taking notes. And if we instead work in the traditional way and focus on the central issues, how can we capture the recurring interstice?

## THE JAWBONE OF AN ANTHROPOLOGIST

In 1849 Dr. Jeffries Wyman, who joined the Anthropological Society of London in March 1867 as the corresponding member from Boston, was required by the coroner to examine remnants of bones discovered in a Boston furnace. He was asked to determine whether they had belonged to one Dr. George Parkman, who had recently disappeared. Wyman produced a "drawing of the human skeleton, exhibiting, by means of various coloring, the parts of the body covered with flesh, the bones found in the furnace, and the absent parts not accounted for." He also took care to reassemble the right lower jaw, and the attorney general believed it matched a mold of Parkman's jaws taken at a time when he had ordered a set of mineral teeth. Parkman's jaw had a conformation "so peculiar, that, unless through some caprice of nature, their precise counterpart could not exist." The resulting inference that this had to be Parkman is still disputed, but it is universally agreed that in so doing, Wyman founded forensic science.[25]

Likewise, modern digital instruments can enable us to collect the remains of characters whose bodies have been burned in the furnace of social editing. Consider, as illustration, the following experiment devoted to Hyde Clarke's appearances in the British press. Hyde Clarke is a useful starting point because he is precisely one of those characters who did not leave proper papers in a college or university, who does not have an official biography or entry in the proper dictionaries, and who thus exists very much as a historical blind spot. And yet, as will be gradually discovered, he is a main character in the anthropological drama. The fact is that Hyde Clarke ("Esq., D.C.L., F.R.H.S., F.S.S., Vice-President of the Anthropological Institute; Honorary Member of the Anthropological Institute, New York; Corresponding Member of the Ethnographic Society of Paris; etc. etc. etc." as one contemporary source has it) is difficult to pin down, and many people saw him in many roles, jointly and severally.[26] Paul Frank Metzaros, the main scholar of the early years of the Corporation of Foreign Bondholders, stumbled on this and, losing patience, called him "the ubiquitous Hyde Clarke."[27]

Suppose we want to calculate Hyde Clarke's coordinates around the time of the *Athenaeum* controversy in the manner Dr. Wyman used for Dr. Parkman's remains. As a rudimentary starting point, we can use two newspaper databases in conjunction: the British Library's British Newspaper Archive and the Times Historical Archive.[28] We can make the request simply for appearances of the phrase "Hyde Clarke" (an exact phrase command was used to get rid of Hyde Park and the many Clarkes) and limit the search to the period between Hyde Clarke's elevation to the Council of the Ethnological Society in May 1868 and the immediate aftermath of the Council of Foreign Bondholders' launch in December 1868, when Hyde Clarke became involved through Isidor Gerstenberg. The *Athenaeum* controversy, started in August, stands in the middle of the request's time frame.

This returns a rather long list of odd results. In May, Hyde Clarke was featured as a promoter of science, taking up the idea of creating a fund to honor Charles Tilstone Beke, whose "scientific knowledge of Eastern Africa" would have permitted the success of a British campaign to free Europeans held hostage in Abyssinia (now Ethiopia) by Abyssinia's ruler, King Tewodros II.[29] Beke was a member of the Geographical Society and had played a role in the hostage crisis, which began in 1863 and eventually led to military intervention in the first half of 1868 to liberate the hostages and remove Tewodros. In June, Hyde Clarke was mentioned for his work with benevolent societies. He was spotted at a meeting of the Society for the Encouragement of the Fine Arts[30] and at the annual gathering of the Newspaper Press Fund, an important charity ostensibly supporting journalists and their widows and children and of which he would later become treasurer.[31] That same month, he was also reported as having launched an Archaeological Society with other scholars.[32]

By August Hyde Clarke was giving policy advice on the subject of consular courts in the Ottoman Empire.[33] The ancient system of capitulations, whereby disputes involving an Englishman or someone from a territory under British authority had to be decided by consular courts, was coming under the strain of an expanding commerce and an expanding empire. Joining a high-profile controversy that had begun in Parliament for the abolition of capitulations, Hyde Clarke wrote to the *Times* stating his opinion on consular courts and claimed that, based on his own knowledge and experience of the region, Turkish and Egyptian government would be "greatly improved by the enforcement of the real responsibility of their officials" should the capitulations be abolished "to-morrow." He concluded that "the 'Capitulations' are very convenient for jobbery, but not essential for the safety or comfort of residents." This letter was sent literally a few days before he sent his famous first letter to the *Athenaeum* about the Anthropological Society of London, which was to lead him to proffer again the accusation of "jobbery," this time against Dr. Hunt and the Anthropological Society's clique.

August proved to be a busy month for Hyde Clarke, who featured prominently in the program of the Norfolk meetings of the British Association for the Advancement of Science with no fewer than three papers in three sections. There was one paper "On The Connection of Historic and Prehistoric Ages in Western Asia" (Geography and Ethnology section),[34] another "On the Progress of Turkey" (Economic Science and Statistics section), and a "Note on the Western Asia-Minor Coal and Iron Basin" (Geology section).[35] In the same month, he was also spotted with Gerstenberg, his future associate at the Corporation of Foreign Bondholders, which was still in the planning stages. The occasion was a report of his efforts to amalgamate two different committees of Venezuelan bondholders, almost at the same time when he was involved in the failed amalgamation of Anthropologicals and Ethnologicals.[36]

In early September, as the *Athenaeum* controversy escalated and a meeting was taking place at the Anthropological Society to consider Hyde Clarke's exclusion, he was occupied with a railway company, the Ottoman Smyrna Aidin Imperial Railway, publishing a denial of his role as leader of a shareholder rebellion. One month later, he was seen at a shareholder meeting of the said company among the said rebels.[37] November came with an announcement of his lecture at the London Institution in Finsbury Circus on the "Examination of Phenomena Common to the History of All Nations." The institution, listed in later guides as a learned society, featured conferences and advertised Hyde Clarke as "correspondent of the Royal Society of Northern Antiquaries of Copenhagen and of the German Oriental Society and the American Oriental Society."[38] Finally, in December, at about the same date when Gerstenberg launched his Council of Foreign Bondholders, Hyde Clarke floated the idea of a Shareholders Protection Association whose stated purpose would have been similar to that of Gerstenberg's bondholders association, except that the shareholders association would have focused on disciplining British company directors rather than foreign governments. Unlike Gerstenberg's project, however, the Shareholders Protection Association never came to be. Not coincidentally, the chairman of the shareholders association was one Henry Brookes, a member of the Anthropological Society who joined forces with Hyde Clarke in the *Athenaeum* controversy.[39] In a digital age, even the past has lost its privacy.

Such are the bone debris from the British press furnace where the remnants of Hyde Clarke are discovered. The exercise helps to reconstruct the location of Hyde Clarke in Victorian society. If we tag each tiny bone with an indication of the layer where it was discovered (be it that of the austere *Times* and its penchant for pontificating and thundering; the *Morning Post*, which rivaled the *Times*; or the *London Standard* and its taste for covering business and foreign issues), and if we factor in the mercurial willingness of their editors to focus on a subject submitted to them, then we ought to be able to start reconstructing the contours, not of the true Hyde Clarke, who probably does not exist, but of the positional Hyde Clarke, the one who matters here. Thus Hyde Clarke is displayed in a number of ways that reflect different information strategies—how he wished to perform in certain media and how others used the media to project an image of him. These can be mapped to a geography of the places where performances were played, for neither the choice of the *Times* to enter the dispute on capitulations nor that of the *Athenaeum* to enter a dispute on a learned society was innocuous.

Another important takeaway from the previous exercise is that it does construct Hyde Clarke's "body" from an apparently heteroclite assortment of items. At first sight, the result may be reminiscent of Borges's Chinese bestiary, which included "Those that belong to the emperor," "Mermaids (or Sirens)," "Fabulous ones," "Et cetera"—not forgetting "Those that, at a distance, resemble flies." But upon closer scrutiny and longer meditation, it is possible to discern a form of consistency in the shape of a space defined by four axes that span charity, science, business, and policy making. The conclusion that such axes define Hyde Clarke is valuable in itself.

Further attention reveals a common narrative in these various features. This includes on the one hand an obsession with good governance as inspired by morality, knowledge, and information. On the other hand, disclosure of information always takes the form of a public performance. Information, exposure, and the cleansing of knavery through public scrutiny and open access as opposed to decisions made by insiders behind closed doors are also features discernible in the *Athenaeum* controversy. If this conclusion is accepted, then the appointment of Hyde Clarke at the Council of Foreign Bondholders may have had a close link with his participation in the *Athenaeum* controversy, permitting, at the very heart of a history of science and culture, the introduction of the materiality of institutions and the ways they tie and bind. Such is my hypothesis, and we are going to use it to help fill in the interstices left by the silent archive.

I think of the social fingerprints approach, which I have just described, as part of the modern investigative ways of history permitted by the implacable serendipity of the Internet search. The technologies that enable these searches are bound to produce transformations because for the first time in the recorded history of the social sciences, the transaction costs of working across fields are reduced. This is a revolution because it allows us to overcome the tendency of individual social sciences and fields and subfields to be self-centered. Writing the history of anthropology—or any other social science—no longer requires first and foremost a familiarity with the field; quite the contrary (although I eventually ended up reading an enormous number of the articles published by the *Anthropological Review* between 1863 and 1871). Locked into the naive belief that anthropology is what anthropologists do, the old approach took language and metaphor a bit too literally, and constructed (or deconstructed) a history of the birth of anthropology from images borrowed and transferred and from the subsequent myths on which the discipline is based. By contrast, as we shall see, the Internet search opens new ways to combine individuals and their deeds and it orients the exploration of archives in new directions. It outlines the importance of the media and its content and the equal importance of how that content was produced. And it brings to the fore the question of value: Material posted and displayed is never randomly posted and displayed.

And thus it is that with the tools available today, we do not need to feel compelled to reproduce the divisions across disciplines as they have been established. In fact, the very search tools that permit us to profile where Hyde Clarke is found and where he is not also give us a quick grasp of how previous research has constructed its own modern silo where it ended up trapped. To take an example, the literature on the history of early anthropology can be itself profiled in a few irreverent clicks rather than after long, silent, and respectful visits to the library—if one was lucky to have a library rich enough to contain the relevant material. A question naturally arises as to what discipline owns this way of doing research. My answer is that, since the material thus collected does not belong to one discipline (as the profiling of the ubiquitous Hyde Clarke suggested), it seems natural to host such reflections within history. History and historical research have been traditionally a generous host for isolated evidence, and the overflow of interstitial material can always be parked there. That is why I am calling this tale that was initially inspired by economic insight—this tale of crime, finance, science, politics, empire, and anthropology—a history, for history is all interstice.

This history book will make it clear that interactions between anthropology and the foreign debt machinery and machinations of the 1860s and early 1870s have nothing to do with coincidence. Just as anthropology and anthropologists are forgotten facets of the Great Foreign Debt Mania, finance and white-collar criminality are an untold part of the Great Anthropological

Epic. Such an outsider perspective obviates the need for mind-blowing dissertations on culture and things Victorian or for doing such things as determining the genuine nature of early anthropology or for trying to disentangle the modern legacy of past anthropological research currents. It also obviates the need for going after the final allegiance of such-and-such anthropologist, either to anthropology or to any of the sister disciplines and societies to which early anthropologists tended to belong simultaneously. Was Lubbock an anthropologist, an ethnologist, an entomologist, an antiquary, or a banker? Did he derive his insights from biology, and what was the influence of his collaboration with Darwin? We do not know, and to an extent, we should never have cared—although of course the findings in this book will inform such secondary questions.

CHAPTER TWO

# Rise of the Cannibals

George W. Stocking's centenary narrative of the making of the Anthropological Institute of Great Britain and Ireland ("What's in a Name? The Origins of the Royal Anthropological Institute, 1837–1871," 1971) inaugurated a Manichean tradition kept alive by the more recent works of Ter Ellingson (*Myth of the Noble Savage*, 2001), Adrian Desmond and James Moore (*Darwin's Sacred Cause*, 2009), and Efram Sera-Shriar (*Making of British Anthropology*, 2013). "What's in a Name?" chose to portray Hunt and his Cannibals as the villains and pitted them against the fair, the gentle, the Darwinian ethnologists of the X-Club. The dispute, Stocking suggested, principally consisted of a conflict of ideas, opposing so-called polygenist anthropology (the Cannibals) and monogenist ethnology (the X-Club). Polygenist anthropologists sought support and legitimation for racist arguments in a theory of the different origins of man while monogenist ethnologists who were anti-slavery activists in the "man and brother" tradition claimed that humanity came from but one thread. As the story got repeated, it also got simplified and morphed into the epic tale of a single combat between the two societies' respective champions—James Hunt for the Anthropologicals and Thomas Huxley for the Ethnologicals, the Cain and Abel of the science of man.[1]

In the same vein, it has been suggested that the twin vices of racism and sexism epitomized the Anthropological Society and circumscribed its intellectual perimeter. Indeed, racism and sexism were in full display in the minutes of the Anthropological Society of London's lectures and meetings. The fact that there was a dispute with the Ethnological lends itself naturally to this interpretation since the Ethnological Society dated to 1842 and is usually described as the scientific offshoot of the Quaker-dominated Aborigines' Protection Society, a humanitarian organization devoted to the protection of indigenous peoples.[2] Against the Ethnologicals' goodwill, we have the foundational text of Anthropological racism in Hunt's infamous *Negro's Place in Nature* (1863), where he legitimized slavery on the basis of anthropological arguments and concluded that "the Negro is more humanized when in his natural subordination to the European, than under any other circumstance."[3]

Sexism found its expression in Hunt's concern to exclude the "fair sex" from anthropological meetings when pure business logic, he claimed, should have encouraged the inclusion of females. They would have swelled the subscriber list and the receipts of the association, but anthropologists insisted they had higher standards than those of a shopkeeper. Hunt scolded Section E (the Geography/Ethnology section), which hosted ethnologists' papers at the annual meetings of the British Association for the Advancement of Science, as the "ladies' section."[4] Given the involvement of women in organizations such as the Aborigines' Protection Society, which anthropologists sought to provoke and antagonize, the gender conflict overlapped with that of race.

This came under the crudest light when Hunt presented his ideas on the "Negro's Place in Nature" as a lecture entitled "On the Mental and Physical Character of the Negro" at the annual meetings of the British Association for the Advancement of Science in 1863. There, in front of a large audience, he was met by William Craft, a runaway slave who had escaped from the United States following the Fugitive Slave Act of 1850 and sought refuge that same year with his wife, Ellen, in London, where they received support from abolitionist campaigners. According to one British newspaper narrating the conference, the "coloured philanthropist" Craft showed up in the fashion of "Banquo's Ghost in Macbeth's Banquet." He then undertook to ridicule Hunt's academic constructions with a few jests. Hunt's speech was greeted by "mingled cheers and hisses" (*London Standard*) or just by hisses (*Caledonian Mercury*).[5] The hisses of course, came from ethnologists and "the ladies." According to anthropologists, this was nothing to be surprised about. They emphasized that their more determined scientific pursuits were "not likely to please the ladies who attended [this] section in such large numbers." And they could find solace and vindication in Paul Broca, the secretary of the Société d'Anthropologie de Paris, who used observations by German anthropologist Rudolf Wagner to show that, regardless of the age, a female brain weighed between 7 and 12 percent less than its male counterpart.[6]

Such is the popular narrative. This chapter will argue that by focusing on the all-too-obvious dimensions of race and gender, previous writers pioneered by Stocking have locked themselves into a territory that is a fraction of that occupied by anthropology. They have mistakenly received at face value the Anthropologicals' own discourse that they were a minority of extremists. This discourse had been articulated by the Anthropologicals themselves in order to emphasize that they were scientific men fighting the domineering humanitarian majority and their army of ethnological Amazons. By focusing on what the Cannibals and their opponents were ostensibly showing them, modern scholars have neglected more important issues having to do with the deeper significance of the dispute between anthropologists and ethnologists. It was a social dispute.

To show this, the chapter will review and criticize key aspects of the conventional narratives of the Anthropological Society: the exclusive emphasis on Hunt, the suggestion that the rise of the Anthropological Society's intellectual identity was shaped by the U.S. Civil War, the simplistic dichotomy between humanitarian ethnologists and racist anthropologists. We will see, as we proceed, that the conventional intellectual distinctions historians have established between the Ethnological and Anthropological Society do not withstand careful discussion. Instead of the Manichean narrative, another significant contrast will emerge, actually anticipated in D. K. Van Keuren's unpublished dissertation: the respective populations of the Anthropological and Ethnological Societies were very different from one another. The first marker of the dispute between the Anthropological and Ethnological societies we identify is that their different "truths" can be mapped into different social statuses. In other words, before an intellectual history of anthropology and ethnology is attempted, a sociological history is needed. This chapter lays its foundations.

## BURTON

Because of the *Negro's Place in Nature* pamphlet and the Craft episode, previous historians of the Anthropological Society have focused on James Hunt. Yet, Hunt's histrionics might have been a smoke screen behind which Burton's hand can be identified. Burton's *Mission to Gelele* (1864) gives clues as to the context of Hunt's essay. It had a chapter, dedicated to Hunt and called "Of 'The Negro's Place in Nature,'" where Burton engaged with the outrage raised by Hunt's eponymous presentation and complimented him with a recollection of Hunt's doorplates ("Esq., Ph.D., F.S.A., etc. etc. etc."). Hunt, a martyr of science, could be proud to have been greeted by "those encouraging sounds, which suggested a mob of Eve's tempers rather than a scientific assembly of her descendants." The compliment was followed by Burton's own views on the subject, which he stated having organized

while staying in Agbome (Abomey, in today's Benin) long before he had seen
Hunt's pamphlet. The parallel with *Negro's Place in Nature* is striking.
Reading Burton's chapter, it is hard to suppress the thought that in one way
or another Hunt's pamphlet had been inspired and perhaps commissioned by
Burton himself.[7]

This suggests that the focus on Hunt has been exaggerated, for Hunt
stuttered after Burton. The limitations of the conventional narrative are best
seen in the manner in which the curious case of Richard F. Burton has been
dealt with—or, rather, ignored—by previous authors, again going back to
Stocking. In "What's in a Name?" Stocking made of Burton a mere masthead, the "shining ornament" of the Anthropological Society as he put it.[8]
In his seminal book *Victorian Anthropology*, he dispatched Burton in two
lines as "a latent homosexual who maintained a virtually asexual relationship with his morally rigid Catholic wife." He then left the matter at that for
subsequent writers to devoutly pilfer.[9] This most curious characterization of
the Victorian explorer appears to have been lifted from an even more curious reading of Fawn McKay Brodie's conclusion in her biography of Burton,
where she discusses Burton's sexuality at length in an attempt to use psycho-
analytic arguments to explain Burton's personality.[10]

In fact, in writing about Burton's allegedly repressed sexuality, Stocking was not so much following Brodie's suggestions but rather restating a
libel that had accompanied Burton during his entire life and dated from an
episode during his time with the East India Company. Working under Major
General Charles James Napier in 1845 as linguist and informant, Burton had
been commissioned to write a confidential report on homosexual brothels
in Karachi. Operating in disguise, he managed to "visit all the *porneia* and
obtained the fullest details, which were duly dispatched to the Government
House." His ethnographic interest and taste for precision led him, according to his own later testimony (for the original report is lost), to provide
such things as insights on the price of different types of prostitutes and the
reasons for such differences. Boys, he found, were more valued than eunuchs,
and he conjectured that this might have to do with the "scrotum of the un-
mutilated boy . . . used as a kind of bridle for directing the movements of the
animal."[11] The report would later be used to spread rumors about Burton and
their brimstone smell would follow him everywhere. Who else but a pervert
could write on such topics? And if found later to marry a Catholic wife, then
Burton was ready for some bargain Freudian sentencing.[12]

When leafing through the pages of the publications of the Anthropological Society, one finds Burton's influence everywhere: in the obsession
with Africa, the focus of many of his explorations; in the emphasis on the

command of language, a precondition for the participant observation that Burton advocated; in recurrent papers or discussions on phallic worship and other subjects belonging today to what is known as the "anthropology of the body," of which Burton was a pioneer (as Fawn M. Brodie actually emphasized, but Stocking chose to ignore); in the anti-clerical tone and the taste for pagan rituals (the articulated skeleton in the society's window bearing testimony among many others); in the penchant for exaggeration and verbal provocation; and in the fetishism that provided the impetus for the collection of artifacts that a number of Burton's connections in Africa and elsewhere sent to the Anthropological Society's budding museum. While many other influences were at work in the evolution of the Anthropological Society, and while this book is not at all about attempting to vindicate Burton's importance in the history of the society, it remains that Burton's excision from conventional narratives is troubling.[13]

The extent to which the Anthropological Society could attract visible supporters owed a lot to the social prominence of the British explorer. It was Burton and not Hunt, only an aspiring man in his early thirties, who had the required standing to attract to the Anthropological Council men such as the English-Australian politician Sir Charles Nicholson (1863–64) and the explorer-politician William Wentworth Fitzwilliam (Viscount Milton, Council member 1865–68). Likewise, a number of members of the Council such as W. Winwood Reade or Colonel Lane Fox were personal friends of Burton. The relevance of the masthead for the society's successful canvassing effort cannot be underestimated, and indeed it must have been Burton who managed to sign in as permanent fellow Edward Stanley, the 15th Earl of Derby and the future foreign secretary in the Conservative government that came to power in 1866.

Once Burton is brought into the picture, things become incommensurably trickier, for it is much less innocuous to disprove a patented Victorian hero, especially when this hero, a linguist and undoubtedly a broker, could change sides and occasionally vituperate against British imperialism and the arrogance of the colonizer in language that would have scandalized the well-wishers who attended the philantropic lectures at Exeter Hall almost as surely as Burton's blasphemies in favor of slavery. "Whenever Madam Britannia is about to break the eighth commandment," he wrote in *Scinde, or the Unhappy Valley*, "she simultaneously displays a lot of piety, much rhapsodizing about the bright dawn of Christianity, the finger of 'Providence,' the spread of civilization, and the infinite benefits conferred upon barbarians by permitting them to become her subjects and pay their rents to her." The Eighth Commandment states: "You shall not bear false witness against your neighbor."[14]

Burton enjoyed fighting and this also imparted one of its distinctive traits to the Anthropological Society. Consider for instance Burton's involvement in a controversial discussion of the annual executions by the king of Dahomey. The episode is significant because it took place around the time when Burton launched the Anthropological Society. The Admiralty, reacting to an 1863 report by Commodore Arthur Eardley Wilmot, who had been on a mission in Dahomey to inspect the slave trade, had let its displeasure towards the manslaughter be known. But Burton, in his *Mission to Gelele* published in 1864, just one year after a parliamentary paper had transcribed the "Madam Britannia" stern view on the subject, explained that the executed individuals were either criminals or war captives and by way of a comparison with violence in Britain suggested that the Admiralty was meddling with the internal affairs of the Kingdom of Dahomey. He noted that the ritual also served a function. It enabled the king to send a message to the land of the dead. He then added, "We can hardly find fault with putting criminals to death, when in the Year of Grace 1864 we hung four murderers upon the same gibbet before 100,000 gaping souls at Liverpool, when we strung five pirates in front of Newgate . . . A Dahoman visiting England but a few years ago would have witnessed customs almost quite as curious as those which raise our bile now . . . The executions are, I believe, performed without cruelty; these negroes have not invented breaking on the wheel or tearing to pieces their victims, as happened to Ravaillac and the half-witted Damiens."[15]

This was not to please the righteous *Athenaeum*, which concluded a review of Burton's book where it made ironic comments on Burton's fascination for Gelele's army of Amazons by asserting that "there is no excuse for murdering captives taken in war . . . There is no excuse for killing a man every time the living prince desires to send a message to his sire in Deadland. Yet Capt. Burton failed to make any impression on either Gelele or his ministers, as regards the necessity of abolishing this custom."[16] Such evidence as this renders somewhat hopeless previous attempts at pinning down Anthropologicals to the board of racism as some kind of exotic butterfly. A flap of wing, and they're gone.

## BANKROLLED BY THE BANKRUPT?

As said, previous authors have unanimously loved to hate the Anthropological Society. This is best illustrated by the recent success of a theory, popularized by historians of anthropology Adrian Desmond and James Moore in their *Darwin's Sacred Cause*, that narrates "How a Hatred of

Slavery Shaped Darwin's Views on Human Evolution." In their book, they argue that Charles Darwin's *Descent of Man, and Selection in Relation to Sex*, published in 1872, was profoundly shaped by contemporary disputes on race. The chapter "Cannibals and the Confederacy in London" provides the usual depiction of the fight between Darwinian-inspired ethnology and devil-inspired anthropology, but with a twist: according to Desmond and Moore, Darwin's theory of the origins of man was in large part motivated by the need to contradict the pro-slavery ideas propagated by the Anthropological Society. The publication of Darwin's *Descent of Man* in 1872 and the taming of the Anthropological Society in 1871 through its merger with ethnology were the two sides of the same coin.[17]

According to Desmond and Moore, the birth of the Anthropological Society owed everything to the U.S. Civil War, which inspired in a few "hard-headed dissidents" to divorce themselves from the Ethnological Society. Moreover, Desmond and Moore argue, if such minority ideas came to being on British soil, it had to be a conspiracy. "It is no coincidence," they write, "that a new society to study mankind was formed during the Civil War, *and even less that the Confederacy had its paid agents inside.*" The smoking gun uncovered by Desmond and Moore is the Anthropological Council, the society's executive board, which they say included permanently one Henry Hotze, a Swiss-Alabamian who was the Confederate States' propagandist for Europe. Hotze, Desmond and Moore emphasize, had been recruited by the Confederate secretary of state and was bankrolled by the government in Richmond, which would have thus channeled money to the Anthropological Society through Hotze. The Confederate defeat in 1865 would have heralded the demise of the Anthropological Society, a claim that has been reproduced subsequently. One suspects that part of the success of this view owes to the fact that it makes the Anthropological Society's racism something that originated abroad. Just like official Anthropological Institute history today boasts its resilience in the face of interwar racism (despite anthropologist George Pitt Rivers's endorsement of Mosley and Hitler), Victorian anthropology might have been infected by Confederate ideas, but briefly and only through money.[18]

On the intrinsic merit of the hypothesis, we may remark that (1) historian Robert E. Bonner has analyzed the intellectual trajectory of Hotze, emphasizing his connection with Mobile, Alabama, anthropologist Josiah Clark Nott, who had hired Hotze in the 1850s to translate "scientific racist," Arthur de Gobineau's *An Essay on the Inequality of the Human Races* (1853–55); (2) Hotze was not permanently on the Council of the Anthropological Society; (3) there was no more Confederacy to bribe the Anthropological

Society after 1865 although six full years lapsed between the end of the Civil War and the end of the Anthropological Society; (4) during this phase, the society experienced several shocks, including the controversy started by Hyde Clarke in 1868 and the death of its leader James Hunt in 1869; and (5) if one goes by the identifiable contributions of Hotze, the amounts are modest.[19]

In fact, even the fascinating material disclosed by J. F. Jameson in 1930 (the source still used by a number of people writing on the Anthropological Society) portrays Hotze as an agent with limited means compared to his Union counterparts, and it shows him reflecting on the art of efficiently bribing British journalists (he talks about the difficulty of "reaching the rich and strong blood"). As Hotze discovered, a more satisfactory way to proceed than outright bribery was to establish some form of reputation for his main organ, a newspaper called the *Index*. In a second stage, working with persons, Hotze hoped to establish himself as a source of understanding of the politics of the Civil War. This evidently does not fully square with the bankrolling view.

If such was Hotze's plan, then we can see the benefits he must have found in association with the Anthropological Society, since it would give his propaganda the endorsement of scientific men who would in large part promote his ideas *for free*. Such was Hotze's stroke of genius if there was one—the Cannibals would do the job for love. Reviewing his successes with the concern to show his contribution in the most favorable light, Hotze stated with evident exaggeration that he had secured the "almost undivided ear of three fourth of the newspaper reading public" in England, and that the Civil War being so "thoroughly understood" was owing to his being "one among many moral and intellectual agencies which jointly contributed to this result." There is no disputing that Hotze portrays himself in a flattering light as the conductor of such agencies, but I do not see that he boasted having *created* them.[20]

All this proves is that Hotze saw the benefits of infiltrating learned societies because he understood something that this book will emphasize: that, for a number of economic, political and sociological reasons, learned sociability was a powerful instrument of propaganda in an elitist society. In the attendance of scientific gatherings, the blue, the rich, and the strong bloods were found. For such an infiltration to have happened, therefore, there must have existed something to infiltrate. Moreover, even if we admit that the Civil War was a catalyst that would have helped the Anthropological Society into being, which is doubtful, it is quite misleading to label the Anthropological Society's venom as foreign in origin. As we saw when discussing Burton and his friends, one must reckon with the fact that the poison was homegrown. After all, it is well known that important interests in Britain supported their country's entry in the Civil War. I have never heard any historian claiming that such interests were merely summoned through bribes.

A more accurate perspective therefore would emphasize that Hotze had observed that a man holding reactionary views on races and expressing them publicly in 1863 could launch a very successful organization. Therefore, science could be one instrument in a set of propaganda tools. On the other hand, this says something about the lack of discrimination (according to modern standards) that characterized an early society such as the Anthropological, bringing us back to the already mentioned question of the "acceptable" or "true" anthropologist. It is easy to figure what had happened: in their efforts to create an organization from scratch, anthropologists were not too picky and all good will was signed in.

Illustrating this, the first set of vice-presidents included for a while one Honoré Pi de Cosprons, the self-styled "Duke of Roussillon" who was admitted on the strength of a philological essay, *Mémoire sur l'origine scythocimmérienne de la langue romane*, published in 1863. Cosprons had been quite effective in penetrating circles of learned sociability, although according to one source, his usurpation of a noble name was eventually exposed by the Royal Society of Literature, the Anthropological Society's neighbor and landlord on St. Martin's Place.[21] Just like Cosprons sought to capture the benefits of social certification from learned societies, Hotze sought to secure affordable propaganda for the cash-strapped Confederacy. One should not read too much in Hotze's membership therefore, and the simplistic story of a *foreign* infiltration of British ethnology through the Trojan horse of anthropology, successfully resisted by the Darwinian antidote, is a good tale for TV shows. In summary, it does not do justice to the racism of the Anthropological Society to make it a narrative of bankrolling, for it had obviously more embarrassing origins than the few pennies received from the soon-to-be defeated and already bankrupt Confederate States of America.

## HUMANITARIAN BLUES

The historical context in which the Anthropological Society was born indeed points towards a resurgent and assertive Victorian racism, commonly described as a backlash from earlier humanitarian attitudes. Burton himself described in this fashion the pendulum of ideas: "1. The popular opinion touching the negro in the pre-Abolitionist times; 2. The general sentiments during that period of violent reaction; 3.The present state of the public mind when it is gradually settling into a middle and rational course." Burton

expressed fear that the "middle and rational" course might not hold, and
the exaggeration of the "unnatural 'man and brother'" could relapse into the
"nigger," "savage," and "semi-gorilla."[22]

This is Burton's own language, but historians have also observed a turning point in attitudes during the first half of the 1860s. According to Freda
Harcourt, this resulted from the "demise of the humanitarianism of earlier
decades" and found its political expression in Prime Minister Disraeli's first
significant tenure in 1866–68, as chancellor of the exchequer then prime minister, when important ideas about the contours of empire were discussed.[23] The
reasons for the backlash have been diversely assessed but usually relegated to
those corners that are less accessible to empirical exploration—mysterious
societal changes, the lack of wisdom of crowds, and a looming domestic crisis,
which would have found a political exit in assertive external expansionism. In
a similar fashion, but this time more narrowly focused on the subject of ethnology and anthropology, scholars have conventionally described the 1850s
as a period of crisis in the science of man. According to anthropologist Ter
Ellingson, in the mid-1850s "ethnology was balanced on the threshold between institutional extinction and transformation with the only alternative
to its death being a radical change of direction to move along the swelling
currents of new developments."[24] And this, he contends, would have partly
motivated the new direction taken under the hateful lead of anthropology.

For our purpose, however, suffice to say that at some point in the late
1850s, racism would have become more strident, and this had perhaps to do
with the Sepoy Mutiny of 1857, the need to rationalize the brutal repression
that followed, and the sexual realignment the mutiny made possible. Early
signs included Scottish writer Thomas Carlyle's "Occasional Discourse on
the Negro Question" (1849), which suggested behind the veil of a trans-
parent anonymity the reintroduction of slavery. African-American history
scholar Thomas C. Holt has described this backlash against earlier humanitarian tendencies and suggested that it may have had to do with feelings
that developed among former slave owners. The more recent work by historian of abolition Nicholas Drapper would encourage us to think of a divide
between large and small slave owners, with the largest ones such as John
Lubbock's father having done a splendid job at capturing the manna the British state distributed for indemnification of slave ownership, while others
accumulated resentment.[25] In the works of specialists in the links between
gender and slavery, we also find indications that women had previously
identified with the slaves and played major roles in abolitionist societies.[26]
Perhaps the Sepoy Mutiny and the atrocities then imagined or committed
provided a context for the aggrandized fantasy of a sexual threat from the

freed dark man against which white males would themselves be aggrandized, now measuring up with predators in the jungle, now raising up as saviors and lords.

Or perhaps the reconfiguring of the political map of Europe, once Britain had sided with the slave-cultivating Ottomans in the Crimean War, required a political adjustment of Britain's official anti-slavery stance. The spread of industrialization and the centrality of cotton in the resulting surge in global trade belonged to the same problem. Britain was the center of the cotton economy, but the bulk of the raw material was secured from the slave system in the southern United States and financed by bills drawn on London, which had to go through the London money market. Credit and commercial links between the City and Charleston were tighter than those between Charleston and New York. The ideologies of the 1850s and 1860s must be replaced in their materialist context. Ever since Eric Williams published his *Capitalism and Slavery*, historians have recognized the possibility of ideological opportunism, and some have accepted the idea that culture might not be the relevant starting point. The expansion of the world economy was bound to throw the supplier and the intermediary into each other's arms. The increasing tension between North and South led the latter to seek a natural alliance with its source of capital while in Britain, investors and speculators welcomed Southern aspirations for political autonomy because they opened the door to profitable combinations.

Last and not least, in a world that was erecting knowledge as a fundamental source of efficiency, the experience of the United States with slavery could be reworked as an object of study. For many contemporary observers it was perceived as a valuable repository of facts from which to draw lessons: economic, political, and anthropological. According to someone like Burton, who valued the fieldwork above everything else, nothing could surpass the firsthand experience white Americans had of the "races" they dominated. As he argued, "the American, totally unlike the Englishman, understands the Negro before leaving his own country; [the American] is a practical man, not a theoretical philanthropist."[27] In this context, despite what we might naturally be inclined to hope, anthropology would operate on the premise that science would lead to more, not less racism—or rather, to "better racism." Once started, the pursuit of efficiency knows no boundary.

## THE MORANT BAY MASSACRE

In October 1865, political unrest among blacks in Jamaica stemming from accumulated abuse whose latest twist was legal chicanery boiled up into a

riot in the small town of Morant Bay. Edward John Eyre, the governor of Jamaica, quelled the riot with ruthless brutality. With the help of maroons—
runaway slaves used as military auxiliaries—his troops killed five hundred
Jamaicans, flogged another six hundred, and executed George William
Gordon, a "mulatto" Jamaican landowner and politician whom Eyre had
considered a source of political opposition. Upon Eyre's instructions, Gordon had been summarily judged by martial law and hanged. This was a patent violation of Gordon's rights to a civil trial and became a *cause célèbre*.
A number of prominent Liberals—among them Thomas Huxley of the Ethnological Society, John Stuart Mill, Charles Darwin, and Herbert Spencer—
established a Jamaica Committee in 1866 and called for Eyre to be tried in
London.[28]

But as the Ethnological Society thus came down stridently against Eyre,
Hunt declared publicly, "We anthropologists have looked on, with intense
admiration, at the conduct of Governor Eyre as that of a man of whom England ought to be (and some day will be) justly proud."[29] Not incidentally,
Eyre was a member of the Anthropological Society. The Anthropological
Society commissioned a study on the causes of the rebellion. The Cannibals
reached a paroxysm of excitement when, on February 1, 1866, the Society
chartered St. James's Hall for Commander Bedford Pim, a discharged Navy
officer, to read a paper on "The Negro and Jamaica." The paper vehemently
defended Eyre against the Jamaica Committee. The event was widely pub-
licized in advance. Applications for tickets were received, and after Pim's
paper had been read and discussed, it received all the honors of the society.[30]
A summary of the meeting was published as a special, stand-alone issue of the
society's new *Popular Magazine of Anthropology*, and the full text of the lecture was printed with publicity for the *Anthropological Review* on the back.[31]
The resulting pamphlet, entitled *The Negro and Jamaica*, was mailed all
over the world, and the online Google version one can inspect, which comes
from Harvard Library, bears a mention of it having been "presented by the
Anthropological Society."

Commander Bedford Pim, one of the important characters in this tale,
certainly belonged to the section of British population that was sensitive to
the Southern states. He was a close friend of the Reverend F. W. Tremlett,
the vicar of St. Peter's in Belsize Park in London's up-and-coming Hampstead district, where the financier, the nouveau riche, and the Trollope character settled. Tremlett was also the leader of the pro-Confederacy Society
for Promoting the Cessation of Hostilities in America. Pim was a member
of that organization and was secretary of the committee formed in London

in 1864 to present a sword to Raphael Semmes, captain of the commerce raider *CSS Alabama* in replacement of the one he lost when the *Alabama* was sunk off Cherbourg by the *USS Kearsarge* earlier that year.[32]

Pim's article encapsulated the pet themes of the Anthropological Society, jargon, and scientific gymnastics. In an ostensible attempt to design a controlled experiment, it looked at the "Negro" first in the "land of his birth" and then "in those parts to which he has been transplanted." It asserted that the rigorous study of dead bodies and the like proved beyond the shadow of reasonable doubt that, although "Negroes" had long been in contact with "advanced peoples," they had failed to take advantage of this opportunity. "In short, after centuries of contact with the Egyptians, the Persians, the Greeks, the Carthaginians, the Arabians, the original characteristics stick to the Negro; he has gained no permanent good; and I commend this fact to the negrophilists."[33] After long digressions that contained a catalogue of commonplace fantasies—including the mandatory stop at sexually charged descriptions of torture and cannibalism—Pim wrapped the argument by stating that he could not see "any hardship to the negro in deferring the claims of the negrophilists for equality on the part of their idol, until he has done what every man amongst us is obliged to do, viz. *prove his title* before he is admitted into fellowship."[34] Pim concluded that he "earnestly hoped that if ever he should find himself in a similar position to that of Governor Eyre, he might have the courage, determination and self-reliance to act precisely as his Excellency had done." As many previous accounts have reported, the meeting concluded with "three cheers . . . loudly called for" and "responded to . . . most enthusiastically."[35]

## ALL CANNIBALS

Previous studies of the Anthropological Society usually stop here, backing away from the horror of Pim's putrid statements and even more so from the enthusiasm of their reception, although anyone who is familiar with the minutes of contemporary shareholder meetings knows what to make of those "Hear Hear" and "three loud cheers" that parse such reports. Yet taking the cheers at face value provided conventional narratives of the eventual merger of anthropology and ethnology in 1871 with a means to give a cathartic quality to the emergence of the Anthropological Institute. Gone were the days of degenerate anthropology. The merger would purify the horrid word and the horrid ways. It would enable the coming of the age of a "decent" ethno-anthropology, one that would have John Lubbock in the armchair.

And the truth is that there was indeed much to purify, much more than conventional accounts are prepared to recognize and much more than the coming of age of the institute could ever achieve. In fact, the Anthropological Society's extreme positions were much less extreme than its leaders succeeded in having historians believe. To take but one example, the Anthropological Society's defense of Eyre was echoed by the Geographical Society's all-powerful president, Sir Roderick Murchison, an adversary of the Anthropological Society in many other battlefields, and as such often portrayed as being on the side of angels. Yet Sir Roderick had no qualms in defending "Governor Eyre, an old medalist of the Society, from what [Murchison] regarded as persecution."[36] Murchison's position was not isolated. To confront Eyre's opponents, a supporting committee was formed that included such important men as Thomas Carlyle and Charles Dickens.

Recently, anthropologist Efram Sera-Shriar, while acknowledging the conventional view of a race-based dispute, has nonetheless emphasized the possibility of a narrative that would stress similarities. According to him, both Hunt and Huxley would have championed the empirical study of man as opposed to the utilitarian philosophy of previous thinkers in the Benthamite tradition. This tradition had imagined man, his desires, inclinations, motivations, incentives, but Hunt and Huxley both understood the merit of the fieldwork. Their societies pioneered modern techniques that included the collection of artifacts, evidenced-based discussion, advocacy of the fieldwork before the term was invented, and the effort to make anthropology an autonomous field of investigation. Commenting on the acrimony between the Anthropological and Ethnological societies, a writer for the *Pall Mall Gazette* called them symptomatic of a "struggle for existence" in science.[37]

Matching the achievements of the two societies actually yields a curious, somewhat embarrassing result, for it does underscore (for want of a better term) the entrepreneurial spirit of the anthropologists. For instance, anthropologists pressed the idea of running a museum with much more determination than the Ethnological Society. In the early 1870s the museum, by then part of Lubbock's Anthropological Institute, would be described in Edward Walford's *Old and New London* as "small but interesting." The Anthropological Society's success at accumulating objects was secured by relying on its members abroad, such as field collector Robert Bruce Napoleon Walker, whose anthropological artifacts the modern curators of the Pitt Rivers Museum of Anthropology (itself the descendant of the merged Anthropological Institute's museum) acknowledge formed part of the museum's founding collection. By contrast, it appears that calls by the Ethnological Society to have members make donations to the museum were not so successful.[38]

The extent to which Ethnologicals were political reformists should not be exaggerated either. For instance, in a very interesting article in which she does not hide her sympathy for the anti-Eyre group, historian Catherine Hall feels compelled to recognize that the wonderful progressives on the Jamaica Committee nonetheless emphasized the "potential for equality rather than an equality which was as yet fully realized. Women, blacks and browns, having been denied opportunities must now have access to them, but it would be necessary for them to learn civilization."[39] Burton could have agreed. In fact, the split on racial issues was not as clearly recognized by contemporaries as it has been by later historians. Augustus Lane Fox (later Pitt Rivers), a member of both the Ethnological and Anthropological societies, declared in his "Primitive Warfare" (with the "Negroes" from Africa in mind) that it happened that "races" experience "arrested growth; and finally, the intellect of the nation fossilizes and becomes stationary for an indefinite period, or until destroyed by being brought again in contact with the leading races in an advanced stage of civilization."[40] Elsewhere in the same essay the gallant colonel stated that "the savage is morally and mentally an unfit instrument for the spread of civilization, except when, like the higher mammalia, he is reduced to a state of slavery."[41] It never shocked the Ethnologicals, who welcomed him.

Likewise, Thomas Huxley, the leader of the fair Ethnologicals, was hardly a radical in both race and gender. Not unlike the Cannibals, he regarded the equality of black and white "so hopelessly absurd as to be unworthy of serious discussion."[42] In the lists of members that the Ethnological Society circulated, a total of two women are found.[43] At the Ethnological Society, women were the passive gender and could only exclaim and clap. They were valuable as important spectators of the Ethnological intercourse, but the male was the active element. In addition, very much like the Cannibals, Huxley would come to exclude women from meetings of the Ethnological Society. Conversely, very much like the Ethnologicals, the Anthropological Society came to invite them in the wider public events and lectures. Admittedly, this token of goodwill on the part of the Cannibals was only given after the death of Hunt when the Cannibals faced a crisis in attractiveness. But it remains that the evolution of the two groups planed down initial differences between them.[44]

In fact, resistance to the admission of women is a lame criterion by which to distinguish the two societies. Beyond some illustrious exceptions, such as Florence Nightingale's membership in the Statistical Society of London in the late 1850s, leading societies of the time resisted stubbornly the admission of women. As historian Evelleen Richards has shown, this was egregious in the society that had the largest membership of all and perhaps the greatest prestige. The Royal Geographical Society, which *tolerated* female explorers, kept excluding women from fellowship and participation in ordinary meetings although they were welcome in larger promotional gatherings.[45]

Likewise, while the Aborigines' Protection Society is a significant connection to define the Ethnological Society, emphasis on religious origins does not take us further than where the spirit of the time and its contradictions lingered. Ethnologists had religious origins at the same time as they boasted their disagreement with religious leaders, and they recruited among the "bishop-eaters." One need look no further than their own leaders for evidence. Known as "Darwin's Bulldog," Huxley had become famous for his 1860 debate during the Oxford meetings of the British Association with Samuel Wilberforce, the bishop of Oxford and a relentless opponent of the idea of evolutionism. No doubt that despite their links with religious currents, the Ethnologists thought of themselves as soldiers in the battle against what Boyd Hilton has called the *Age of Atonement*.[46]

A similar conflicting relation with religion was noticeable with the Cannibals. There too there were links with the religious establishment and a dispute. Anthropologicals included among their ranks more than one reverend with more than one doorplate (M.A., D.D., F.G.C.H., M.R.A.S., F.E.S., D.C.L., F.S.A., F.G.S., Ph.D., and so on). Hyde Clarke, in his letter against the Anthropological Society, hinted that they suffered from a religious, anti-scientific bias, and he complained about the presence of clergymen from rural counties among members. But there too, the leaders of the Anthropological Society could be called "bishop-eaters." One would have expected this much from a society whose intellectual reference, Richard F. Burton, was indeed a professed miscreant who loved to call religion a superstition. A contemporary paper from Burton's friend Winwood Reade on the "Efforts of Missionaries among Savages" published in the *Anthropological Review* and scolding the role of the church in Africa, illustrates where the majority of the Anthropological Society's Council stood. Reade's text was an article of faith of an anti-religious kind. And when Reade quipped in this paper that "anthropology is the most catholic and the most comprehensive [of all the sciences]" he implied no endorsement of Catholicism of course, only that anthropology was, as Catholicism, all-encompassing. Reade's "Efforts of Missionaries among Savages" created controversy by drawing a response from the lord bishop of Natal, John Colenso, a social reformer, defender of the Zulus, believer in human equality, and a man of science. Thus had the Cannibals managed to stage their own Huxley-Wilberforce controversy.[47]

## AN ANTHROPOLOGICAL BUBBLE?

A more promising perspective than the one that emphasizes the dispute on race, gender, or religion might be to put to the fore what is pedantically referred to as the "strategies of science." Hunt and Huxley were academic entrepreneurs striving for academic domination, and the story of anthropology can be told from this vantage point as one of competition rather than opposition. As Efram Sera-Shriar noted, Hunt, a former secretary of the Ethnological Society in the 1850s, who thus had experience in the management of a learned society, could have believed that the Anthropological Society, which he owned and controlled since the *Review* and the *Journal* were both his personal property, would be a useful instrument of academic power, one he would never have been able to secure if he had tried only to rise through the ranks of the Ethnological Society.[48]

Focus on the strategies of science, if taken seriously, means a study of its institutions, organization, and governance. The rise of anthropological science cannot be separated from the broader trend of science and its expansion through the learned society, its vehicle at the time. When the Anthropological Society was launched in 1863, the growth of learned societies, like the expansion of navigation, commerce, and finance, was not a new phenomenon. Like navigation, commerce, and finance, it had origins in the seventeenth and eighteenth centuries. But as in navigation, commerce, and finance, a dramatic quantitative transformation was taking place. The phenomenon was striking enough to attract the attention of one contemporary statistician who made it the subject of a paper entitled "On the Progress of Learned Societies" presented during the 1868 meetings of the British Association for the Advancement of Science in Norwich.

The statistician, Leone Levi, estimated there were at that date "upwards of one hundred and twenty" learned societies in the United Kingdom, having in the aggregate "upward of 60,000 members." This was three times more than the number 30 years earlier, when total membership had stood around 20,000 only.[49] If society membership is an indicator of the size of the scientific marketplace, then the market for science was expanding dramatically. But there was another important finding in the Levi study. It consisted in the membership edge the Anthropological Society had by 1868 taken over the Ethnological Society. Levi put the Ethnological Society's membership in 1867 at 219 members and its annual income at £300, while the respective numbers for the Anthropological Society were 1,031 members and £1,327. In both membership and earning power, Anthropologicals dominated.

These estimates require caution, and would be later criticized. There were different forms of membership, and Levi's numbers went by a narrow definition for membership in the Ethnological Society but broad for the Anthropological Society. Yet if we try to correct for such biases and limit ourselves to both members who paid their annual subscription fees and life members who had compounded them with one large lump-sum payment, we get, for the end of the year 1866, 550 members at the Anthropological Society (700 if we include late payments) against 210 for the Ethnological Society (219 with late payers included). (See the "Supplements" at the end of the book for details on computations and sources.) The money-raising capacity of the two societies reflected these trends, even after controlling for creative accounting. At around £900 to £1,000 between 1865 and 1867, the annual cash flow of the Anthropological Society was about three times bigger than that of the Ethnological Society, which stood from £280 to £300. This was truly remarkable, considering that the former society had just been created in 1863. These amounts were substantially less than, for instance, those of the venerable Royal Geographical Society, with its 2,102 members and £5,981 income at about the same date. On the other hand, they were ahead of the increasingly important Statistical Society with its 371 members and £779 of income.

Thus the Anthropological Society began auspiciously. A few years later, even the society's more ferocious adversaries would acknowledge this, although they would depict this in negative terms, speaking of a bubble. The nightmare of an inflating anthropology was expressed in the "Anthro—that horrid word" letter Ellen Lubbock sent to Emma Darwin in 1873. She wrote about this "mushroom society with no good men in it." The quote reveals Ellen's appreciation of the might of the Anthropological Society, although the reference to the "mushroom" (which, like bubbles, grows overnight) questioned the quality of anthropological inflation. In sum, a statistical comparison between the Anthropological and the Ethnological Society reinforces the impression of the greater dynamism of the former, especially in view of the stagnating income and membership of the latter. In fact, this remarkable success of the racists may explain the eventual triumph of the word *anthropology* once the amalgamation took place, a triumph which has puzzled earlier historians. Fundamental for a proper assessment of what happened in the 1860s is to recognize that the proper starting point is not the failure of the anthropologists to meet up with modern standards of political correctness but their brilliant practical results. Desmond and Moore's portrayal of the Anthropological Society as a den of the Ethnological Society's dissidents subsidized by a detestable foreign power is a case of the tail wagging the dog.

Society leaders had worked hard to produce the desired outcome. Since apart from donations, membership fees were the main source of revenues, membership increases became a hallmark of success.[50] They were an integral part of the machinery whereby sciences were being "advanced" and "emancipated," to use the language of Cannibals, who appropriated the language of slavery and freedom in their disputes with ethnologists.[51] This resulting free scientific citizenship, although not unique, was an innovation compared to earlier times, when learned societies and clubs typically put themselves in the red and relied on wealthy patronage for writing off their debts.

The Anthropological Society grew like a medium does. On the one hand, it made important outlays so as to ensure visibility. The first years of the society show publicity budgets and promotional expenditures, including the copies of the anthropological literature it showered on potential or confirmed members. On the other hand, it made strenuous efforts at canvassing readers. An initial target of 500 members was set, achieved in early 1865, then raised to 1,000. In 1866 Assistant Secretary J. Frederick Collingwood prepared a list of "8,000 names of gentlemen who [he] consider[ed] may be likely" to become subscribers.[52] Both the published minutes and the archive of the Anthropological Society bear witness to substantial expenditures on efforts made to achieve these goals. In early years, there were discussions to increase the assistant secretary's wage in case membership should exceed 500. The more detailed accounts for the year 1868 show, under the secretary's wages, an entry for "Commission on subscriptions" for about £30.[53] Reaching membership targets was an occasion for celebration. When the society claimed to have signed in its five hundredth member in early 1865, a promotional dinner was given to commemorate the event (and honor Burton, who was to take his new assignment as Her Majesty's consul in Santos).[54] Lord Stanley, son of the leader of the Conservative opposition, and as we saw a fellow of the Society, presided over the dinner.[55]

## VARIETIES OF CANNIBALS

As in any other learned societies, the members of the Anthropological Society came in a variety of guises. First was the executive body of the society, the Council, which was elected annually during the anniversary meeting. The Anthropological Society's Council was principally peopled by the individuals who formed the Cannibal Club, those whom Clarke would call "Dr. Hunt's clique." This group remained consistent over time and included veterans such as J. Beddoe, R. S. Charnock, Owen Pike, Charles Robert des Ruffières,

W. S. W. Vaux, R. King, Berthold Seemann, Barnard Davis, and W. Bollaert, all of whom served for extended periods of time. In 1867 the Council was enlarged to welcome two eager promoters of anthropology, Captain Bedford Pim and Colonel Augustus Henry Lane Fox.

While mostly Britons and only British residents comprised the Council, the "honorary fellows" who formed the other high-visibility group of the society consisted of a collection of generally old and generally Continental luminaries. The official narrative of the Anthropological Society's development claimed that the Continent had taken a lead in matters of anthropology (primarily Broca's Société d'Anthropologie de Paris). The society thus sought the endorsement of members from Continental societies and got it by including men from the top brass of French anthropology, such as Paul Broca and Jacques Boucher de Perthes. Other famous names included the Swiss-born professor of natural history Louis Agassiz, at that time teaching at Harvard University; the American anthropologist Ephraim George Squier; and the Geneva-based German professor Karl Vogt, whom Darwin praised in the opening pages of his *Descent of Man*. In this group, comparatively fewer Britons were found. Among them, we find John Crawfurd (at one point president of the Ethnological Society), Sir Charles Lyell, and Charles Darwin.[56]

If the relationship with honorary members brought to the headquarters a flow of credibility, that with the society's "corresponding members" brought a flow of information. Corresponding members were entrusted with supplying details on the latest developments of the science of man in their place of residence. The network of corresponding members was geographically eclectic, more than that of honorary members. It tapped into established clusters of anthropological discussion, and as a result, corresponding members were mostly found in Continental Europe. Among these, Parisian anthropologists were well represented, and we note the presence of members of the Copenhagen school, which pioneered anthropological archaeology. In 1866 only four correspondents were located beyond Europe: one in Canada, one in Peru, and two in Argentina. Interestingly while the Anthropological and Ethnological societies shared many honorary members, the public of corresponding members did not overlap at all, and they tapped into regions that were wholly distinct. Unlike their Anthropological counterparts, corresponding members of the Ethnological Society were principally located in the British Empire or in places under more or less exclusive British influence such as New Zealand, Natal, Burma, and China.

But this is not to say that the Anthropological Society forgot about the broader world. After all, its creator, Richard Burton, was a British consul. In such places, the Anthropological Society created a special type of membership that was without counterpart in the Ethnological Society and indeed in many other societies—the local secretaries. In 1866 the Anthropological Society's network of local secretaries spread widely across the world: in Africa (Gabon, Nigeria, and Natal); along the Mediterranean coast (Algeria, Egypt, and Turkey); in Asia and Oceania (India, China, and Australia); in the Americas and the Caribbean (Canada, the United States, Mexico, Nicaragua, Ecuador, Jamaica, Venezuela, Brazil, and Argentina). Local secretaries were to have an important role in the history of the Anthropological Society. In principle, they reported on local matters of anthropological interest or fed the society with photographs, skulls, and others. The supply of such studies or artifacts, however, was but one projection of their local presence. In the empire of anthropology, they played the role that British consuls, with their ill-defined duty to advance the trading interests of Britain in ways they deemed fit, played for the British Empire—they promoted the interests of the Anthropological Society in ways they deemed fit.[57] Their goodwill stirred information and anthropological enterprise. It was as local secretaries that key characters of the subsequent drama such as Hyde Clarke and Bedford Pim secured access to the Anthropological Society.

## A BRIEF SOCIOLOGY OF THE CANNIBALS

Essential to a proper understanding of the Anthropological Society is the fact that the Cannibals were a different crowd from the Ethnologicals. This fact was observed at the Council level but also at the level of regular membership. The overlap between the two societies' leadership was reduced to a handful of individuals—John Beddoe, Charles Robert des Ruffières, Richard King, Augustus Lane Fox, and Viscount Milton—who occupied seats in the respective executive bodies simultaneously or in succession. Likewise, comparing membership for the late 1860s gives 29 common members among 636 Anthropological and 230 Ethnological fellows respectively. This limited overlap in fact helps to explain the attraction of the merger. Fewer than 5 percent of the members of societies dealing with the science of man duplicated their subscription, so that the loss in foregone fees would be minimal, while general expenses would be pooled.

In what way were the Cannibals different? To date, the most complete attempt at pinning down the sociology of the Cannibals and of their rivals is provided by an unpublished dissertation by D. K. Van Keuren, who sought to determine the overlap between the respective members of the two societies' governing bodies on the one hand and significant circles in Victorian society on the other.[58] For instance, he explored connections between the two councils and such social institutions as the Athenaeum Club. The Athenaeum Club was and still is a cluster of literary and scientific leaders (incidentally, there is no relation between the *Athenaeum* journal and the Athenaeum Club). Created in 1824 with the purpose of accepting "individuals known for their scientific or literary attainments, artists of eminence in any of the Fine Arts and Gentlemen distinguished as liberal patrons of Science, Literature and the Arts," the Athenaeum Club became identified over time with the most prestigious learned societies and the highest circles of the British academic system (the current director of the Royal Anthropological Institute, David Shankland, is a member of the Athenaeum Club).[59] But as Van Keuren discovered, while the percentage of Athenaeum Club members in the Ethnological Council was about 50 percent, the comparable ratio for the Anthropological was only about 10 percent.[60]

As I discovered, these sociological traits Van Keuren identified for the leadership are also visible for regular membership. An intriguing feature of the membership lists published by the two societies is the frequent indication of the club affiliations of the members. This was a signal of social position since these gentlemen's clubs typically restricted admission. Thus societies splashed out the names of their members, and they in turn splashed out their social affiliations. Using these lists to identify club membership outlines the differences between anthropology and ethnology. When one pores over the Ethnological Society's members, one is again struck by the fact that the prestigious Athenaeum Club comes up frequently, while for the Anthropological Society we find instead the Reform, Garrick, Carlton, Conservative, Oriental, and United Service clubs. These clubs were all important and venerable institutions of Victorian Britain but were undoubtedly less rarefied than the Athenaeum, to which even Disraeli struggled to secure admission.[61]

Using a published list of Athenaeum Club members, I was able to quantify the relationship between the Athenaeum and the two rival societies. I found that a total of only twelve members of the Anthropological Society, or less than 2 percent, belonged to the Athenaeum Club around 1867 against sixty-two members of the Ethnological Society, or about 30 percent.[62] This evidence cuts a long story short. The contrast between the Anthropological and the Ethnological societies was one of social standing. Ethnologists were the ultimate insiders of British society. This impression of interlocking circles is reinforced by other interesting computations Van Keuren undertook showing the overlap between Ethnologists and the Royal Society and the number of senior positions members of the Ethnological Council occupied in other learned or scientific societies. Likewise, F. Galton, W. Spottiswoode, and Clements R. Markham were simultaneously the editors of the *Proceedings of the Royal Geographical Society of London* and important members of the Ethnological Society. Sir Roderick Murchison, the Royal Geographical Society's president and a powerful member of the British Association for the Advancement of Science, sat on the Ethnological Council at one point. This favorable position of Ethnologists was unlike that of Anthropologists, who had fewer connections to begin with, although they tried hard and hailed their successes at penetrating such places as the Royal Society.

All these indicators of greater social and political connectedness of the Ethnological Society with respect to the Anthropological Society do point to a single conclusion. Compared to the standard *modus operandi* of the governance of British science, which rested on interlocking councils and cooperation within a blue-blooded oligarchy of scientists (an "alliance of interests" as Van Keuren decently puts it), the Anthropological project innovated by seeking to establish itself from scratch and without the endorsement of the inner circle and governing institutions of British science embodied by the Royal Society, the Athenaeum Club, and the British Association for the Advancement of Science. This explains the Anthropologicals' emphasis on their foreign origins in Paris (through Broca), Boston (through Agassiz), and Geneva (through Vogt), which others have emphasized but not analyzed. International connections were used to remedy the deficiency of domestic endorsement. Given this, the limited overlap in the membership of the two societies is no surprise. The point is that the two societies had radically different sociological hunting grounds.

Thus, behind the scientific dispute between anthropology and ethnology and its all too evident references to ideological disagreement on the subject of race, there was a social dispute. This renders even more intriguing the fact that scholars writing on the subject have engaged in normative judgments and supported the ethnologists without realizing that the "Lubbock circle" squared with "those inside." The dispute, therefore, was about the social status of disciplines. Ethnologists were social insiders. Anthropologists were less so. This qualifies many previous historical assessments.

For instance, Van Keuren is critical of the Anthropological Society's efforts at aggressive canvassing and making itself heard through a variety of public outlets such as the short-lived *Popular Magazine of Anthropology*, comparing it unfavorably to what he called the Ethnological Society's "low profile" and its soft-spoken recruiting of new adepts. But this simply

rested on people from the same milieu talking one another in. Unlike the
Anthropological Society, discretion and elegance was the tacit rule at the
Ethnological Society, and the society, in Van Keuren's phrase, "stayed out
of the public eyes," thus "eschewing public spectacles." But this was only
because the Ethnological Society owned an entry in the big public displays
such as those provided annually by the British Association for the Advancement of Science of which the Ethnological Society was a stakeholder. More
adequately, perhaps, we should say that the Anthropological Society, being
an outsider—sociologically and politically—of British science, had to organize its own circuses.

## CHAPTER THREE

# Anthropologists without Qualities

The problem of the quality of anthropologists, which the previous chapter has raised, was to form the bulk of the accusations Hyde Clarke raised against the Cannibals in the summer of 1868. One was that the society's fellows had been "canvassed for among general practitioners and clergymen in the country." For the contemporary public, the censure boiled down to stating, at a time when the adoption of the second Reform Bill had stirred minds about the question of the proper extension of democracy, that anthropologists were not proper men in that they would never have been accepted in the best clubs and societies. Yet inspection of contemporary lists shows nothing of the sort, certainly nothing as simplistic, and in particular not nearly as many clergymen as Hyde Clarke implied. As far as general practitioners were concerned, it is true that the consumers of anthropology tended to be professional men.

Excluding the very few members who did simultaneously subscribe to the Ethnological Society, we find that the rank and file of the Anthropological Society included an enormous number of surgeons or physicians, some with a distinguished pedigree such as Holmes Coote, who had been in charge of the wounded from the Crimean War at Smyrna Civil Hospital and wrote after his return *A Report on Some of the More Important Points Connected with the Treatment of Syphilis*; and William Nathan Chipperfield, a professor of anatomy and physiology at Madras Medical College. There were engineers as well, such as Henry James Castle, the first man to hold the professorship of surveying at King's College, Cambridge. And there was an enormous number of technicians from the army and navy. While many members lived within the United Kingdom, many also lived abroad within the British Empire and dominions, in particular in British India.

Naturally, Anthropological fellows abroad had been recruited in places where the promoters of the society already had connections. At the time the Anthropological Society was launched, Richard Burton was frantically trying to escape the fevers of Fernando Po, an islet off the shores of West Africa, and he managed to sign in Henry Stanhope Freeman, governor of Lagos Colony, and his successor, John Hawley Glover, who would later organize the indigenous against the Ashanti in the War of 1873 and be made K.C.M.G. (Knight Commander of the Order of St. Michael and St. George for the uninitiated).[1] Unsurprisingly given that Burton himself was a consul, the Anthropological Society had a number of consuls and vice-consuls as members, several that it shared with the Ethnological Society but several others of its own. Among the consuls with dual memberships, for instance, were Charles Duncan Cameron in Massowah in modern-day Eritrea; T. J. Hutchinson in Argentina, who would join the last Anthropological Society Council before the merger; and Thomas F. Wade, the sinologist and influential secretary of the Beijing Legation. Members that the Anthropological Society could claim as its own included George Thorne Ricketts (consul, Manila); P. Henderson (vice consul, Benghazi); C. Treasure Jones (consul, Shanghai); D. B. Robertson (consul, Canton); J. H. Skene (consul, Aleppo); J. W. Studart (consul, Ceara, Brazil); and Whitock Darby (consul, Puerto Rico).

The examples above suggest resisting the temptation to join the game of comparisons invented by Hyde Clarke; instead, we should investigate the reason why this game was invented. The whole contemporary dispute about the qualities of members of the Anthropological Society, just like the modern dispute about what constitutes a "real" anthropologist, is exactly where Hyde Clarke and the ethnologists wanted the debate to go for reasons that the previous chapter has clarified. Hyde Clarke was successful in setting the tone. The problem with anthropology, as the dominant literature from Stocking to Desmond and Moore subsequently asserted, was that the Anthropological Society was an unworthy institution, and as a result, they have meticulously sought to disprove anthropologists (as Ellingson did, establishing the vacuity of Hunt's ideas) or "expose" that Anthropologicals were puppets in the hands of the abhorred Confederacy.[2]

But a more interesting perspective, and incidentally a less jingoist one, would be to reflect upon the mindset that produced the urge to compare various groups of scientists. It is evidently this mindset that produced the conventional contrast between the anthropological clique and the ethnological circle, between the anthropological dilettantes and the ethnological specialists.[3] Holding on to the notion of the superiority of ethnologists could only happen in a country where academic success is still a road to knighthood, and where learned society door-plates obtained in mysterious ways are still worn proudly as hallmarks of social status. Yet recalling that the Anthropological Society catered to laymen while the Ethnological Society lists included noble laymen should cast doubt on the intellectual solidity of the implicit claim about anthropologists and ethnologists. Noble laymen were not primarily dignified by academic achievement but rather by their social position and the possibility that their status conferred on them the opportunity to mingle with scientists.

Thus it was that the preexisting Victorian hierarchy produced the hierarchy between the twin sciences of man—the bluer-blooded ethnologists claimed more elevated truths, but the redder-blooded anthropologists insisted it was the other way around. Anthropologicals sought to reverse the scales and insisted that they were the only genuine scientists and that the blue-blooded ethnologists were the dilettantes. If it was a race to prestige and status, and if status were with the incumbent and *they* were the outsiders, then the apparent paradox of their numerical superiority and emphasis on elite science is resolved.

In sum, the two societies were not unlike the aristocracy and the middle class with respect to each other, and the way they discussed the subject of race bore the stamps of this social struggle. On the one hand were the rulers of an inner circle of old wealth. On the other hand were middle classes still roving the world with hungry bellies, still voracious, and still eager to transform the world into a commodity. This conclusion suggests a different *entrée* to the history of the Anthropological Society than has been taken thus far—one that would focus on those aspiring Englishmen who could identify with Burton. They longed for their political voices to be heard and for the right to loot "this magnificent world of ours." Such were the Cannibals in all their complexity, and the reason for my belief that erudite discussion of the content of the controversy between anthropology and ethnology is of limited use, after a while.

## HUMAN GEOGRAPHY

One missing block of the conventional story of the science of man in the 1860s is the pivotal role the Royal Geographical Society (RGS) played in it. The RGS was one of the most prominent and active societies of Victorian Britain, and its membership had steadily increased in the 1850s and 1860s on the back of expanding British and international commerce. According to the table of leading scientific societies supplied by Levi, there were 607 members of the Royal Geographical Society in 1850 (ranking fifth) and 1,354 in 1860 (ranking third), surpassed only by the United Service In-stitution, which signed in every military officer, and the Royal Agricultural Society, which signed in every landowner and cultivator.

The link between anthropology and geography is a fundamental one and can be spied through the many chinks of the conventional story. The nature of Victorian geography as the science of empire par excellence has been em-phasized before, although its role as a hotbed for anthropology has not been so forcefully emphasized. But as appears from inspection of early presidential addresses at the Geographical Society, successive presidents put the science of man squarely within the territory of geography. In the language of W. R. Hamilton, its president in 1841, geography would assist with Britain's "re-sponsibilities" of "civilizing the yet benighted portions of the globe, and for bearing her part in forwarding and *directing the destinies of mankind*."[4] And thus the leaders of this powerful organization did not look favorably upon the exodus of what was for them a subdivision of their field of competence.

As a result, the history of the Royal Geographical Society and that of the Anthropological Society are enmeshed. First, geography and the science of man had been linked to one another since 1851 in the British Association for the Advancement of Science's Section E (Geography/Ethnography sec-tion), not incidentally through the intervention of Sir Roderick Murchison, then already a member of the British Association's higher echelons.[5] In this capacity, Sir Roderick became the nemesis of anthropologists, who would describe him as the "destroying angel" who had slain an early subsection devoted to the topic. Hunt and his followers deplored Sir Roderick's "lack of compunction toward the science of man," and his obstruction in conjunc-tion with the Lubbock group inside Section E toward letting anthropology gain or regain "sovereignty."[6] Various arrangements, always intended to di-lute the power of anthropology, were subsequently opposed to the Anthropo-logical Society's requests for a separate section. This enraged anthropologists whose sheer numbers, they felt, secured for them the means to dominate Huxley, Lubbock, and the rest had a science of man section been opened. Indeed, the antagonism between anthropology and the Geographical Soci-ety is such a vital part of the story that the letter by Hyde Clarke in the *Athenaeum* started with a complaint about the "lampoons" that Hunt had printed not only against Huxley of the Ethnological Society but also against Sir Roderick Murchison.[7]

Another element that demonstrates the importance of the geographi-cal origins of the Anthropological Society is the number of Anthropologi-cal Society members who were also members of the Geographical Society. More than one prominent Cannibal was a current or former important member of the Geographical Society, beginning with Burton, who Clements R. Markham, member of the Ethnological Society and a long-term secretary of the Geographical Society, tells us belonged to the list of famous explorers Sir Roderick's society supported "morally and materially."[8] Another prominent geographer turned Cannibal was William Bollaert. This steady member of subsequent Anthropological Society councils was a regular of discussions held at the Geographical. Poring over the *Proceedings of the Royal Geographical Society of London* in early 1863 when the Anthropological Society was launched, one quickly comes across Anthropologicals. Future Anthropological Society's Council member Winwood Reade is found discussing his African travels. Likewise there are discussion of reports by Paul B. du Chaillu, Antonio Raimondy, and Arminius Vambery who became local secretaries in the west coast of Africa, Lima, and Pesth respectively. There is also evidence that the Anthropological Society canvassed some of its converts inside the Geographical Society, probably through the agency of Burton's own network. Here the best example is Edward Eyre, the governor of Jamaica who joined in 1864 and was soon to become the author of the merciless repression against the Jamaican Rebellion in 1865. Eyre had been an explorer of Australasia and a medalist of the Royal Geographical Society, which he had joined in 1857.[9]

It is no surprise that the anthropologists' canvassing efforts turned to subscribers of the Geographical Society rather than to general practitioners and clergymen in the country, as Hyde Clarke would have it, or to the Ethnological Society. The large membership of the Geographical Society formed a much better hunting ground from which to poach than the more limited membership of the Ethnological Society. In sum, the story of the rise of the Anthropological Society might be told as an anthropological secession within a highly networked science. The problem of emancipation of a group of scientists trumped that of the subject. They were prepared to rework the contours of the field to achieve sovereignty. It is notable that when recollecting the creation of the Anthropological Society in 1865, Burton emphasized a problem of access—the impossibility to publish on certain subjects. What was at stake was a matter of political sovereignty and the design of science against the insufficient openness of British institutions. The precise scientific purpose was almost secondary.[10]

## THE SOURCES OF ANTHROPOLOGY

Pointing to the same direction of a secession inspired by Burton, the launching of the Anthropological Society in 1863 coincided precisely with the deterioration in the relations between the Geographical Society and Burton. This was a period Burton's wife Isabel called the "Geographical disagreeables." A few years earlier, the quest for the source of the Nile River led the Geographical Society to sponsor a mission by Burton and his co-explorer John Hanning Speke to find the large lakes reported to be in the center of the African continent. In February 1858 they reached Lake Tanganyika, and in May 1858, Speke alone (for the sick Burton was grounded in Kazeh) reached Lake Victoria. A protracted quarrel erupted between Burton and Speke when the latter used the facilities of the Royal Geographical Society upon his return in early May 1859 to claim he had discovered the source of the Nile.

This happened while Burton convalesced in Aden and, according to Burton, despite the fact that both explorers agreed in Aden to appear together at the Geographical Society. Burton was outraged to learn in late May 1859 when he arrived in London that Speke had already claimed ownership of the discovery that Lake Victoria was the source of the Nile and now turned against the theory. He was incensed by the fact that Sir Roderick Murchison had given his blessing, proved so quick to accept Speke's narrative, and was already sending him back on a second exploration. This was to lead to an increasingly bitter controversy, culminating with the scheduled discussion at the British Association's meetings scheduled in Bath in September 1864. The meeting never occurred because Speke died the day before from what was described as a hunting accident.[11]

Burton had interpreted his nomination in 1861 as Her Majesty's consul at Fernando Po as a deliberate measure against him. "They want me to die," Burton said bitterly, "but I intend to live, just to spite the devils." He organized his survival in this "white man's grave." Upon learning that the Spanish had just established a sanitarium at 400 meters above sea level, he built one of his own even higher up. In the meantime, his wife, Isabel, managed to secure four months' leave, and Burton landed at Liverpool in December 1862. One of the first things he did was launch the Anthropological Society a few weeks later.[12]

But Burton was to return to his post, although he did his best to transform his time at Fernando Po into an exploration of the kingdom of Dahomey, the famous "mission to Gelele," and returned again to London for the British Association meetings in 1864. Meanwhile, Isabel had all but worn out the patience of the Foreign Office. In a letter, John Russell declared to her: "I know that the climate in which your husband is working so zealously and so well is an unhealthy one, but it is not true to say that he is the smallest of consuls in the worst part of the world." Russell nonetheless reassured Isabel that he would keep an eye on possible vacancies. Finally, in early 1865 Burton accepted a new assignment in Santos, Brazil.

Some aspects of the rhetoric of the Anthropological Society, such as its persistent ranting against Murchison, its taste for lashing out at various British imperial administrations such as the Admiralty, or the enthusiasm that the Anthropological Society would later display in joining in a press campaign against Prime Minister John Russell's political and ethnological incompetence, may therefore be put in relation with the underlying conflicts between Burton and the Geographical Society on the one hand and the Foreign Office on the other. According to the hypothesis I am laying out, the creation of the Anthropological Society was therefore prompted by an issue over the ownership of a discovery and Burton's subsequent impression that he needed his own institution.[13]

This conjecture is reinforced by the very way in which Burton handled the dispute with Speke. Unhappy to have lost ownership of a knowledge he felt he had originated, he sought to be nonetheless rewarded for his discovery of Lake Tanganyika when he appeared before the Royal Geographical Society. He attempted to redefine his contribution by delineating Speke's *geographical* achievements from his own *ethnographic* advances: "To Captain J. H. Speke are due those geographical results to which [Sir Roderick Murchison has] alluded in such flattering terms. Whilst I undertook the history and ethnography, the languages and peculiarities of the people, to Captain Speke fell the arduous task of delineating an exact topography, and of laying down our positions by astronomical observations—a labour to which at times even the undaunted Livingstone found himself unequal."[14]

To understand Burton's tactical move, it is important to recast the Royal Geographical Society in the context of the 1850s as a learned society undergoing tremendous expansion encouraged by underlying mercantile trends.[15] This expansion had been managed by Murchison's long rule, permitting him to influence durably the society's evolution. Among the many things he did was staging popular events such as the famous search for Captain Sir John Franklin's lost expedition for the Arctic. Franklin had departed from Britain in 1845 and had never been heard from again. The inaction of the Admiralty provided an opportunity for Murchison and the Geographical Society to pose as gallant scientific gentlemen who supported the widow's desperate plea. Expeditions were finally launched, and the Royal Geographical Society benefited. The sales pitch had all the elements of a blockbuster—the adventurous explorer Sir John Franklin, his wife in despair of finding a body to mourn, the reluctant Admiralty, whose arm had to be twisted, Murchison insisted, by the Geographical Society. The twists of events created further sensation. In the early 1850s, remnants of the expedition were gradually discovered, and later accounts were collected by the Inuits, bringing to the

fore the interest of anthropological investigations. Though the number of
Geographical fellows had stalled in the late 1840s, it more than doubled
during the 1850s.

As the interest for the ices of Franklin's expeditions began melting in
the late 1850s, the heat of Africa took over. After the travels of Burton and
Speke and the fallout between the two explorers, the new season now focused on the African travels of Speke and James Augustus Grant between
1860 and 1863, ostensibly to test the hypothesis Speke had articulated that
Lake Victoria was the source of the Nile. It was yet another blockbuster, and
through it the subject of race made a triumphant debut. Publishers raced to
secure the rights to those stories. The House of Blackwood purchased the
rights for Speke's memoirs for a hefty £2,000 only to discover Speke's dismal
grammar and hunt for the right ghostwriter who ended up rewriting Speke's
manuscript into a popular ethnography of "Dark Africa." *The Sources of
the Nile* became a major success with 7,600 copies sold the first year in
three editions, and rights sold to France, Germany, and the United States.
As shown by historian David Finkelstein, this twist toward focusing on natives had been inspired by the publisher himself, who sniffed opportunity
and commissioned from the ghostwriter a book that would realize "savages and savage life in a way that nothing else ever did." This was ironical,
given how Speke and Burton had divided territory, but this also provides perspective on the logic pursued at the time by the Anthropological Society of
London.[16]

The Anthropological Society, therefore, might be usefully described as
the studious pupil of the Geographical Society. There is indeed a powerful parallel between the two societies. In making the fracas it has been reproached for causing, the Anthropological Society was essentially copying
those techniques Murchison had perfected for the promotion of the Geographical Society. It is under this light that the episode of Bedford Pim's
lecture on "Negro and Jamaica" is to be told. The Morant Bay Rebellion had
brought the headlines toward the peculiar domain on which the Anthropological Society sought to rule, and it was natural that the society should
take advantage of it. Ter Ellingson has underlined the role of P. T. Barnum's
"Grand Ethnological Congress of Nations" as an inspirer of the Anthropological Society's project and has portrayed Hunt as "the Prince of Humbugs"
of anthropology.[17]

Hunt was one such prince, but definitely not the only one, and traveling this route, there were many P. T. Barnums in the British science of the
time. The Pim lecture was just another show to be put on a list of successful

social-scientific events that included narrating the Franklin expedition and the controversy on the sources of the Nile. These shows were further epitomized by the "Barnums" of the sessions at the annual meetings of the British Association for the Advancement of Science, such as Huxley's debate with Wilberforce.[18] In fact, while they drummed in spectators with Pim's show, the great wizards of anthropology engaged in smoke and mirrors by giving tremendous publicity to their rejection of publicity. By the same trick, they were making known the Anthropological Society's insistence on avoiding the Ethnological Society's errors of involving women and degrading science by popularizing it beyond proper bounds.

And indeed, following the "Negro and Jamaica" lecture, subscriptions to the Anthropological Society kept coming in unabated. There were a couple resignations such as Anthropological Council member Hugh J. C. Beavan, a polygenist and the translator of French polygenist George Pouchet, and more prominently, Charles Buxton, MP, a leading figure of the Ethnological Society and an antislavery activist.[19] Buxton mentioned the "ridicule" with which the Pim lecture was spoken of among "scientific men," but what should we make of the fact that he had felt so little of the ridicule of Hunt's earlier *Negro's Place in Nature* in 1863 that he had risked compromising the name of his father, the masthead of abolitionism, by compounding his fee so as to become a *permanent fellow* in 1864? These isolated moves have to be set against letters of congratulations for organizing this session, which were directed to the Anthropological Society. It is no surprise that, in what was primarily a mercantile society, race and gender were turned into assets and liabilities.

That all this was just a matter of business should never be forgotten. Indeed, it was important to try and prevent things from getting too personal. And thus, as the Anthropological secession proceeded, Burton never gave up either his membership in the Ethnological and Geographical societies or his social connections such as having the occasional dinner with Murchison and showing up at the Geographical Society (although he was dissatisfied with meetings there and claimed they were "slow").[20] Beyond the split, we can discern the explorer's efforts to leverage his own position and increase his bargaining power toward the establishment, which he felt had ill-treated him. In the hands of Burton, the Anthropological Society meant resources, a public, possible financial security, and the capacity to pressure British authorities and secure the right consular assignments, all without forgetting the society's potential usefulness as an instrument to own future discoveries. In fact, the micro-politics of the Anthropological Society outline the contours of that social nexus to which Burton aspired to secure access and in which he wanted to remain.

## BLACKBALLED

After the aggressive growth and vibrant success of the first years, difficulties began to surface for the Anthropological Society in the shape of a widening imbalance between the members signed in and the money that actually rolled in from membership fees. There were a number of reasons for this phenomenon. First, given the scattered nature of the society's membership, there was of course a great difficulty in securing prompt payment—what could be done if a Madras surgeon was late by a few months for instance? However, while the society's leaders went by this theory and as a result entered late payments as a credit that accumulated against late-paying fellows, there was another phenomenon at work, the importance of which became increasingly clear over time—the zeal of previous canvassing efforts had led to signing in more subscribers than were actually willing to pay. In effect, the high point in membership reached in late 1866 would never be surpassed in the following two years. Instead of continuing with the early trends, membership and revenues leveled off at about their 1866 figures. The demand for anthropology was saturated.

The language of annual reports, which had been enthusiastic and fairly detailed, became unclear, hailing the new converts but glossing over the losses. Illustrating this, the report for 1867 remarked on the "number of fellows steadily increasing during the past year, the new admissions more than counterbalancing the losses by withdrawals and deaths, 152 having been elected."[21] Yet the gap between the number of members signed in and the actual paying members grew bigger. In 1867, when the society reported to Leone Levi up to 1,000 members, the revenue figures implied a number of subscriptions that, once we add the life subscribers, does put the society around perhaps 600, not implausibly 700 if one was generously counting late payers, still well below the 1,000 mark.

There is no doubt about Hunt and Burton's remarkable entrepreneurial achievement. They had constructed a society's subscription list, which remained a solid asset despite all its fledgling members. Even by the revised measure I am suggesting and even without counting the late payers, the Anthropological Society was still three times the size of the Ethnological Society in both membership and revenues and still ranked above other important societies such as the Statistical Society. Its size was a third of the Royal Geographical Society, a society that had been around for years, had a natural access to the higher levels of the Royal Navy, enjoyed a centrality in circles of British science, and, as already said, had provided British society with epic narratives.

However, with its aggressive campaigning, the Anthropological Society had put itself into debt. The society distributed its publications generously, for instance, without looking too much into whether subscriptions rolled in right away. Hunt himself came to admit that making sure that signed-in fellows would pay their dues proved challenging.[22] The debt was substantial but not crushing, nor was the society in 1867 on the verge of bankruptcy despite subsequent statements by quantitative light-heads. Indeed, debts equaled roughly one year of revenues, and they seem to have been carried by the publisher, the house of Trübner, which was happy to let the debt grow since this was a successful society that organized dinners with political heavyweights. Besides, the printer held the society's main tangible asset, its stock of books composed of such things as the translations of Vogt and Broca. If the situation had to be stabilized, then stabilization was by no means beyond reach. But this was a fragility the society's adversaries did not fail to exploit. In the middle of the failed merger discussion of 1868, Hyde Clarke declared to the Anthropological Society delegates that they had better consent to the terms offered by the Ethnological Society: "Your Society are in the position of toads under the harrow" (a contemporary Anglicism meaning to find oneself in a delicate situation).[23]

It was precisely in 1868, when the overextension or leverage of the Anthropological Society was becoming most visible, that Hyde Clarke, a member of both societies at the time, started the *Athenaeum* controversy. The mismatch was then largest between the number of members the society claimed and the number of members visible in the balance sheet through the fees collected. Significantly, the attack focused precisely, if not exclusively, on the sincerity of the society's accounts. In his letter to the *Athenaeum*, Hyde Clarke charged that the Anthropological Society's membership was suffering a hemorrhage, itself resulting from the bad governance habits of the Cannibals. Members, he claimed, had left "by the hundreds"; how otherwise to reconcile the numbers in the balance sheet and the membership figures that the Anthropological Society had boasted imprudently to Levi?

There was a self-fulfilling element in Hyde Clarke's accusations, for membership began to decline afterward.[24] His articles encouraged members to resign and late payers to forfeit their debts. New recruits became harder to win over. As a result of the letter and the ensuing dispute, the Anthropological Society was now finding itself in the position of "toads under the harrow" with its growth prospects severely hampered. On the one hand was the invective Hyde Clarke threw at the Anthropological Society and the effect it had on membership and the ability to raise credit. On the other was the difficulty for the Anthropological Society to expand beyond the size it had already reached, especially as long as access to the sacrosanct institutions of British science such as the British Association was closed off. But this treatment by the British Association and Murchison was fully justified, critics of the Anthropological Society could insist, given the ill-treatment— the "lampoons" as Hyde Clarke had it—that Huxley (leader of the Ethnological Society) and Murchison (leader of the Geographical Society) had suffered from Hunt. The Anthropological Society's "growth strategy" was petering out.

With Murchison's resistance unabated, Hunt's repeated attempts at establishing anthropology as a sovereign or emancipated citizen in the parliament of science were thwarted. During the Exeter meetings of the British Association in the summer of 1869 (one year after the *Athenaeum* controversy had begun), anthropologists were instructed to submit their papers to Section D (Biology), whose council was dominated by several ethnologists such as Busk, Lubbock, and Balfour. Die-hard Cannibals Beddoe, Pike, Charnock, Hunt, and Carter Blake withdrew their papers in protest.[25] This seems to have been the last blow to Hunt, whose health was fragile, as many earlier statements in the *Journal of the Anthropological Society of London* suggest. Reporting on Hunt's death, the *Western Mercury* wrote that he was "so annoyed and mortified at the treatment anthropology received at the British Association" in Exeter that he was seized by an excitement that brought a "brain fever" and fell in the middle of the street. He was sent home to Hastings in charge of a friend. His next action had been to summon the Cannibals to the Half-Moon Hotel for one last council of war to consider the course pursued toward the science of man by the British Association, but his health never rallied. Hunt died on August 29, 1869, literally when the British Association meetings were closing.

It must be written somewhere that the man would always be tracked by academic controversy, for death did not release him. On September 29, 1869, J. Frederick Collingwood, secretary of the Anthropological Society, published a rebuttal of the *Western Mercury* allegations in the *British Medical Journal*, stating that the "illness was induced by sun-stroke" (in Exeter?) rather than by the obstruction that anthropology had met at the British Association. "Those who knew Dr. Hunt were well aware of his ability to sustain opposition," Collingwood added. The *British Medical Journal* in turn wondered whether Hunt belonged to a new class of "martyrs of science" and reflected that "recent occurrences would suggest that, in the present day, the pursuit of science is more exciting than that of theology."[26]

Without Hunt, the society was deprived of its energetic leader, and membership kept falling. It was also deprived of the writing energy of Hunt, who had been a driving force of the *Anthropological Review*. The *Anthropological Review*'s arrangement with Trübner was terminated in 1868, and at the same date Trübner became the publisher of the Ethnological Society's journal, taking it over from the powerful John Murray. Given the existence of the Anthropological Society's debt towards Trübner, it is hard to believe that the move was coincidental. The most likely interpretation is that Trübner was receiving some financial guarantee from the Ethnologicals and collecting its reward.

In 1869 another publisher, Asher and Co., was found for the *Anthropological Review*. The last issue, dated April 1870, contained the text of the paper Hunt had intended to read during the fateful Exeter meeting of the British Association. Heralding a change of style, it reproduced exchanges that had taken place between Huxley and the Cannibals on the subject of the Celt and the Saxon. The tone was much more dispassionate than had been the case before. In 1870 the *Review* was replaced by a *Journal of Anthropology*, now published by Longmans, Green and Co. To describe "the aim and scope of anthropology" in the opening article of the new *Journal* in July 1870, veteran Cannibals displayed one Charles Staniland Wake, a young anthropologist who had entered the Anthropological Council in early 1868 upon Charnock's recommendation on the strength of his *Chapters on Man*, an essay in comparative anthropological psychology. Wake had not participated in the previous unpleasant exchanges. He was not even a supporter of polygenism. The essay, which was preoccupied with the re-branding of the Cannibals, argued that the origin of man was in fact not in the orbit of anthropology. Breaking with previous habits, he insisted on maintaining a low profile.[27]

And thus the Cannibals were sending signals of civilization, and some of their leaders showed their better side in exchange for due indemnifications.[28] In the last months of 1868, Burton, while traveling in Latin America, had received the coveted Damascus consulate from Lord Derby. Although the new Liberal government elected in late 1868 reconsidered several Tory consular appointments, this one was maintained. In the next years, beginning with council member W. S. W. Vaux in 1870, some Cannibals started being admitted to the Athenaeum Club.[29] The disappearance of major protagonists alleviated acrimony. Sir Roderick, at seventy-eight, was slowing down and would die in October 1871. When the British Association met in Liverpool in 1870, a separate department of ethnology and anthropology was created within Section D (Biology), of which Dr. Beddoe, a long-time Ethnologist and Anthropologist who had become president of the Anthropological Society following Hunt's death, was made a vice-president. It was further decided that Beddoe would join the Council of the British Association. With peace made with the British Association, the time was ripe for the amalgamation with the Ethnological Society, which occurred in early 1871 and gave birth to the Anthropological Institute of Great Britain and Ireland. Its *Journal of the Anthropological Institute* was carried by Trübner. The status of the Anthropological Institute reached a new high in 1907 when it became the Royal Anthropological Institute, the name under which it still exists today.[30]

## THE DESCENT OF THE ANTHROPOLOGICAL INSTITUTE

In describing the last moments of the Anthropological Society, conventional accounts generally suggest that Anthropologicals were essentially bankrupt and that this was the reason for the merger. Mark Patton, for instance, writes, "by the beginning of 1871, with the Confederate funds having dried up, and amid allegations of fraud, the Anthropological Society was bankrupt." What is unclear in this interpretation, however, is why Lubbock and Busk, beyond their splendid manners and gentlemanly intentions, proceeded the way they did. If the Anthropological Society was almost bankrupt in 1871, then why did Lubbock, the banker informed by Hyde Clarke, the careful auditor of scientific bottom lines, settle for a merger that gave Ethnologicals only half of the seats and saw the triumph of the abhorred word "anthropology" now used for the common organization? Why didn't they just let it go rather than perform this curious Ethnological bailout of Anthropological debts?[31]

    In fact, the triumph of the word *anthropology* bears witness to the power relationship between the two parties. Despite the difficulties of the Anthropological Society, the brand was still perceived as attractive. Moreover, at the time of the merger, and despite the bloodletting that the Anthropological Society's membership had experienced, there were still many more anthropologists than ethnologists. This meant that the ability of the Anthropological Society to raise cash through subscriptions was still superior to that of the Ethnological Society. Even when subjected to attacks on its credibility, even when decapitated, even when suffering from financial strain, the Anthropological Society was still imposing. When Leone Levi updated his statistics for scientific societies in 1879, he forgot altogether the

Ethnological origins of the Anthropological Institute and construed the Anthropological Society as the sole ancestor of the Anthropological Institute.

Rather than recognizing this, Lubbock and his friends organized a much-repeated story—that of the "debts of the Anthropological." The matter came to the fore in late 1873 and early 1874, when it was decided to raise a special redemption fund with the ostensible purpose of freeing the Anthropological Institute from the debts it had inherited from its predecessor institutions—chiefly, it was repeated, from the Anthropological Society. Following the adroit suggestion of Busk and Lubbock, a subscription was opened "so as to relieve [the Institute] of its load of debt." The subscription became a way to give a tangible shape to the previous misconduct of the Anthropological Society. The price of atonement was set at £700, the amount of the "Anthropological Society's debt." When one reads modern authors, one almost comes under the impression that the £700 was the salary of racism. The list of contributors was printed with the names of Busk and Lubbock opening and closing it respectively.[32]

Yet the extent to which the debts of the Anthropological Society had been entirely the result of anthropological misconduct, as has been subsequently claimed, may be disputed. First and foremost, the debt had been incurred in large part to canvass members who were now part of the merged institute. These members represented a large earning capacity for the merged institute. Because of the initial predominance of ex-anthropologists in the rank and file of the new society, a more rigorous computation would adjust the debts of the Anthropological Institute by its assets, with the effect of writing its overall contribution in black rather than in red. Second, as shown in figure 2, the merger was followed by a hemorrhage of fellows, the result either of little outward promotional effort by the new institution or of the continuing bickering that took place inside, especially in 1872–73 as each clique tried to pack meetings and outmaneuver the other.[33] In other words, the extent to which existing debts were sustainable or not hinged on policies adopted after the death of Hunt—after the Anthropological Society's accounting books had been closed.

Thus available elements suggest an alternative interpretation of the descent of the Anthropological Institute in 1871, one that has Lubbock and his allies taking over the Anthropological Society. In this alternative story, the Ethnological Society, distanced in the 1860s in the race for membership, would have drawn the greatest benefits from the *Athenaeum* controversy, making it understandable that the man who led the attack was a member of the Ethnological Society's Council. This had dented the Anthropological

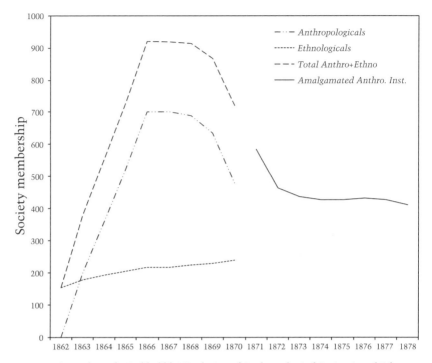

FIGURE 2. An anthropological bubble? Evolution of Anthropological Society's and Ethnological Society's and Anthropological Institute's membership (1862–1878). *Source*: Author,
from material discussed in supplements.

Society, crippled its money earning capacity, and eventually permitted a
merger on equal terms. Next, the financial difficulties that resulted from
the post-merger period gave ethnologists a decisive edge. Their superior social connections gave them access to pools of charity money on a typically
larger scale than their competitors. Indeed, subscriptions to the redemption
fund show the average ethnological contribution at about twice the size of
the average anthropological contribution. A financial squeeze gave an edge
to wealthy individuals over the middle class, and to the logic of charity over
an income-generating, profit-oriented approach.

   In other words, everything happened as if the Ethnological Society had
purchased, cheaply and without deploying all of Hunt's efforts, the goodwill
of the Anthropological Society. Leaving aside the subject of the Ethnological leaders' intentions, the result of the successive actions could be likened
to vulture capitalists rocking companies in order to take advantage of them.
For £700, the Anthropological Society's nuisance factor had been acquired

and put to rest. Following this line of reasoning, it may not be so surprising that anthropology subsequently entered a period previous scholars have characterized as somewhat dull—according to some, the age of "armchair anthropology." But should we be surprised given that competition had given way to a monopoly, and that pompous men had taken over the scoundrels?

In contrast with conventional accounts, which argue that ethnological common sense prevailed against the racism and sexism of anthropology and that this moral economy must have had some counterpart in financial returns, this chapter has argued that balance-sheet numbers are equivocal at best and that the successes of the Anthropological Society ought not to be underestimated. If we take seriously the financial dispute that developed at this point and formulate it in the language of business rather than that of a BBC educational program, then we discover that a twofold meaning of credit was at work: credit scientific and credit financial, the two sides of the same coin in the attack by Hyde Clarke.

In lieu of the modern scholarship that emphasizes the moral high ground of ethnology, Darwin's sacred cause, and the ethically logical triumph of ethnology, an alternative story emerges that puts social conflict and the creation and destruction of credit, understood in a financial way, at the heart of what happened. To summarize, the new, richer narrative would include the initial success of the Anthropological Society, vibrant and unparalleled; the significance of the Royal Geographical Society; the understanding by Burton, a hero invented by the Geographical Society, that selling Africa after the Arctic would be a hit; the way he and Hunt took advantage of the controversies of the time to spell out their sales pitch; the challenge Anthropological success posed to the scientific establishment of the time, and to that ridgepole of interlocking directorates of learned societies, the British Association; the resulting confrontation with Murchison and the Geographical Society; the aborted Ethno-Anthropological merger in 1868, shunned by the arrogant anthropologists because they held the high ground; the Anthropological Society's problems with overextended finances; the successive shocks experienced through Hyde Clarke's attacks on the society's financial credit; the death of Hunt; the merger of 1871, which gave ethnologists a foothold in anthropology; and finally the continuing tension and membership decline that provided Lubbock's circle (or clique or faction, for it's all one) with almost complete control over the Anthropological Institute.

Interestingly, the history of anthropology has the distinct quality of a financial undertaking where it's "speculation if you lose" according to the formula. Bubble-shaped anthropology is visible in how rising and falling membership numbers fit the pattern, and it also transpires in the language of contemporaries—perhaps most characteristically, the language of opponents such as Ellen Lubbock, who described anthropology as a dismal and bloated mushroom, a bubble whose eventual puncturing had been needed. This takes me to the ambition of this book—to tell the story of a science with the language of speculation. In history, bubbles have taken place in such heteroclite markets as tulips, South Seas companies, and homes. There is nothing really eccentric in the endeavor apart from its having never been attempted before.

Such is the conjecture this book will seek to elaborate upon. It will recognize the importance of the material motives that underpinned the question of anthropological and ethnological knowledge; it will tell the adventures of the science of man as if they were a subject of political economy; it will discuss the success of the Anthropological Society and probe whether it was genuine or a bubble and mushroom; it will assess the durability of a learned society and ask whether it is a question of truth or credit; and it will explore the history of anthropology and wonder whether it is best informed by a discussion about the identity of science or by the tools normally used to discuss mergers and acquisitions.

Note that it is already striking how language and metaphor bring us inexorably beyond the taboo that separates the interested and the disinterested and toward their interconnections. Those writers who have pictured the Anthropological Society as a conspiracy inspired and bankrolled by the Confederacy were not entirely wrong in that they rightly heard that a matter of business was lurking. But their notion of an Anthropological Society manipulated by a foreign power, itself possibly motivated by their concern to purify the origins of anthropology, required them to detain the Cannibals in a zoological garden. I want to set the Cannibals free with traceable devices tacked on their backs and follow them on my Internet and archival radar as they disappear into the wilds of money and science.

CHAPTER FOUR

# The Ogre of Foreign Loans

"Fee-Fo-Fi-Fum! I smell the coin of an Englishman! / Yes I am the ogre of Foreign Loans / I'm a curious sort of beast / I fatten on widows' and orphans' bones / And revel and shout when I hear their groans / I'd rather hear them than the sweetest of tones while making my usual feats." Thus roared the moneybag-bellied "Terrible Monster" in a caricature the Dalziels splashed in *Fun* magazine on April 10, 1875 (figure 3). Once again, the disgraceful foreign loan ogre had gobbled up the savings of investors. His line was adapted from "Jack and the Beanstalk," but in the more explicitly cannibal original rhyme, the smell of blood rather than the smell of coin gives the Englishman away: "Fee-fi-fo-fum / I smell the blood of an Englishman / Be he live, or be he dead / I'll grind his bones to make my bread." But then, as Marx had noted, wasn't the coin the true lifeblood of the Victorian Englishman?[1]

As the cartoon shows, the ogre was not allowed to perpetrate mischief. On the bottom left corner of the drawing, we can see the tiny figure of the Liberal M.P. Sir Henry James in the black robe of the lawyer, exposing the feasting beast with a kerosene lantern (and holding a slender, less-than-threatening blade). For introduction, picture James as a kind of Victorian Ferdinand Pecora, if such a typically American hero can ever have had a counterpart in Victorian Britain. Ferdinand Pecora (1882–1971) was the Italian-American lawyer who became famous in the 1930s as chief counsel to the United States Senate Committee on Banking and Currency for the way he led the examination of Wall Street banking and stock brokerage practices and exposed corporate wrongdoings. In a like fashion, James, attorney general for the Liberal government of William Ewart Gladstone during its last year of rule before the triumph of Disraeli in 1874, was to the general public the mastermind of the Select Committee on Loans to Foreign States, the parliamentary inquiry into corporate wrongdoing signified by the lantern in the

77

FIGURE 3. Another man-eater: "The Ogre of Foreign Loans." *Source*: *Fun* magazine, April 10, 1875.

cartoon. As the poem concludes: "Thanks to the gallant Sir Henry James / . . . / The ogre must soon bid adieu to his games."[2]

Or not. The ogre of foreign loans had fed on the mercantile expansion of the 1850s and 1860s when booming shipping and trade motivated investment in improved transportation infrastructures: railways, canals, and telegraphs.[3] Roads served to connect markets, and the rise of a world economy brought the rise in the demand for connection. Roads also served to shift relative political values. Land that used to be bought for a song became highly priced and ready to welcome the flow of migrants who were leaving Europe for the New World. Roads finally served as instruments of external influence. Foreign ownership of a road created means of cooperation with, and control of, local elites and thus became an integral part of programs intended to subvert local opposition and redesign the world.[4] This project stimulated the creativity of advertisers. It was an age, one contemporary would write in 1875, "which has seen electric cables stretched across the ocean, a passage for ships opened between the Mediterranean and the Red Sea across the Isthmus of Suez, locomotives traversing the Alps after the perforation of an immense tunnel in Mount Cenis, and works prepared for opening another tunnel from England to France under the Channel."[5] To what extent did purses open automatically at the barker's pitch?

Many did and do exaggerate the legendary naïveté of the investor. On the other end of the rope was the dishonesty of the borrowing government, itself no less legendary. In the language of the time, many borrowing governments lacked "character." As the century advanced, a story evolved that mapped credit and creditworthiness to the traits of borrowers such as culture, mental habits, and irrational behavior. This was a capital market story, one of trustworthiness. The language of banking and finance had prepared the ground for the cultural narrative. Just like financiers warning one another against certain borrowers, character was routinely spoken of in the press— not only in the City article—and character led to race and race thus became naturally the concept of choice to discuss creditworthiness. The resulting financial racism provided a natural highway whereby representations could arrive that criminalized the underdeveloped. They dot nineteenth-century written productions, always hinting at the superior credit of the imperial or colonial arrangement. Thus, for instance, the *Times*, commenting on a Paraguayan loan in 1872 that would be defaulted upon and included within the scope of the select committee's investigations, wrote: "If the outlay is to be controlled by Paraguayan politicians, persons acquainted with the region will not be sanguine as to its results. If, on the other hand the works and their costs are to be regulated by a really responsible London committee, the possibility of great advantages may be admitted."[6]

Forty years later, a prominent British administrator in Egypt who also happened to come from a leading banking family would declare of the Oriental borrower, "the Western financier will act wisely if, casting aside some portion of his Western habit of thought, he recognizes the facts with which he has to deal, and if, fully appreciating the intimate connection between finance and politics in an Eastern country, he endeavors, so far as is possible, to temper the clean-cut science of his fiscal measures in such a manner as to suit the customs and intellectual standard of the subject race with which he has to deal."[7] The subject race, on the other hand, was to be governed by men of character, and historian P. J. Cain has spotted the language of character used to justify the possession of empire in the mouths or under the pens of what he calls the "ultra-imperialists" of the last part of the nineteenth century. I am unsure why Cain has spoken of "ultra-imperialists," however. For such turns of phrase were hardly limited to the "ultra"—they belonged to the language of capital and capitalism.[8] The parallel between this and the modern theory of the serial defaulter, according to which some countries just have lower political morality, is striking. This theory promoted by economists Carmen Reinhart and Kenneth Rogoff builds on the "observation" that certain countries and parts of the world exhibit greater incidences of default. In its historical variant, it goes back to a treatise by lawyer Robert Dudley Baxter, a member of the Statistical Society whose study of national debt dynamics suggested to him confirmations on "current hypotheses" on the "decline of the Latin race and ascending march of Anglo-Saxon and Teutonic races." The concept could be generalized and map shades of skin into debt dynamics.[9]

This link between foreign lending and Victorian prejudice regarding the character of certain borrowers is important: It turns out to have been an indirect (and probably unknown) source of inspiration for literary theorist Edward Said's popularization of the concept of "orientalism" as a way to characterize stigma and the fantasizing of the oriental. In his *Orientalism*, published in 1978, Said, who had studied writer Joseph Conrad in an earlier book drawn from his dissertation, now dissected in detail the process of stereotyping he had observed at work in Conrad's *Nostromo*. *Nostromo* is a novel that could be defined as a tragedy of character, for character and reputation are indeed the main obsession and tragedy of the novel's title character. This gives background to Said's discussion of the process whereby the untrustworthiness of the Latin American was meticulously stigmatized.

As few people realize however, one of Conrad's sources of information had been a book written by one Edward B. Eastwick, *Venezuela; or, Sketches of Life in a South American Republic, with a History of the Loan of 1864*, published in 1868. Eastwick typifies many of the characters in this tale. He had begun a military career in India in the 1830s when his rapid mastery of Indian languages led the East India Company to employ him in a political capacity. Back in Britain in the 1840s, he was appointed professor of Hindustani at Haileybury College, was elected a fellow of the Royal Society in 1851, and became a scholar-diplomat as secretary to the British Legation in Persia (1860–63). Upon his return in 1864, he was employed by Samuel Laing's General Credit and Finance Company (GCFC) to monitor a loan Laing had just issued in partnership with the Frankfurt House of Oppenheim to the benefit of the new

government of Venezuela of General Juan Crisostómo Falcón. The loan—at
10 percent—had been shunned by Barings who, in terms of trustworthiness,
preferred the recently toppled caudillo José Antonio Páez. As far as the punctual payment of the new debt was concerned, Barings was soon found to have
been right. Through Eastwick's narrative, this inspired Conrad's remarks on
the imaginary Latin American republic of Costaguana as being a "bottomless
pit of 10 percent loans and other fool investments."[10]

Among the reasons for his involvement in this sovereign debt interlude,
Eastwick the orientalist-turned-financier mentioned the pleasure of seeing a
new country, of learning a new language, and the experience with financial
transactions he should gain in such a mission. Besides, Eastwick notes, he
had just spoken with Richard Cobden, the patron saint of all men of many
doorplates, and Cobden had told Eastwick: "Why don't you go to the City?
They will treat you better there." It is probable that the General Credit and Finance Company had purchased Eastwick's social capital. Soon enough, however, financial troubles began, and Eastwick was back in Europe busy with
other matters. Prompted by Charles Dickens, Eastwick's *Sketches* were half
political ethnography, half controversial revelation on the loan, augmented
by a tribute to the policies of Disraeli's Conservative government applaud-
ing its more assertive, manly—in one word, British—policy toward defaulting races. The book was dedicated to Sir Philip Rose, Disraeli's right hand
and a senior partner in the firm of Baxter, Rose, Norton and Co. of which "racist" statistician Robert Dudley Baxter was also a partner. The *Sketches* were
released right in the middle of the political and scientific controversies on
which this book focuses, before finding their way into *Nostromo* and as a result, through this tortuous chain of events, becoming an element of background for Said's critique of orientalism and tying it to the raptures of international finance.[11]

## FRAUD AS SOCIABILITY

As the select committee's investigations would discover, the foreign loan
ogre fed not so much on the ignorance of widows and children in the coun-
tryside or on the depravity of indigenous governments in the jungles as it did
on the shady machinations of various intermediaries in the London Stock
Exchange. The aforementioned Venezuelan loan offers a case in point. It
ended up in a trial in which one contractor (Oppenheim) sued the other (the
General Credit and Finance Company) over the subject of appropriation of
underwriting fees (Oppenheim alleged that the GCFC would not have been

sharing the fees in a sincere way).[12] Such trials, which became increasingly
frequent, brought publicity to the operation of the stock exchange and more
generally to company fraud. These revelations added to the earlier narratives of a new genre by journalists such as Morier Evans's *Facts, Failures and
Frauds* published in 1859.

Along the way grew a significant literature on the subject. On the side of
fiction, Charles Dickens, Robert Louis Stevenson, and of course Anthony Trollope tried their hands at it. Fiction and nonfiction mingled. In 1865 Malcolm
Ronald Laing-Meason published in Dickens's *All the Year Round* a series of articles detailing the techniques of financial abuse. In anticipation of Trollope's
railway from Salt Lake City to Vera Cruz, Laing-Meason cast the story as one
of an imaginary Rio Grande railway and a Patagonia bank. The articles were
combined in a volume published that year as *The Bubbles of Finance* under the
authorship of "a city man." In the aftermath of the Overend, Gurney Panic of
1866, Laing-Meason came out with a book suggestively called *The Profits of
Panics, Showing How Financial Storms Arise, Who Make Money by Them,
Who Are the Losers, and Other Revelations*. As Laing-Meason turned to investigating turf frauds, Dickens continued to cover the topic with the help of
Eastwick, the former employee of the General Credit and Finance Company
whose manager, Samuel Laing, appears to have been related to Laing-Meason.[13]

As indicated, the report of the select committee focused on the role of
the London Stock Exchange. In the collection of propositions offered for sale,
the stock exchange had not been able to discriminate. A collection of fraudulent schemes had made their way in the pile. Contemporary fraud had its
patterns, style, and technologies, and real-life models provided counterparts
to the stories told by Laing-Meason and others. After the select committee's
investigations, abuse became, so to speak, officially recognized. The typical
railway machination relied on three-way partnerships between the agents
for the borrowing government, the underwriting bankers, and the engineering and construction companies that had conspired to exploit both the investors and the taxpayers from borrowing countries. In other words, the loans
may have been made for railways in Honduras, Costa Rica, Paraguay, Liberia, and Egypt, but the swindles were nonetheless concocted in the financial
headquarters of the capitalist world. In the Dalziel cartoon, the name tagged
on the belly-bag where the purloined coin-of-the-Englishman is stored—"Le
Swindle, Cheatham & Co"—hints of a French ingredient, an indication of
the lack of sincerity that characterized the cosmopolitan crowd of this global
market.

Key in the process was what is known today as the "road show," a term
employed to designate the series of presentations that investment bankers

use to promote a financial investment. In the nineteenth century, the foreign debt road show made extensive use of existing social platforms or created social events to launch the loan. Respectability was an essential condition for raising money, giving rise to "new forms of ingenious crime, which could only be perpetrated by business men, and by large, prominent, wealthy or at least credit-worthy business men at that," as historian Harold Perkin has described it. The growth of the mercantile economy raised the value of names as social signals of trustworthiness and thus, by the same token, created incentives to compromise those very names as reflected in the increased use of dummy directors—aristocrats or well-known individuals sitting for a fee on the boards of more or less dubious schemes. The result was what Martin Joel Wiener identified as a tight association between respectability and criminality. This explains the importance of clubs, societies, and other instruments of social exchange and validation set in motion by the contractors—or originators, to use the modern expression—of foreign loans.[14]

Henry Drummond Wolff, a diplomat and Conservative politician who had been involved in several bondholders' committees operating during this period, has left an ironical account of the role of sociabilities and social technologies in the launching of a new loan:

Before the issue of the loan, a dinner is often given in the name of the financier, but at the cost of the Syndicate. It fairly comes within the item "market expenses." Here our financier presides, the diplomatic representative of the country on his right, a relative of the Finance Minister upon his left. Near them are some retired Indian and Colonial officials beribboned, a few Admirals, a Member of a late Government—all anxious to become Directors of Companies, all counting upon selling the new loan at a premium. Near them are capitalists, leading stockbrokers, some newspaper-writers, a politician who has perhaps made a hobby in Parliament of the country in question, and a clergyman whose face is well known to most of the guests as a dabbler in Turks and Lombards, and as an eager though silent attendant at meetings of discontented bondholders. Dinner over, and the speculative divine having said grace, speeches are made by the M.P. and the financier, who look forward to this loan as a new pledge of alliance between the two countries. The prospectus-maker airs the adjectives cut out of his prospectus by the solicitor. The diplomatic representative makes speeches in a language utterly unknown to his fellow-guests. The financier, not the better for his dinner, confides in English equally unintelligible to the Finance Minister's relative that the loan, if successful, will be entirely owing to his—the financier's—

abilities; that if unsuccessful failure will only be attributable to the rotten and bankrupt state of the borrowing country.[15]

And thus it is that despite all the prejudices members of the select committee may have nurtured about the character of the Latin American, the final assessment they provided of the foreign debt mania depicted it as a London-centered problem of financial predation rather than one of endemic Southern delinquency. The source of the trouble examined in 1875 was not with the foreign government, although it had done its share, but with British society. As they found, one had to give up the pretense that the curse of foreign loans was the dishonesty of foreign states because only a small portion of the money had reached the borrowers. This was carefully documented in the case of Honduras, whose investigation absorbed the biggest part of the energy of the committee. As it found, "enormous sums had been abstracted . . . and appropriated among those who were entrusted with its management." The experience manifested the existence of a peculiar class of traffickers whose intention was to exploit both ends—borrower *and* investor. The government or railway loan was an excuse for white-collar crime. In the final assessment, the select committee reported being in agreement with the Honduras Legation that "the fault of the failure falls with equal force upon all who have interests, rights, claims, complaints, or any participation whatever in these matters. It is a kind of *original sin*, which reached even the most innocent who have anything to do with this undertaking."[16] In other words, the real cannibals were in London. The Dalziel brothers agreed on this too when they portrayed the ogre as distinctly white-faced, wearing the checkered trousers of the London Stock Exchange jobber.

Contemporary pundits offered their insights on why this had happened. According to Walter Bagehot, the editor of the Liberal-leaning newspaper the *Economist* (and one of the fellows of the Ethnological Society), the reason "Why the Stock Exchange Is Likely to Have More and Greater Frauds in It Than Any Other Market" is that financial schemes do not abide by the laws of supply and demand. They have the magic property of knowing no other limit than the imagination of the projectors. In the case of a loan to a new state or railway or distant mine, the "article is worth just what the public can be made to believe that it is worth." The effect was that the stock exchange presided over what Bagehot called "the 'inception' of the value." Price, in the end, was a "fancy" (a claim that Karl Marx, with his theory of fictitious capital, would not have disowned).[17] In sum, the sacrosanct scarcity of goods on which the price system relied to perform its allocative role just did not apply.[18] In a society that put the Darwinian sauce on every plate (and to be fair, Darwin confessed having put the *Economist*'s editor sauce on his own when he published his *Descent of Man*), the result was described by Bagehot as a predator-prey relationship where the "unskilled outsiders" (the widow and orphan) were being devoured by the "insiders" and thus kept "alive the group of skilled machinators at head-quarters."[19] Such would have been the ecology of the foreign loan ogre. And in the end, the sustainability of the ogre hinged critically on the ability to create an impression of truth, something that required more than gifted salesmanship. It required organization and, as we shall see in this chapter, the techniques of science of which learned societies were masters and depositaries.

## BANKING DELINQUENCIES

The select committee exposed to the wider public the tricks used to deceive investors. Despite all that is usually said about the stupidity of human nature and its inability to learn the facts, nothing was in truth more difficult than selling a bad loan, and the low prices at which they were issued, offering returns above 9 percent when British government bonds yielded 3.5 percent, were not always enough to convince buyers. Once the loans had been issued, and since many bonds remained unsold, the schemers thus resorted to various tactics intending to create a fictitious market, which would hopefully lead investors to lose their minds by persuading them that others were making money. Speaking success is showing success: the most popular trick, which the select committee documented in detail, was to bait the investors through the simultaneous buying and selling of the new security at a premium in order to create the impression of easy financial gains. One broker would be instructed by the sponsoring contractor to sell at, say, a 5 percent premium over issue price while another broker would receive instructions to buy at the same price. This was especially done before the security would be fully subscribed or allotted at a time when the market lacked transparency because a large share of the issue was still in the hands of the sponsor.

These insincere prices resulted from dubious transactions that used stock exchange brokers but were not properly speaking taking place *on* the exchange and thus not reported on the official stock exchange list but in a few newspapers whose goodwill had been purchased in one way or another. One contemporary judge described the technique as essentially going into the market and pretending to buy shares through "a person whom you put forward to buy them . . . in order that they may be quoted in the public papers as bearing a premium."[20] The scandal and much of the surrounding evidence discussed in courts brought to the fore the pervasive practice of rigging the market. With such artifices, which became the principal focus of the examinations of the select committee, a price increase in early trading or "run-up" would be recorded in the media and by salesmen who could point to this premium as an argument for the loan. (Technically, the run-up is the price difference between the price at which a security trades in early markets and the issue price. A high run-up is suggestive of a successful issue.)

This explains why journalists and opinion makers were so closely involved in the process of launching securities. As the select committee emphasized, the editors of the "City article" in the more austere newspapers had succumbed to the temptation of reporting so-called pre-allotment prices and in effect helped puffers. Revelations on the lavish lifestyle of Marmaduke B. Sampson, the City article editor of the *Times* who had purchased a mansion in Hampton Court in 1871 and was reported to have accumulated £400,000 from his practice as City editor, created much controversy. Sampson had trained as a committee secretary of the Bank of England in the 1840s. He also lectured on criminal phrenology and had an interest in Central American transit.[21] His name suddenly came up when a former associate of "Baron" Albert Grant (the real-life model for Trollope's Melmotte), a man with the improbable name of Rubery, sued Grant, alleging that he had threatened him in the course of a dispute of using against him the pen of Marmaduke B. Sampson. Marmaduke B. Sampson was, Grant countered, "an intimate friend of his, and a very useful man to the commercial community." On cross-examination, Grant reiterated his allegations, admitting that "various large sums of money, in one year amounting £5,000, had passed between him and Mr. Sampson." He had also, he explained, "allotted shares to Mr. Sampson" and went on to defend this as the "usual custom," not only very good but "laudable." Why, Mr. Sampson had taken a "great deal of trouble to obtain information," and it was only right that "his labors should be recognized."[22] Sampson was forced to resign, causing much commentary.[23] "Signs of the Times," *Fun* jested.[24]

Of course, it was difficult for an investor in Glasgow to know that the transaction was not real and that, through the agency of the underwriter or contractor, the same person was buying and selling. As the report of the select committee put the case, the public, having "no means of learning that the contractor was principal to both transactions," was led to believe that the fancy price was rising. This was the mechanism used to try and jump-start the market and the speculation in new shares and bonds.[25] In the mind of contemporaries, the techniques used for puffing securities relied on the opacity of the market and were thus described as jobbing since the jobber was the individual who went on the market and actually passed the orders. The select committee put it in more stilted parliamentary paper style: They talked of "flagrantly deceptive techniques."[26]

The other aspect of baiting involved the telling of a development story, usually a Great Transportation Epic, a narrative of progress and profit— continents domesticated, oceans connected, and 9 or even 12 percent. The report of the select committee would outline the misrepresentation of the borrowers' true situation made in financial prospectuses (*Fun's* ogre cartoon has a perplexed bespectacled top-hatted man poring over the "PROSPEC- TUS"). They contained "exaggerated statements" of the "material wealth" of the contracting states. Just like the decoy of mock transaction, such tech- niques had a long tradition. At the beginning of the century, the wealth story was originated in the London Stock Exchange along with the loan on which it was grafted. Bankers and brokers sought the help of journalists who had become experts in the puffing business through their promotions of literature. The largely uncharted nature of foreign prospects left much room for Bagehot's "fancy." Contractors typically resorted to the hired pam- phlet or journal article or produced the journal and pamphlet as docile in- struments of puffing.

A now-famous early nineteenth-century example was the association in puffing between publisher John Murray, whose house was tightly related to British science, and Colombian debt and foreign mines promoter John Dis- ton Powles. Murray had employed the writing talent of the young Benjamin Disraeli to argue in the middle of the 1825 London Stock Exchange bubble that the market could not possibly go down. And the way Disraeli wrote scathingly about the incantation of the counter-puffers—"bubble! bubble!"— reminds one of the incantation of the witches in Macbeth: "Double! Dou- ble!"[27] (A few years later, Disraeli would remember the experience and de- scribe Diston Powles in *Vivian Grey* under the traits of one "Mr. Premium.") Another well-known case is that of an enterprising Scot called Gregor Mc- Gregor who had sold the bonds of the imaginary kingdom of Poyais along the Coast of Mosquito, where he also managed to send settlers. Poyais had its geographer, one Thomas "Strangeways," "KGC," captain of the First Na- tive Poyer Regiment and aide-de-camp to his Highness Gregor Cazique of Poyais who wrote a *Sketch of the Mosquito Shore* (it is believed today that Strangeways was no one else than McGregor himself).[28]

These fascinating but poorly understood times often lead to incorrect characterizations of which modern financial media are fond.[29] As suggested in a later chapter and for a later episode involving the Mosquito Coast, it is not quite right to say that the kingdom of Poyais did not exist. Rather, it was a relevant, if singular, British protectorate—singular because the protection took the form of the story of an old alliance between the indigenous king ("the King of Mosquito") and the British monarchy. And this narrative had tangible effect, as explained by historian Robert A. Naylor, who shows how it had been relied upon by British policy makers and traders to further their interests at the expense of Spain.[30]

But beyond that, and somehow systematizing the principle upon which Strangeway's "geography" rested, many publications existed that sought to provide investors with some information on foreign securities traded on the exchange. They were typically produced by stock exchange brokers who sold the concept by emphasizing the need for exhaustive and accurate information. From the 1820s onward, the subsequent editions of market compendia such as Thomas Fortune's *Epitome of the Stocks and Public Funds*, Carey's *Every man His Own Stock-Broker*, Fenn's *Compendium of the English and Foreign Funds* or some occasional volumes such as Bernard Cohen's *Compendium of Finance: Containing an Account of the Origin, Progress and Present State of the Public Debts, Revenue, Expenditure Etc.* began documenting the position of foreign countries in detail. Safety was sought in numbers, and importantly, this demand for figures grew independently of the colonial state or the colonial bureaucracy. Rather, it was a part of, and a result of, the desires of the stock exchange. This was how a statistical charting of the world, a statistical mode of knowing, came to be created at the impulse of the stock exchange. Numbers were incomplete, data imperfect, and even figures for the good borrower were fragmentary at best, but this did not reduce the urge, much to the contrary. The will to know and the gaping holes in figures conspired to invite other forms of imagining that would fill the void. A kind of primitive anthropography was invented that substituted for the deficit in order to gauge the creditworthiness of the borrowing state—considerations about the character of people, the natural offspring of financial engineering. Consider the following entry from the 1833 edition of *Fortune's Epitome*: "Oh Spain! Who hast bartered thy former heroic valour and chivalric prowess for beads relics and pilgrimage, where are now thy gains? Where is the noble Castilian blood that once flowed in thy veins?"[31]

The appetite the stock exchange displayed for data grew with the appetite for loans, and they fed one another. Significantly, the promoters generally supplied the information themselves as in the case of McGregor's geography of Poyais. The connection between the publications mentioned above and the sell side of the market leaves little doubt since as indicated many of the originators of these publications were themselves more or less eminent brokers.[32] As Herbert Spencer (from whom I borrow the title for this section) explained in his "Morals of Trade": "But it is scarcely to be expected that [trade circulars put out by the sellers] should be quite honest. Those who issue them, being in most cases interested in the prices of the commodities referred to in their circulars, are swayed by their interests in the representations they make respecting the probabilities of the future. Far-seeing retailers are on their guard against this."[33]

As a result, specialists of information provision, if they wanted to be believed, had to arrange displays that suggested they were dealing at arm's length with the sellers of financial products. An interesting example that reflected this evolution was the Nashes. R. L. Nash Sr. is known for pioneering accounting techniques to sort out the performance of railways during the railway mania of the 1840s, and he did it under the authority of the London Stock Exchange, leading in the late 1840s to violent disputes over his interestedness. His son, R. L. Nash Jr., subsequently took care of editing the updates of *Fenn's Compendium* after 1867 when the foreign debt mania started. According to Andrew Odlyzko, a University of Minnesota mathematician and amateur historian of accounting, the younger Nash might have worked simultaneously for the *Investors Monthly Manual*, an influential magazine for investors started by the *Economist* in the critical juncture of the 1860s. A sign of the times, the son's link with the market was no longer as evident as had been the case for his father.

Suggestively, we see him moving toward further scientific posturing while simultaneously involving himself in foreign debt monitoring. Logically for an individual interested in scientifically probing the character of countries, he joined the Anthropological Society of London in 1868.[34] In the 1880s, he would conceive a statistical ratio, which he called the "trade test," that was meant to predict the likelihood of a foreign government default.[35] In other words, the producer of information and financial opinion looked increasingly emancipated from the sell side of the market. Emancipation coincided with both increased assertiveness displayed by the press (reference to the fourth estate was not yet overused when in 1850 Frederick Knight Hunt had published the first volume of his *The Fourth Estate: Contributions Towards a History of Newspapers, and of the Liberty of the Press*) and the political triumphs of the leaders of the press-agentry industry such as Richard Cobden, whose campaigns against newspaper stamp duties—the "knowledge tax"—succeeded in 1854. As we shall see in this book, this reinvention of the media man from sinner employed by the tradesman to independent preacher of moral truths is crucial to the story of the relationship between anthropology and the stock exchange.

## BONDHOLDING DECONSTRUCTED

Opposite the conventional image of the destitute investor losing it all in foreign debt markets, bondholders of the Victorian era are better described as an industry, and their leaders were powerful, extremely wealthy men.[36] Their business was started after the foreign debt bubble of the 1820s. While on the one hand, loans of the more solid borrowers, mostly from western Europe and originated by the House of Rothschild, sold well and delivered good performances, the other countries' bonds—essentially new countries that included the already mentioned examples of Colombia and Poyais— had been originated by a variety of smaller merchant banks and failed almost instantaneously. In most cases, they did not even sell out and were quickly repudiated by their governments. There were some for which the London minister of the borrowing country had found it profitable just to issue the loan so as to generate a commission but forgot about asking permission from his government.[37] Following this debacle, and under attacks from newspapers about the lack of commercial faith, efforts at demonstrating goodwill were often made by those who had originated the securities, such as John Diston Powles, the distributor of a Colombian debt, and Samson Ricardo, a brother of economist David Ricardo and a stock exchange broker who had participated in the issue of a Greek independence loan. These financiers took it upon themselves to organize the bondholders.

They created committees that drew legitimacy by claiming they were representatives of a society of bondholders. They also proclaimed themselves bondholders although the conditions in which they had acquired the bonds were shady. Given the confused way in which securities had been distributed, it was unclear whether theirs were securities seriously acquired at the time of the issue, whether they had somehow picked the bonds at bargain prices from the few investors who had been imprudent enough to subscribe, or whether such securities consisted in the enormous piles of leftovers, which as bankers they had been unable to distribute. This is how bondholder committees led by what amounted to vulture investors were put together during the 1830s and 1840s. There were the self-styled Mexican bondholders and the Greek bondholders, and the Colombian bondholders. These groups had a loose governance with a permanent committee leading the group, calling meetings, harassing the Foreign Office, putting announcements in the press, and bribing the underpaid British consul into pressing the local government with more or less vague threats.

Those who got involved in defaulted foreign government debt were not poor individuals. Beyond Powles (*Vivian Grey*'s "Mr. Premium") and

Ricardo (a prominent London Stock Exchange brokerage firm), we find extremely rich men such as Richard Thornton, said to have been one of the wealthiest Englishmen of the nineteenth century.[38] Of him, David Morier Evans, this fine connoisseur of City men, wrote in 1852 that "next to the Rothschilds and the Barings," Thornton stood "A1 in point of wealth and connection with foreign countries" and was said to "*have accumulated a considerable part of his property by his successful operations in this article.*"[39] Evans continued, "He is a very large holder of Portuguese and Spanish stock, which the public may suppose, by finding him frequently placed in the chair at the meetings of the bondholders and great service he has done their cause." Another name that often occurs in sources documenting Victorian bondholders' committees is that of the Haslewood dynasty. Lewis Harrop Haslewood, the father, had provided the capital, in partnership with merchant banker Hambro, to underwrite one of the risky loans Italy made in 1851 to finance its war of independence against Austria (which Rothschilds opposed). His sons Edward and Clement Alfred were regulars of bondholders meetings and the ultimate insiders of the foreign debt market. The conventional image of the poor bondholder evidently fades away in favor of the savvy capitalist.[40]

The rich and powerful ordinarily do not take the defense of the poor if there is no good reason to do so. What attracted the foreign debt vultures was controlling the flow of information generated by participation in foreign debt trading. The chronic situation of default and permanent debt restructuring afflicting many borrowing governments created valuable opportunities for insider trading. The bondholders wanted bond prices to go up and wanted defaulting foreign governments to resume payments of their coupons. But for the insiders in the committee who had access to all the technologies of the stock exchange—enabling both upward and downward speculation through the forward market—every piece of information, good or bad, meant a trading opportunity. Information was always disclosed to the committee-men first, enabling them to trade on it before releasing it to their constituency. And there was no reason for jobbing to stop there, since after all the committee-men not only obtained information but made it. Through deft statements astutely released in the right journal, they could take advantage of, say, some foreign official's visit to London to let it be understood that a certain favorable or unfavorable development might happen, putting the uninformed bondholder in scramble or panic mode and exploiting the resulting market movement.[41]

A fascinating article by historian W. M. Mathew recounts such an episode during a Peruvian debt restructuring in 1848 when Peru's government found itself bargaining with a Peruvian bondholders' committee that included a number of veteran foreign debt specialists such as G. R. Robinson, an M.P. who was also a representative of the Jamaica lobby in Parliament, and John Diston Powles.[42] The correspondence of Peruvian authorities shows they had no illusions as to their bargaining adversaries. The country's creditors, the Peruvian foreign minister wrote in 1848, "are not, almost without exception the original ones, but having bought the bonds they now hold at low prices, do not seek in a settlement of the loan an indemnification for delay, of which the lenders alone could complain, but for an augmentation of positive profits." As Mathew explains, the price of Peruvian bonds began a sudden rally during the negotiations with Peruvian authorities in December 1848. Early that month they were at about £37 and in about three weeks leaped forward to £49 (a 33 percent advance). Members of the committee were taking advantage of the information they had of an impending settlement. The increase in bond prices was fully factored in when the agreement between Peru and its bondholders was drawn up and made public in late January 1849. To ordinary bondholders who complained in a meeting that there had been insider trading, the committee-men provided a "most perfect denial of anything like an unfair use . . . of the information," but the *Times* wrote for its part that there had been some "gross breach of confidence, apparently for speculative end."[43]

To the extent various groups of bondholder leaders competed with each other, information about the underhanded market operations of the foreign debt vultures was bound to surface, resulting in disputes between rival factions within committees. Conversely, this meant that industry leaders were encouraged to develop some kind of organization, which secured their position and enabled them, undisturbed, to extract value from those outside. To this end, they had developed a kind of federal set structure, the Spanish American Bondholders Committee, an entity created in the 1830s that united the bondholder committees in Spanish-speaking America from Mexico to Buenos Aires (then a sovereign state). Other forms of inter-committee cooperation included kinship (as illustrated by the Haslewood, Ricardo, and Mocatta Stock Exchange dynasties) as well as joint participation in a variety of mostly foreign companies.

The indications we have suggest that, over time, bondholder organizations became increasingly powerful. Because of a rule at the London Stock Exchange that prevented defaulting governments from making fresh issues if they had not made good with a majority of bondholders—the foreign debt vultures—who not surprisingly comprised a number of brokers, in effect controlled access to fresh capital. The vultures formed an insurmountable roadblock for any attempt to promote a new foreign loan. An analysis of the interlocking committee system as it prevailed between 1855 and 1865 reveals the centrality of old John Diston Powles, the godfather of the foreign debt vultures industry. He was surrounded by a handful of chosen allies, cronies, and rivals with whom he cooperated and occasionally competed. One was Isidor Gerstenberg, who was to take advantage of Powles's death in 1867 to take the lead of the industry, which he decided to remodel. His alliance with Philip Rose gave birth to the Council of Foreign Bondholders in 1868 and eventually, in 1873, to the Corporation of Foreign Bondholders of which John Lubbock would be a vice president and Hyde Clarke secretary. Another member of the clique was Henry Brinsley Sheridan, a Liberal M.P. who was chairman of the Mexican Foreign Bondholders Committee in the late 1860s. This fellow of the Royal Geographical Society joined the Anthropological Society in May 1864. In February 1869 he became a member of Gerstenberg's first Foreign Bondholders Council.[44]

## STOCK EXCHANGE SUPREMACISTS

At the time this story begins, one scheme among the endless machinations bondholders were permanently hatching received special attention because of the pervasive push for land-grabbing and colonization, a result of the transformation of the cotton economy brought about by global economic expansion. This was shaking the world over, and in effect, many characters in this tale, whether they be found in Central America or in Abyssinia, had in one way or another a concern with cotton.[45] Within this general trend, which started in the 1840s and 1850s, plans to increase the scope of the cotton economy beyond the American South were articulated by Southern states. As recently described by Susanna Hecht, at the time the Civil War broke out Southern states saw the Gulf of Mexico as Confederate waters and Central America and the uplands of the Amazon as territory that could be "disciplined and developed as part of a new American 'Confederacy.' " Geographical knowhow was accumulated, the figurehead here being Matthew Fontaine Maury, "the most decorated American man of science of the 19th century" and the promoter of a "Southern Manifest Destiny."[46]

With the outbreak of the Civil War, the plans for the Greater Confederacy's projects lacked financial means and political priority. The subject naturally ended up in the hands of British investors, at the same time when Matthew Fontaine Maury and a number of other prominent Confederates reached London for safety. There they met and mingled with the foreign debt specialists who were pushing their own agendas for colonization. The Confederates had the knowledge and the maps, the London Stock Exchange had the money. This rationale met yet another line of thought. In the City, the idea to collect unpaid claims in the shape of land grants from defaulting governments had gained momentum in the 1850s and 1860s and had become a major focus of the vulture industry.

A pamphlet written in 1863 by Edward Haslewood discussed the business logic.[47] He first reviewed his own vitae and explained how, after having been a "member of the Committee of Spanish American Bondholders, and the Chairman of the Committee of Greek Bondholders," he had become an expert in foreign government extortion. Or rather, because he did not put it in those words, how it had fallen upon himself to exercise the "high duty" to "weigh in nations in the balance, to study their wants and their difficulties." He was thus a credible proponent of an arrangement with a number of Latin American defaulting states that would permit the purchase of a huge patch of land to start new English-speaking colonies in the "uplands of the Amazon," a territory "about the size of one-sixth of Europe." This territory, he explained, would be obtained from its current owners—"Ecuador, New Granada [Colombia] and Venezuela."[48] These were, he stated, countries with immense money needs but whose "credit is—nearly NIL." In one scenario for funding, Haslewood declared that recruiting British government money was feasible. Such was a country that had not hesitated to pay "£20,000,000 sterling to free the negroes," a reference to the indemnity raised in the 1830s to compensate former slave owners, underscoring the role which the emancipation bounty played in the imagination of traffickers. Most importantly, Haslewood put his scheme in the perspective of a long series of similar acquisitions by the United States in particular: Louisiana from France, Florida from Spain, and California from Mexico. The latter experience in particular, with the simultaneous and opportune gold discoveries and the sudden birth of the city of San Francisco, was especially stimulating. Thus was the spirit of Haslewood's project. By arriving at the right time in a lending and defaulting cycle, one could buy land for a penny-the-hundred-thousand-acre, organize settlement and the development of transportation infrastructures, and sell back the land at a multiple of the original price.

These were not mere castles in the air. Of all the actual schemes the vultures promoted, the most notable was perhaps Isidor Gerstenberg's Ecuador Land Company, which had been set up in the 1850s along with John Field, another prominent member of the bondholding clique who had taken over the publication of *Fortune's Epitome* in the 1830s. The company sought to promote German immigration in lands that "British bondholders" (meaning Gerstenberg and his friends) had received "in discharge of obligations" from the Ecuadorian government. The prospectus of the Ecuador Land Company indeed splashed the California dream as sales pitch: "When it is considered, that the site of . . . San Francisco some ten years since was desert lands, . . . it seems not unreasonable to anticipate that the Pailón will rapidly rise to the position of an important town, having gold fields in its immediate vicinity, equal to those of California."[49]

This obsession with California we find again in the memorandum Gerstenberg published in 1866 with the purpose of launching his idea of a Council of Foreign Bondholders. The pamphlet gave the loss of California to the United States as the ultimate reason for organizing the bondholders in his corporation, stating that "Britain had suffered from the lack of determination by British bondholders," and as a result failed to grab the Golden State, which Mexico was prepared to give away. Essential indeed in the background of this stock exchange view of the world was the idea of an unavoidable economic confrontation with the United States in which Britain, with its unlimited access to finance and migration, was to have the upper hand.

Consistent with this interpretation, supporters of the default-colonization nexus invariably expressed their hatred of the Monroe Doctrine, and in the 1860s, they thought and proclaimed (as Haslewood did in his pamphlet) that the Civil War offered a rare opportunity to reverse existing trends whereby successive British administrations had retreated in the face of Yankee pressure. However, under the banner of free trade that "Brother Jonathan" (a popular way to refer to the United States of America) could not entirely dispute, vast patches of land could be made agreeable to British interests. What the United States had done in California could be replicated in a place where the Union Jack would be planted. And unsurprisingly, the same group of bondholders was looking with amity to the government in Richmond, possibly the first bridgehead of a resurgent British empire in North America.

But even if all went according to plan, there remained pending issues after concessions were granted and land surrendered to the creditors. The sort of land defaulters were prepared to give away was hardly the best. The Ecuador Land Company had reportedly good land near the ocean, but as Haslewood recognized, one could not expect to receive the most settled territory, but rather, land subject to "neighbour disputes," "waste lands," or "large tracts of wilderness" without a "road" or any sign of "civilization." This was a time when the ability of Latin American governments to control their own territory was uncertain, and there were indeed vast indigenous enclaves. Between good land near the capital and land somewhere in the jungle infested with mosquitos or filled with hostile natives, it is easy to figure out which land a Latin American government would award the bond-holders. As Haslewood noted in his pamphlet, the whole success of his business idea depended on whether the title and ownership of the acquired land would be regarded as "indisputable, though the Indians dispute the authority" of the existing governments.[50] Consistently, this particular group of buccaneers had a natural tendency to dislike Indians, blacks, and more generally anyone who stood in the way of their schemes, and this antipathy was blended with the notion that only the Anglo-Saxon race would be able to generate value from such land—the case of California again given as the archetypical example.

An illustration of the resulting economic racism may be found in Haslewood's pamphlet, in which he stated that British Prime Minister George Canning overestimated the subsequent development of Latin America in the 1820s because he "forgot the sluggish current which runs in the veins of the descendants of Spain when mixed with that of Indian and negroes. They certainly will not labour as the pure Anglo-Saxon race toils and labours."[51] If dealt with by the Anglo-Saxon race, Haslewood ventured, these lands could experience a "ten-fold" increase in value. This gives perspective to his indictment of emancipation for evidence showed that emancipated slaves "have never done anything since of any benefit to mankind." This was the classic argument, which had been regularly used to legitimize both the Western colonization of occupied land and the exploitation of people. Earlier references to "unused emptiness" to justify external claims were reworked into one about the under-exploited nature of the soil—the feeble human density that so-called primitive life permitted.[52] The next stage of Haslewood's argument thus insisted that externally developed, superior modes of production had a right to displace local inhabitants in the name of efficiency. It is not surprising that this line of reasoning was set for an encounter with anthropology. To the extent that it was now recognized that there were inhabitants (since they were studied), the next step was to ask the experts of native studies what science's take on the abilities of the locals was. Were they bound for extinction or could they somehow be made productive?

With the mention of Confederate explorer Matthew Fontaine Maury, we have spotted the scientific origins of Central America and Amazonian colonization schemes. Now wrapped in the language of business, they traveled the Atlantic Ocean to London where they were unpacked by prominent stockbrokers and then repackaged in the language of science. Reflecting this, we have the chapters contributed by Anthropological Society Council member and botanist Berthold Seemann to a book he published in 1869 with Bedford Pim—but this time clothed as science. Tropical America, Seemann asserted, was *the* "field of colonization of the future." To such individuals, debates on natural selection and adaptation provided an endless source of vocabulary and metaphors. Seemann quoted at length a paper by Joseph Dalton Hooker (the grand priest of botanic Darwinism and son of Seemann's mentor William Jackson Hooker), "On the Struggle for Existence Amongst Plants," noting that as in nature, where "new arrivals have always the advantage over the old," one was witnessing the "rapid spread of European species in New Zealand and the displacement of the indigenous."[53] He wrote of showing the sights in London to a Nicaraguan politician who, after touring "those noble monuments of architecture, the great bridges of London," declared to Seemann "mournfully" that his country would never have such works of modern art until "after all [his] country men have passed away and [Seemann's] taken possession." Seemann then went on explaining why Central American elites were resisting the entry of Anglo-Saxon immigration. First, they had "neither the bodily nor mental power to hold their own against such rivals [as ourselves]" and second "they have seen enough to understand that [immigration] would be the making of their countries . . . but also they feel instinctively that it would be their own 'unmaking.'" And a few pages down: "Foreigners have already done a great deal for these countries; and if they should but arrive in sufficient numbers, would doubtless regenerate them."[54]

Bedford Pim's own chapters in the same book developed yet another variant of this racist political economy of settlement when he discussed the relative performance of post-colonial North and South America, concluding with a table that illustrated the more modest demographic growth of the Latin element—"a tabular statement," he explained, of what the descendants of Spanish colonization have "since accomplished with the possession of more liberty and equality than they know what to do with."[55] Nor did the catalogue of clichés from a "struggle for life" view of colonization stop there. Just like Haslewood's, Pim's contribution contained the classic ranting against the Monroe Doctrine and a poignant call for Britain to strive "to civilize and reclaim Central America" or else, "the [Yankee] Americans [who from] the first adopted the bolder policy, [are] therefore, to use a phrase of their own, 'bound to win.'"[56] The struggle for life, it would appear, took place with heightened ferocity in the upper echelons of Anglo-Saxon racial superiority.

We should be careful not to ascribe this view to some extremist mind. The mercurial Marmaduke B. Sampson, editor of the *Times'* "City Article" did not differ much in his views from the aforementioned stock exchange supremacists when he lamented on the occasion of the Paraguayan "Public Works" loan of 1872 about "Paraguayan politicians," summarizing the problem in the following fashion: "The country is one of the finest in the world, and, could an intelligent and honest population be placed upon it, its progress might be more rapid than any yet witnessed anywhere in South America."[57] Stated in simple terms, the great stock exchange question of the time was how to make money from expropriating the tenants of the less efficient parts of the world. That they should first send anthropologists to study the tenants was of the essence.

## CALLING AT THE LEARNED SOCIETY

The scramble for foreign land required capital, and those who did not have enough resources sought to raise it. But to persuade investors that the tenfold profit was around the corner, promoters of land companies needed to show some evidence. Publicity could only get you so far, although it was never neglected. In his joint book about Central and Latin America with Pim, Seemann stated that "after the northern parts of the New World, Australia, and New Zealand shall have become fully peopled, our millions will pour into this long-neglected region, and found thriving colonies and happy homes along the magnificent mountain ranges and on the splendid table lands, while busy steamers will ascend the mighty rivers, railroads break in upon the stillness of the rising forests, and silent telegraphs flash along intelligence, telling of the great deeds of mankind, and giving the latest account of the pulsation of the world."[58] This was a beautiful image, but promoters needed to secure some form of certification that the territory was malaria- and cannibal-free.

This is how the learned society came to be sucked into the maelstrom of financial machination, and we'll discover later how McGregor's scheme would be reinvented and refined by a member of the Council of the Anthropological Society of London with more powerful tools at his disposal than McGregor's apocryphal Captain Strangeways. An illustration of the logic at work can be given in the shape of the accreditation that Gerstenberg's Ecuador Land Company sought to secure from the Geographical Society in 1859 when the company was set up by Isidor Gerstenberg with Louis Levinsohn and a group of individuals that included William Bollaert, at that time a member of the Royal Geographical Society and a specialist of Maya grammar who later became a member of the Council of the Anthropological Society and the Cannibal Club. The Ecuador Land Company was selling to aspiring settlers the acres it had acquired from Ecuador as an indemnity for its defaulted debts. With this partnership between Gerstenberg and Bollaert, the significance of the alliance between the stock exchange and anthropology or geography becomes clear. According to the work of historian Ulrike Kirchberger, Gerstenberg had been in constant correspondence with the Royal Geographical Society. In 1858, following a lecture by Bollaert, a general discussion on the proposed company had taken place at the society, which was incorporated the next year.[59]

The first three names on top of the Ecuador Land Company's incorporation documents (Gerstenberg, Levinsohn, and Bollaert) are all found in a contemporary list of fellows of the Royal Geographical Society, Bollaert since 1850 and the two others joining in 1859. The complete list of initial subscribers has twenty-four names, the majority of them declaring themselves as bankers, stockbrokers, or simply marking "stock exchange."[60] A letter in the National Archive at Kew shows the assistant secretary of the Ecuador Land Company writing to Admiral Fitzroy (the captain of *HMS Beagle* during Charles Darwin's famous voyage) at the Meteorological Department of the Board of Trade, with the ostensible purpose of advertising the excellent scientific intentions of this company, "constituted of holders of bonds of Ecuadoran debt having accepted large grants of land from the Government of that Republic in liquidation of arrears." A yacht had been purchased, and an exploration party was about to sail. "The Company," the secretary wrote, "[wished] to render the labours of its officers as available to scientific research as possible." When the prospectus of the company was issued in 1859, it told the usual story of successful colonization, gold, tobacco, quinine, and fertile provinces, and did not forget to emphasize that these "favourable anticipations" had been "fully confirmed by the careful examinations of scientific men."[61]

In other words, facts crucial to the process of "inception of the value" as Bagehot called it, were being adjudicated in a forum of science making the learned society a valuable part of the stock market mechanism. And the learned societies did not seem to resist the pull. Instead, as we saw, they were casting their net across the world and boasted, as the Anthropological Society did, a vast network of local secretaries "in different parts of the world, to collect information." And thus it is, as the rest of the book will show, that the sciences that dealt with individual cogs of the long-distance infrastructure project became involved in this complicated foreign debt food chain, which this chapter has sought to unpack—geography, geology, public finance, and of course, ethnology and anthropology, whose societies became essential places for sorting out the jumble of value assessment, places essential for the preparation of the stock exchange auction. In one word, the financial road show had gotten used to calling at the learned society.[62]

Such was the ecology of the foreign loan ogre in the 1860s—a far more subtle system for transferring and appropriating wealth than accounts of the delinquent southern state would have it. The increasing complexity of the stock exchange machination engulfed and produced more interests than ever before. Earlier scheming had contented itself with the promoter showering the press with gifts, but the foreign debt boom that took place at this point saw the involvement, alongside the financier, of an array of scientific interests carried out by learned societies and their members. Thus my conjecture is that learned societies became increasingly valuable in the process of capital export. They attracted increasing attention as illustrated by the rising membership in anthropology, geography, and geology, which took place in tune with economic, commercial, and financial expansions. It is indeed a striking feature for any student of the pulsations of the global economy that the mid-nineteenth-century acceleration in global trade coincided with the takeoff of membership in a number of learned societies. This transformation had several reasons, but one prominent factor was that they enabled establishment of "truths" that permitted other individuals to transact—a sine qua non of financial markets. A patch of land, once its human and material surroundings had been sorted out, could be sold and bought, and as a result of the geographer's or anthropologist's structuring of the world, financial products could be devised. Anybody who has read a geographical treatise of that time cannot help being struck by the constant emphasis on local conditions from the vantage point of colonization, with reference to the healthy air, mentions of temperature, and the incidence of disease everywhere. As a result of this, a novelty of the period was the emergence of agents who, as specialists of the country in question, straddled science and finance. They had contributed a book or pamphlet, knew about the country's character—its finances, geography, population, language, and history—and they had a long-distance project. Some of those agents even attempted to out-compete the bankers, suggesting that they were the owners and specialists of new symbols of the information economy. Some financiers understood it, and soon harnessed them, as was the case with Edward B. Eastwick and the General Credit and Finance Company. In other cases, the specialists became their own masters under the guise of "agents," and the loan prospectuses bore their names. Among these new entrepreneurs, we may single out individuals such as Horatio Nelson Lay, who was involved in originating early Meiji Japan foreign securities in such creative ways that the Japanese government ended up suing him. Other prominent scientific entrepreneurs more directly related to the tale in this book were George Earl Church, who dealt with Bolivian debts and navigation companies, and Bedford Clapperton Pim, who came to be involved in the infamous Honduran debts studied by the select committee. All three were members of the Royal Geographical Society, and Pim, as already said, was a prominent Cannibal at James Hunt's Anthropological Council. All would leave a track record in tribunals. I conclude that the contours of Victorian knowledge must have been shaped by a combination of mercantile, scientific, and criminal concerns. As already indicated, colonialism and imperialism have always been known to generate their forms of knowledge. In the mid-nineteenth-century variant, the London Stock Exchange would have a heavy bearing upon emerging modes of knowing and in particular upon the design of anthropological science.

## CHAPTER FIVE

# The Learned Society in the Foreign Debt Food Chain

That the importance of the learned society to the operation of the stock exchange has been overlooked in previous research is quite natural since the role of science in what Bagehot called the inception of the value tends to be forgotten. Such an oblivion results from the heuristic and seemingly relevant opposition made by previous historians of capitalism between two successive sources of inspiration for the management of wealth: on the one hand, contingent, personal, social, and relational considerations guiding investment and on the other hand a later approach where local determinants would be superseded by universal, researched, impersonal, and in the end quantitative investment strategies. Because the learned society produced concepts used by the second approach but in its essence operated with techniques borrowed from the first, it has naturally become a blind spot. The comfortable contrast between scientific investment and personalized knowledge assumes away the personalization of the institutions of scientific advice.

It is ironic that this dichotomy should have been first articulated by a sociologist—Max Weber—especially since Weber was at one point himself a student of the stock exchange. Weber described late nineteenth-century modernity as a process of rationalization and bureaucratization that drove out the personal. His perspective still has much appeal, and modern historians of the capital market describe the evolution of the technologies of investment as one where investment trusts and other sophisticated investing bodies replaced individual investors—where EBITDA, the acronym for earnings before interest, taxes, depreciation, and amortization, an indication of profitability, substituted for kinship as investment advice. To quote a leading historian of the British capital market, "the importance of non-financial forces in investment . . . steadily diminished in importance" over time and led to a changeover to "impersonal investment."[1]

In Weber's theory, the process of scientific rationalization triumphs because it has rational legitimacy. Weber never quite explained where this legitimacy came from except in the tautological way in which he postulated that rationalization bred consent without specifying *whose* consent. It is true that sellers of financial advice, even when they have not read Weber, have always advertised themselves as producers of rational knowledge, vaunting the merits of scientific investment, and warning their clientele against the risks it ran from dealing with the competition—material, personal, social, and relational advice. The overlap between the legitimation of science and the sales pitch for scientific investment is obvious. "Today, you need not guess," read an ad by rating agency Standard Statistics a few weeks before the crash of 1929.

Likewise, because narratives about the learned society typically emphasize its kinship with the professional-technical, it has seemed natural to think of scientific institutions as part of a process of extirpation from the social context. But it remains that this is little more than prestidigitation, one whereby reference to rationality and rationalization serves to divert attention from the fact that they are both highly personalized processes from which the social is never absent. Reflecting the ambiguity, we have the scientific and the social embedded in the very phrase "learned society."

This chapter is devoted to exploring the role of the learned society, a social institution, in the process of foreign capital export in the London stock market. To illustrate my point, I will simply follow the tribulations of one scientist who ended up originating a conglomerate of Bolivian transportation companies and raised for this purpose a £1.5 million loan in London before a substantial part of the money evaporated into thin air. (This is £118 million modern pounds in purchasing power according to retail prices). The individual in question was George Earl Church—"Colonel" Church—a multifaceted explorer: he defined himself in the paperwork of the railway company he incorporated in Britain before issuing his Bolivian loan as an engineer, but he was also the author of *Aborigines of South America*, a book that salaciously dwelled on the many ways used by Indians to cook and eat their enemies or relatives or make flutes out of their tibias.

Because smugglers are the ultimate experts when it comes to boundaries, Church's interloping activities will help us to understand the emergence of the contours of anthropological science. As did many characters from the early story of anthropology, Church straddled many universes. Church, a U.S. citizen, would become the first foreigner to be vice-president for an English learned society (the Royal Geographical Society, which he joined in

1872). He was never formally a member of the Anthropological or Ethnological societies, but he is recognized as, among other things, an early anthropologist, as evidenced by the books he wrote or read: the Church collection at Brown University is a remarkable library for students of the anthropology of Latin America in the nineteenth century. Church was also closely related to several characters of the anthropological saga. In particular, he was a close friend of Clements R. Markham, a veteran of the Geographical Society who participated in the Franklin recovery expeditions and explored Peru during the 1850s. Markham was the perennial secretary of the Royal Geographical Society under Murchison, whose kingdom he inherited. Markham's human geography credentials are impeccable. He joined the Ethnological Society in 1864 and is acknowledged by Chapman among the significant contributors of the Pitt Rivers Museum of Anthropology in Oxford.[2] He would join the Council of the amalgamated Anthropological Institute in 1872 and has been recently called by Susanna B. Hecht (author of *Scramble for the Amazon*, 2013) a typical "bio-pirate" from the way he organized and supervised the British accumulation of foreign and colonial artifacts.[3]

During the 1860s, Markham was not a member of the Anthropologicals (as befitted a devoted servant of Murchison), yet he nonetheless gave a lecture there at least once, dissertating in December 1863 "On Quartz Cutting Instruments of the Inhabitants of Chanduy (near Guayaquil in South America)." This was during the session that immediately followed Hunt's "Negro's Place in Nature," where Hunt's ideas on the Negro's "natural subordination" had been articulated.[4] Markham was a central member of the Geographical/Ethnological galaxy, and we'll see that Church's network can likewise be traced to this clique. Crucially, Markham would always vouch for Church. His preface to Church's *Aborigines of South America*, which was published in 1912 following the author's death two years earlier, lamented the loss of an "eminent authority on South America" and praised the "amount of information collected" in this book devoted to the subject of the "noble Indians of South America." Therefore, not only was Church one of the strange creatures who roamed the Geographical/Ethnological galaxy, but he was also a character who was *underwritten* by prominent leaders of this galaxy.

## PROVIDENCE

George Earl Church did not want his 3,500-volume library devoted to Latin America and its explorations dismembered after his death. His will insisted that the beneficiary should keep the books together, and since Harvard wouldn't, Brown University now holds the Church Collection as part of Hay Library. The scope of the collection underlines the complementary character of geography and the science of man underscored earlier: if Church's library were to be classified according to the sections of the British Society for the Advancement of Sciences in the 1860s it would come squarely under Section E (Geography/Ethnology Section). But it does not stop there, and if many researchers who call at the Church Collection are interested by its enormous wealth of geological, geographical, ethnological, and anthropological literature, I came across it initially as part of my investigations of the foreign debt vulture industry. When I called at the Church Collection in the autumn of 2011 and was welcomed by its amicable curator, it was because I had discovered that it held one of the most complete collections of British bondholders literature, comparable in breadth and scope with that held by the British Library in London.

That this bondholder material was immersed in a rich collection of ethnological and geographical volumes I initially regarded as essentially a curious fact. Reassuring indications why this should have been the case were provided by the two main sources on Church, Neville B. Craig's *Recollections of an Ill-Fated Expedition* (1907) and a biographical notice that introduces Church's *magnum opus* on Latin American Indians, the latter having most probably been prepared under the supervision of Markham given that he wrote the preface. These sources state that Church made several travels to Ecuador in the late 1870s and Costa Rica in the 1890s on behalf of the Corporation of Foreign Bondholders.[5] Technically, therefore, Church in those later missions was reporting to the management of the corporation, and thus to John Lubbock and Hyde Clarke, thus associating Church with the institutions of British anthropology in yet another, indirect way.

The biographical entries devoted to Church, however, are silent on earlier entanglements between Church and bondholding, a curious fact because many of the volumes in the collection refer to early productions, covering periods (the 1850s) and countries (Spain) beyond the time when Church was technically involved in bondholding. Yet as evidenced from several volumes held in the collection, the links between Church and bondholding were deeper and more intriguing. Consider for instance the volume that contains the minutes of two Spanish bondholders' meetings held in 1851. Its title page shows a handwritten inscription "[To?] Edward Haslewood Esquire, with Joseph Tasker's Best Compliments, August 30th 1853."[6] We have already come across Haslewood and will meet him again repeatedly until the end of this book. This is the Haslewood described in the previous chapter as a stock exchange supremacist who had authored *New Colonies on the Uplands of the Amazon*, the pamphlet vaunting the merits of colonization. Tasker was another prominent leader of the bondholding vulture industry dominated by John Diston Powles in the 1840s, 1850s, and 1860s.[7]

The dedication is intriguing for many reasons. First and foremost, the minutes of this Spanish bondholders meeting can be read as yet another instance of bondholder machination similar to the Peruvian episode narrated in the previous chapter, when members of a Peruvian foreign debt committee, at the expense of the investors they were supposed to represent, took advantage of insider information about an impending restructuring they were themselves negotiating. This time, the scheme takes the form of a racket with Rothschilds being the prey. The minutes of the first meeting show members of the bondholding vulture clique making threats of bad publicity in an attempt to interfere with Rothschilds's efforts at restructuring a new Spanish loan. The minutes of the second meeting show Rothschilds and the bondholders happily reconciled, implying that a settlement must have taken place behind the scenes. It is hard to resist the conclusion that the dedication to Haslewood was a piece of self-congratulation by the vultures. Second, the presence of the volume in the hands of Church establishes the existence of a connection between the ringleaders of the bondholding industry and Church himself.

How and when could the connection have taken place? The most natural suggestion is that Church somehow acquired or inherited the volume from its owner (Haslewood). To contribute possible answers, we can get back to the biographical notices on Church, which, despite their hagiographic tone and unreliability, may give some clues. Rather than discussing bondholding, the notices dwell on Church's past as a railway engineer in Iowa in the 1850s or as a member of the scientific commission sent by the government of Buenos Aires to explore the southwestern border of the country. There Church would have fought two pitched battles against Indians who charged at midnight "naked and mounted bareback on their splendid horses" and stole the reserves of his party. There is also an evocation of Church as an American Civil War hero (this is when he would have become a colonel), as a U.S. adventurer mercenary (or agent?) participating in the expelling of French forces from Zacatecas in Mexico in 1867 and pursued by the entire imperialist army (again by men on horseback, but this time all dressed boot to cap in the Mexican way), and finally as member of the *New York Herald* editorial staff upon his subsequent return to New York.

There he met with General Quentin Quevedo, himself just arrived from Mexico. Quevedo was a prominent member of Bolivia's diplomatic corps and a former explorer of the Madeira River, whose "appeal" to Church's energy would be the reason why Church launched his great railway adventure in the confines of the upper Amazon in Brazil and Bolivia.[8] The plan that was articulated then, followed on the heels of two significant diplomatic developments—first, Brazil's decision to open the Amazon to navigation by all nations (an issue that was an obsessive theme of colonization companies such as Gerstenberg's, as indicated by his insistence to bring the matter up for discussion at Royal Geographical Society meetings);[9] and second, the Ayacucho Treaty "of friendship and navigation" between Brazil and Bolivia, which ensured that the wealth of Bolivia could be exploited through the Amazon water system to the east, in exchange for substantial land concessions by Bolivia.[10] Previous explorations had revealed that it was possible to navigate the Amazon, then into its tributary the Madeira River and join the Mamoré River farther up in the heartland of Bolivia. The only serious obstacle came from the nineteen falls and rapids extending about 250 miles between the Guajará Falls and San Antonio. Somehow the idea emerged gradually that a canal or railway through the territory of Brazil could be devised, permitting in effect the exploitation of a route 2,000 miles long from Vinchuta in Bolivia to the Brazilian state of Para on the Atlantic mouth of the Amazon.[11]

According to Neville B. Craig, after Church decided to undertake the promotion of Bolivia's railway loan, which required going to La Paz to secure the concessions and make necessary arrangements, he took a rather circuitous route. From New York, he went to London first. Since you would go to Paris for pleasure and to London for business, it is thus possible that it was at this juncture that he met with Haslewood, a stock exchange broker with a professed interest in the colonization of Latin America, in particular the Brazilian border (although the area of interest to Church on the southern border of the Amazon was distinct from Haslewood's golden rectangle on the river's northern border). It may have been there that the connection between the two men was established, materialized by the bondholder pamphlet Haslewood had received from Tasker that was to end up in Church's possession. Indeed, Haslewood later appears as one of the initial promoters of Church's Railway Company application to the Board of Trade when it was created in 1871 and subsequently as a shareholder of the company in 1872.[12] It is also tempting to speculate, but more uncertain, that it was at this time or at least during a subsequent visit to London that Church entered in contact with scientists such as Markham who had, from his own earlier explorations, an interest in the region. In any case, the link between Church and Markham was sealed before 1872, for at that date there is primary evidence that the two men were on friendly terms. Such was the geometry of Church within the intersection between the bondholding ellipse, centered on John Diston Powles, and the ethno-geography ellipse, centered on Murchison and Markham.

Then, from London, Church sailed to Buenos Aires and traveled overland to La Paz. Once in the Bolivian capital, Church secured for himself the rights and land grants for the construction of a canal to avoid the rapids.[13] Note that when he reached La Paz and secured the concession, he curiously had not yet explored a mile of the intended project, instead relying for background documentation on a number of other accounts such as one by the German-Brazilian engineers José and Franz Keller, who had explored the possibility of improving transportation along the Madeira River between San Antonio and the Guajará Falls near the mouth of the Mamoré River in 1867–68. On the occasion, the Kellers, father and son, had given a "picturesque description of the savage Indians, some of them cannibals."[14]

It was thus on the basis of limited explorations that Church was to venture inside the capital market jungle, simply equipped with secondhand knowledge, before he had approached the rapids at San Antonio. But carefully exploring the feasibility of a waterway was perhaps a less essential part of the promotion of distant infrastructures than the task of promoting group interests. This is illustrated by yet another contemporary document from the Church Collection underscoring the technologies that characterized the universe in which Church operated: a volume containing a printed list of Ecuador Bondholders committee members with handwritten inscriptions across individual names noting "dead" or "gone" (this one is harder to decipher). At the bottom is a subtraction: "13—8 = 5 alive."[15] Thirteen minus eight, "and then they were five"—the subtraction hums a lullaby from an Agatha Christie novel. The rhyme, it will be recalled in its politically correct version, has the "Five little soldiers going in for law / One got in Chancery and then there were four." Little soldier Church's grand project to unlock Bolivia and connect it to the Atlantic Ocean via the Amazon was to propel him from the rapids of the Mamoré and Madeira Rivers to British courts of justice.[16]

## STRUCTURING THE MADEIRA-MAMORÉ RAILWAY

Once the concession had been secured, frantic arrangements followed, made slower by the long distances and various complications. Neville B. Craig argues that Church went to New York and Europe after obtaining the concession, and it was there, in Europe, that it was decided that the right way to go was a government loan. A private company alone was not sufficient security, Church was told. Also, the canal gave way to a railroad. Thus arranged, the

Bolivian loan started looking like many of the schemes that were floated at this time in the London Stock Exchange. Church returned to La Paz in late 1869 to modify the concession. While a Bolivian National Navigation Company, which was only national by name, held the privilege of transportation of freight and passengers to and from Bolivia via the Amazon as well as the right to collect tolls from other boats, a company to be called the Madeira and Mamoré Railway Company would hold railway rights as well as a number of mining and land grants. Last, since the railway project was cut through the Brazilian state of Rondonia, there was also a Brazilian government concession that was secured in 1870.[17]

Financial engineering started before railway engineering. The National Bolivian Navigation Company was incorporated in New York by an act of Congress in June 1870. Then in March 1871, Church incorporated the Madeira and Mamoré Railway Company Ltd. in London. Beyond George Earl Church himself and the inevitable Edward Haslewood, "member of the stock exchange," the archive of the Registrar of Joint Stock Companies shows the following among the seven signatures necessary to start a company: Juan Francisco Velarde, the Bolivian chargé d'affaires in London; George Hopkins, a civil engineer; and Morton Coates Fisher, a partner in a London based Anglo-American merchant banking house who also became a top subscriber with Church (his business address—58, Threadneedle Street—was used as domicile for the railway company). Not incidentally, Fisher had joined the Ethnological Society of London in 1870 during this society's phase of more aggressive outreach, at the time of the Anthropological Society's collapse in membership. When he joined, he had already contributed to the society's Transactions an 1869 article about Indian tribes in North America's southern plains entitled "On the Arapahoes, Kiowas, and Comanches."[18]

The structuring reveals a remarkable command by Church and his sponsors and associates of Victorian techniques of financial engineering—and a perfect mastery of the art of puffing. On paper, both companies displayed an impressive capital: 2.5 million gold dollars for the navigation company and 7.5 million gold dollars for the railway company. But this was imaginary money. Of the $2.5 million in shares of the navigation company, $2 million were simply created to pay Church for the navigation rights he had transferred to the company. Remaining shares were then distributed to associates, the biggest allotments going to Morton C. Fisher and Colonel William Henry Reynolds, a Civil War acquaintance of Church.[19] Of the remaining half-million, $50,000 was given as commission to one Julius Beer a wealthy London merchant and the editor of the London newspaper the Observer.[20] How much this imaginary money was really worth we can figure

from the fact that one Irwin Davis, who subscribed in 1874 most of the remaining capital of the Navigation Company (about $430,000), paid 20 cents
on the dollar—and he probably overpaid.[21]

Application to the Board of Trade for incorporating the Madeira and Mamoré Railway Company was made on March 1, 1871. By an overwhelming
margin, the main shareholder of the railway company was the navigation
company (meaning Church, who controlled it), with some additional shares
showered on the usual suspects such as Haslewood and Reynolds for no
money at all. Among the recipients of gifts were also new, interesting additions. The first available report in May 1872 shows two notable allotments.
One was to Leopold Markbreit, a German-American Republican politician
from Cincinnati, Ohio, minister resident of the United States in Bolivia at
the time of Church's visit to La Paz in 1871, who would assist in Church's
efforts. The other was to Señora Lindaura Anzoategui de Campero, a Bolivian novelist (author of La Mujer Nerviosa) and the wife of the Bolivian minister in London, General Narciso Campero, whose name appeared on the
bottom of the Bolivian loan prospectus.[22] As more individuals were co-opted
in the fold and received gifts, other significant names appeared in the list of
shareholders such as William Scully, an Irish-Brazilian who received in 1875
a large endowment of free shares from the Madeira and Mamoré Railway
Company. Scully was the owner and editor of the influential Anglo Brazilian Times, a founder of the Sociedade Internacional de Imigraçao (which was
intended to stimulate immigration to Brazil), and a man considered by contemporaries and historians to have been at one point a close advisor to the
emperor of Brazil.[23]

All these shares were declared as fully paid up, but no money was ever
forthcoming (apart from the Irwin Davis purchase). They were gifts to allies
and cronies among whom we find many individuals able to reach out in
social circles. A paper project was being transformed by publicity and social outreach into something material while an elaborate structure hid the
money trail. One of the judges who was later called to examine the structuring of Church's companies found these were "two shams and fictions created
for the sole purpose . . . of deluding a credulous public into subscribing to a
loan."[24] According to another: "I cannot see any reality in that company [the
railway company]. If there is any shareholder in the Railway Company at all,
it seems to me that that shareholder is the Navigation Company." But that
company, in turn, had "no power at all" because it had "no capital and no
credit."[25]

Craig writes favorably that "by newspaper and magazine articles [Church]
made known the great natural wealth of Bolivia and her need for an adequate

commercial outlet."[26] This serious-sounding language is amusing when it is realized that, as far we are concerned, the information campaign about the great natural wealth of Bolivia had been manufactured in large part by the distribution of stocks to journalists and social intermediaries including one ethnologist, men such as Julius Beer and Morton C. Fisher, and by securing the support of Clements R. Markham, the influential secretary of the Geographical Society. Markham's support had not been purchased through outright distribution of stocks but rather cultivated. Smelling the money, several professional puffers got their cut. For instance, in the immediate aftermath of the subsequent Bolivian government loan issue, Marmaduke B. Sampson (the *Times* City editor who was to fall victim to a scandal narrated in an earlier chapter) became consul general of the Republic of Bolivia. That must have been his just reward for supporting the loan through positive comment or lack of criticism—or did he get the job because he threatened Church with adverse reports?[27]

The last part of the financial engineering was to structure the railway as a government loan.[28] By involving the Bolivian government and offering to reward a number of officials (as the presence of Juan Francisco Velarde among the first signatories of the railway company proves), Church was rendering it even more impossible to read an already complex chain of responsibilities. And thus on May 18, 1871, Church, as "special agent" for Bolivia, signed a loan contract for £1.7 million in nominal capital. The coupon was 6 percent and the bond was issued at 68, providing a yield of 8.8 percent. (See figure 4.) The underwriter was the House of Erlanger. Erlangers was an aggressive Franco-German merchant bank that had originated the Southern Confederacy's infamous Cotton Bond during the Civil War in 1863 (already defaulted upon) and the first Japanese foreign bond, which led to bitter litigation between the government and its agent, Horatio Nelson Lay, specialist of the Orient and a member of the Geographical Society. In London and Paris, the firm was run by Emile Erlanger, stepbrother of Victor, Baron von Erlanger as the family was known in Wiesbaden. In 1868, Victor had become a correspondent of the Anthropological Society and subsequently of the amalgamated Anthropological Institute.[29]

The Public Works Construction Company, an engineering firm, pledged to construct a railway line in the malaria- and cannibal-infested jungles of Rondonia for £600,000. A contract between the railway company and this firm was signed on May 18, 1871, the same day the contract for the government loan was signed. Court's language later distinguished between the two parts of the scheme by referring to the "works contract" and the "security contract" respectively. It is worth emphasizing again that at this advanced

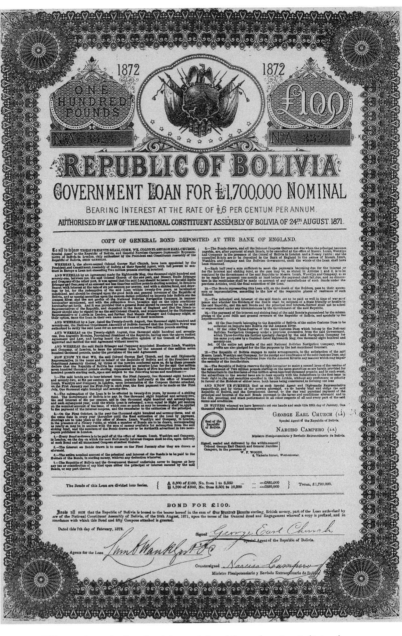

FIGURE 4. What anthropologists do: A bearer's bond from the 1872 Bolivian loan. *Note*: The printed signatures of G. E. Church and N. Campero are visible. *Source*: Mitu Gulati's private collection.

stage, the only material evidence that served as basis for these contracts was a series of reports and travel memoirs that Church had collected and which he would later circulate as evidence against allegations that the railway was a sham. In June 1871, however, the company dispatched two engineers to survey the route. What they exactly did remains unclear. Church accompanied them because there had been a coup in Bolivia and it was necessary that the new president, Agustín Morales, should also agree to the loan, or else the rules of the stock exchange would prevent its issue in London, since the right had been granted by an authority that was no longer competent.

After the new president and Congress gave their approval, Church then proceeded with an exploration party to survey the rapids, and on November 1 the railway terminus was established in the presence of the entire party "consisting of mainly half-clad Indian boatmen, in the heart of a vast tropical wilderness, fifteen hundred miles from civilization, where the interior country was absolutely unexplored and except by primitive savages uninhabited, etc." According to Craig, "animated by the faith of Columbus [Church] firmly believed and fondly hoped [he would] open to immigration and to the commerce of the world this country unsurpassed in latent wealth by any unoccupied territory of equal extent on the face of the globe." As Joseph Conrad put it in *Lord Jim*: "But do you notice how, three hundred miles beyond the end of telegraph cables and mail-boat lines, the haggard utilitarian lies of our civilization wither and die, to be replaced by pure exercises of imagination, that have the futility, often the charm, and sometimes the deep truthfulness, of works of art?"[30]

## PIRATES: SCIENTIFIC AND COMMERCIAL

The loan was announced a few weeks later. It was calculated, and advertised in the prospectus, that the new route, set to 150 miles, would "reduce the distance of the trade centers of Bolivia from Europe and the United States from about 180 days to about 30 days, and the cost of freight from an average of £55 per ton to about £15."[31] Erlanger began the placement on January 20, 1872, with Lumb, Wanklyn & Company acting as trustees and receiving subscriptions. Reliance on a formal trusteeship arrangement was unusual since in general, the banker himself more often received the money. But this innovation would have enormous impact on the rest of the adventure because it effectively locked the money away from a number of stakeholders and created procedures for the release of funds that protected investors.

Most probably, involvement of trustees was conceived by the promoters as a signal. According to the historian of the law firm Baxter, Rose, Norton and Co., Philip Rose, Disraeli's personal lawyer, was a close associate of Erlanger at the time the loan was issued, and the Madeira and Mamoré Railway Company would have been one of the their joint projects. It seems natural to assume that Rose's legal ingenuity had some role in the idea of bringing in trustees. Displaying trustees on the front window was a claim of serious behavior, for the trustees would see to it that monies were properly dispensed with. This provided some security against the risk of money being simply embezzled. As the lord chancellor Earl Cairns would declare later: "I think it obvious that if the money had not been placed in the hands of trustees the loan would not have been obtained."[32]

Subscriptions were closed one week after they had opened, and the loan was reported to have completely sold out. It is true as well that this Bolivian loan provided a substantial return when British consols were at 3 percent, but perhaps more fundamental was the fact that the issue had been clothed in the proper social fabric. A few years later, as Church's schemes became the focus of intense litigation culminating with *Wilson v. Church*, a request filed by bondholders to secure the release of the money held in trust, a majority of judges remained adamant in their contention that investors had not been abused. Brett, L.J., stated so much, arguing in the matter of *Wilson v. Church* that "unless people were reckless (but we have no right to assume such foolish recklessness), they must have inquired into the constitution of the [navigation company] and they must have found that it was a company which could not obtain any credit."[33]

This underscores the significance of the process of social promotion that backed Church's scheme. While brokers under the leadership of Erlanger and Julius Beer pushed up the loan in the London Stock Exchange, while Marmaduke Sampson at the *Times* remained agnostic, Markham pushed up Church in the Geographical Society: literally two days after subscriptions to the Bolivian loan had been opened, Church was voted in a fellow of the Royal Geographical Society.[34] The puffing of prices went hand-in-hand with the puffing of Church. That way we find, in the spring of 1872, favorable press reports regarding the premium on the Bolivian loan at the same time Church was receiving an invitation from Markham to the dinner of the Royal Geographical Society because members wished further "geographical information about Bolivian rivers."[35]

Once in, the money was engulfed by the pockets of the various sponsors. Based on the information later released in tribunals, it is possible to produce a rough account. The loan nominally of £1.7 million was issued at 68 percent of its face value. This produced £1.15 million in effective money (£92 million in today's money using the retail price index). Of this, £187,000 was set aside to pay the first three dividends and "amortization" (the annual reimbursement of bondholders). This left £969,000, from which the banker took his fee and issue expenses for a total of about £114,000 (thus more than 10 percent of the effective capital). Of the remaining £855,000, the Bolivian government, as agreed in contract, got 17 percent, or about £145,000. This left 83 percent of the money, or £710,000 in the hands of the navigation company. Next, £600,000 was put in trust, earmarked to pay the Public Works Construction Company, which received outright a down payment of £75,000 while the remaining £525,000 was invested in United States securities. In the end therefore, about £110,000 ended up with the New York-incorporated navigation company, which Church controlled single-handedly, and was never seen again. In rough numbers therefore, the banker's cut and the explorers' cut were comparable at about £110,000.[36] This is £8.5 million in today's money if the retail price index is used for conversion, and much more with other plausible indices.[37]

It was, strangely, at this point that the Public Works Construction Company woke up, took a serious look at the construction project, and declared it didn't like what it saw. The company discovered that the railway line would have to be 180 miles long or more (later figures would mention 200 or even 230) rather than 150. The contract, they now believed, had been imposed upon them by misrepresentation, and they wanted it rescinded and some of the money in the trust for indemnification. In July 1873 they filed a bill in Chancery to this effect against the navigation company and Church. It gives perspective to state at this point that, as some exploration reveals, Emile Erlanger and Julius Beer were among the chief sponsors of the Public Works Construction Company, which had been incorporated just one month before the loan contract of 1871. Again it appears that Philip Rose had been involved with Erlanger in putting together the Public Works Construction Company.[38]

In August, Church's Navigation Company opened an action at common law against the Public Works Construction Company for breach of contract. The same month, it also filed a bill in Chancery against the Public Works Construction Company and the trustees, claiming that since the company had repudiated the contract, then the trustees were to release the money to the Navigation Company so that it could pay another construction firm. In September 1873 the Navigation Company contracted with Dorsey and Caldwell to build the railroad. The price per mile was £6,000, so that if the length was 200 miles, then the £525,000 left in the trust did not even meet half of the needs. Evidently the goal was not so much the construction of the road as the milking of the trust fund, and for this, one breach of engineering contract after another was taking place.

These events, which signaled that the prospects of the Bolivian project were deteriorating, heralded a decline in the price of Bolivian debt, inviting further scheming. This was especially true because the money in trust was rising in value at the same time, the mechanical effect of Church's initial decision to invest the fund in United States securities. As these experienced a rally during the 1870s with the stabilization of U.S. government finances the value of the trust soared. The situation invited speculators. Informed vultures knew that the initial contract with the Bolivian government had stated that the project was to be completed within two years, otherwise the concession could be revoked. The name of the game became acquiring Bolivian debt from panicked investors and asking for the money in trust to be released. And thus in January 1874, exactly two years after the Bolivian loan had been issued, a group of "bondholders" filed a bill in Chancery against the trustees, the Navigation Company, the Railway Company, the Republic of Bolivia, and Church. If the construction of the railway was impracticable, they asserted, then the fund was to be divided among the bondholders.[39]

One might have expected that, at this particularly tender time, Church would have been frantically trying to ensure that his project would deliver. Instead, on March 9, 1874, we spot him perfectly relaxed with his friend Markham at yet another meeting of the Royal Geographical Society, pontificating over Latin American railways. Markham was reading a paper on a railway completed in southern Peru between Mollendo and Puño as well as steam navigation on Lake Titicaca and praised the "geographical attainments" of Peru's President Don Manuel Pardo, shortly to be proposed as a corresponding member of the society but also the architect of the defensive alliance of Peru and Bolivia against "Chilean imperialism." As was said during the discussion to which Markham, Pardo, Church, and Consul Hutchinson (member of both the Ethnological Society and the last Anthropological Council) participated, "the simple fact that Don Manuel Pardo, an able financier and friend of science, was now president of Peru, was the best guarantee of the future prosperity of the Republic." In his remarks "Colonel Church" pointed out that this railway across the Andes would eventually reach a transportation network that would link to the headwaters of the Amazon to "form a chain of communication between the Pacific and Atlantic."[40] Some years later, following the assassination of Pardo and the outbreak of the War of the Pacific with disastrous consequences for Peru and Bolivia, Markham wrote the first book on the war to be published in England, and it was vigorously anti-Chilean.[41]

A few weeks after his appearance at the Royal Geographical Society, Church's first daughter, Blanche, was born in Paris. This was on May 18, 1874, the anniversary of the contract that had made Church a rich man. The happy mother was the Peruvian musician Olivia Sconzia, singer and songwriter. The title of one of her songs catches the eye: it was called "Down the Stream." That same month, the Republic of Bolivia joined the frenzy in "this crop of litigation, which does not often occur even in the affairs of such a company as this," as one judge of the Church case later recognized. Bolivian authorities stated their opinion that, since construction had ground to a halt, *they* now owned the money. They were also reported to be purchasing their own debt on the market, just like the foreign debt vultures (an operation known today as debt buybacks). This made sense because they could repurchase their debt at a fraction of the money owed: lack of credit-worthiness has its perks.

The last blow to the fortunes of the Bolivian loan came when the money kept to service the debt dried up, and amortization was interrupted in January 1875. Technically, Bolivia was now in default. But because the remaining money was sitting in the trust, the interests of the bondholders and of the Bolivian government became increasingly consistent with one another. Bolivian authorities in La Paz had all reasons to pacify the bondholders, who could block their access to the London capital market, and try to join the scramble for the money in trust. The spectacle of a country scavenging its own securities in combination with its very creditors was an interesting sight.

In 1876 Bolivian authorities formally withdrew Church's concession. In 1877 a deal was cut between Bolivia and the committee of bondholders. They now joined forces against the navigation and railway companies, leading a besieged Church to pull out new tricks. First, he managed to get a mini-railway constructed, reportedly shuttling in the middle of nowhere, but useful as a token of goodwill and diligence to be shown in the courts. Second, he now claimed he had the support of the Brazilian government, which would pledge its credit in order to effect a further loan to cover the difference between the money in trust and the amount needed to actually build the way. This was not long after the Irish-Brazilian confidant to the emperor William Scully had been showered with free stocks.

The first legal battles were lost in 1876 and 1878 by the government and bondholders' coalition, but they eventually won in the Court of Appeal in 1879. They won again in the House of Lords in 1880, when Church's

companies and a dissident bondholder planted there to create further confusion appealed the 1879 decision. The Lords had been unimpressed by the
Brazilian guarantee. The lord chancellor declared, "I do not find any evidence which satisfies me either that the Brazilian Government is bound or
has definitely agreed, to raise a further loan."[42] Likewise, judges found that
the few miles of railroad in the middle of the jungle were not "a railway as
a link in a chain of communication" but an "isolated railway between two
points."[43] The money was released and the story had an almost happy end.
Owing to the bull market in United States securities during the 1870s, the
amount in trust at the time the fund was wound up was nearing £800,000.
This meant limited capital loss for those investors who had bought Bolivian
bonds and subsequently held on to them until the trust fund was released.
Of course, this also meant huge gains for the vultures who had purchased
them in the doldrums of 1874 and carried out the litigation.

## CHURCH REWRITTEN

Telling the story of George E. Church as we have done, emphasizing machinations worthy of Trollope, presents us with one difficulty. On the one
hand, it is consistent with contemporary commentary. When the *Examiner*
of October 31, 1874, discussed Church's railway promotion, it spoke of "the
Bolivian loan, whose history it is difficult to detail without libeling somebody." Safe from the risks of lawsuit, my own narrative has been conducted
in the spirit of what Herbert Spencer called the "often-told tale of adul-
terations." The contemporary "lack of consciousness," Spencer explained,
would have shown "itself in the mixing of starch with cocoa, in the dilution
of butter with lard, in the colouring of confectionary with chromate of lead
and arsenite copper" and was observed with the cloth maker, the manufac-
turer, the silk merchant, and the banker.[44] In the temple of Justice, there
were a few judges who also gave Spencerian overtones to their discussion
of Church. Using the evidence he had before his eyes, Henry Cotton, one
of the judges in the Court of Appeal's panel that ordered the money in trust
to be released in 1879, concluded "an unwillingness, or want of power, to
make the railway."[45] William Milbourne James, another judge from same
panel, was more outspoken, declaring:

> It appears to me that the atmosphere of the temple of Justice is polluted
> by the presence of such things as these companies . . . The [railway com-
> pany] has not a shilling to pay engineer, lawyer, or secretary, or for any

advice or assistance . . . It has not a penny to pay for postage stamp or letter unless, indeed, it levies *baksheesh* or *pots de vins* from the new contractor, e.g. selling worthless shares at par price. Nor would the company as company, if it were a real and substantial one, have the remotest interest in spending the bondholders' money. If the whole scheme were realized, is any person so credulous as to believe that it would ever pay interest accumulated and accumulating, and then the capital sum . . . ? The motives and interests of the individuals acting in the name of the company are of course very different and easy to be imagined. It would be very strange indeed if persons having to expend £800,000 to make the contracts and supervise the work did not find some money adhering to their hands. In this state of things, as it seems to me, the case is at an end.[46]

The strong language used suggests that James would have been comfortable with my rendering of Church. Yet his view was by no means unanimous, even when his conclusion that the money had to be released was endorsed by a majority of other judges in this panel, including William Brett, who came up with a quite different interpretation of Church:

I must confess to such an abhorrence of fraud in business that I am always most unwilling to come to a conclusion that a fraud has been committed, and I have very strong views with regard to what is the legal definition of fraud. It seems to me that no recklessness of speculation, however great, and that no extortion, however enormous, is fraud. It seems to me that no man ought to be found guilty of fraud, unless you can say he had a fraudulent mind and an intention to deceive. Now, speaking for myself, I cannot decide this case upon any ground of fraud. It seems to me that Colonel Church was a person in a peculiar position—in a position which would almost necessarily lead him to reckless speculation. But reckless speculation is not fraud. I am not satisfied that there has been fraud in any part of this transaction; and, so far as I am individually concerned, I must absolve Colonel Church from any charge of fraud, though I think that he cannot be relieved from charges of recklessness and extortionate requirements.[47]

This tone was echoed and amplified in opinions proffered by the judges in the House of Lords. Most characteristically perhaps, William Wood (Lord Hatherley), a lord chancellor under Gladstone's first ministry, rejected the idea that Church had had fraudulent intentions. He declared:

I am not myself impressed, at this stage of the transaction at least, with any approach to *mala fides*. I think that all parties meant to make the railway, and tried their best to make it, although they might have found difficulties in providing ways and means. The Government very much desired to make it. A contract was entered into with the Public Works Construction Company to achieve those works which, after survey, were considered necessary and proper. The works were to be executed within a limited time, which time has unfortunately long since expired.[48]

Hatherley found himself in agreement with the lords justices in the Court of Appeal, inasmuch as they instructed the trustees to release the fund, but he felt it necessary to emphasize his rejection of some of the wordings that had been used there:

I think therefore my Lords, that the conclusion of the Lord Justices is correct, although I do not bind myself to every word and sentence in that conclusion with reference to the conduct of the parties. My opinion is that in the present case there has been a common misfortune, the work has proved to be beyond the strength of those who undertook to perform it, there was a mistake on the part of those who engaged to enter into it, in regard to the supply of the necessary means by which alone the money could be raised, and by which alone the contract made with the public, as contained in the prospectus, could be carried into effect.[49]

This almost neurotic fear of referring to bad faith was vested, given that the House of Lords panel included the current and past lord chancellors, with enormous political significance. The construction of the Madeira and Mamoré swindle as a piece of bad luck from which *mala fides* (bad faith) were absent had the effect of cleaning up Church's credit. And indeed, this endorsement coincided with the beginning of a reinvention of Church, which then went on to further embellish the story of a mishap into one that made of Church a far-sighted discoverer whose great project had fallen victim to a "body of commercial pirates" that others would tell later. These pirates had "purchased Bolivian bonds at a mere nominal price and with the deliberate purpose of wrecking the whole scheme which Colonel Church had matured after many years of laborious effort."[50] There was no question that the Bolivian bondholders had behaved as the vultures they were, but using this to reassess Church's machination was pushing it far.

The signs of redemption that dotted Church's new, virtuous path were of two distinct orders yet enmeshed with one another. On the one hand were manifestations of his rallying the cause of the proper institutions of the defense of investors, in the shape of his becoming an agent of the Corporation of Foreign Bondholders itself. As the biography supervised by Markham put it, Church's first job (Ecuador, with a mission to restructure the debt) occurred "a few months after the wreck of his great enterprise."[51] The transition is remarkable. In the second half of the 1870s, the Bolivian bondholders who were suing Church got increasingly closer to the Corporation of Foreign Bondholders and in effect began working in cooperation with it, using its facilities and lawyer Thomas W. Snagge. Sending Church to Ecuador on behalf of a bondholder association that had been involved in suing him heralded his transformation from adversary into employee.

Other railway surveying jobs on behalf of British investors followed. In 1888–89, Church secured a contract to construct a line in Argentina for the Villa Maria and Rufino Railway Company. As Markham emphasizes and the evidence shows, Church was able to deliver in two years.[52] There was also a mission to Costa Rica in 1895, which required settling complicated diplomatic issues between Britain and the United States and for which the British bondholders employed Church. The apex of this new career in the service of British virtue, however, was reached when Church married Anna Marion Chapman in 1907, a woman who, Markham's notice tells us, was the daughter of Sir Robert Palmer Harding, a pioneer in British accountancy and the first vice president of the new Royal Institute of Chartered Accountants in 1880.[53] Church the swindler had married the daughter of the accountant.

On the other hand, and in parallel, Church's academic star kept rising, leading him in the late 1890s to accumulate prestigious plates such as the presidency of the geographical section of the British Association for the Advancement of Science in 1898 and the vice-presidency of the Royal Geographical Society. He was a member of the Royal Historical Society and of the Hakluyt Society, a club devoted to the promotion of records of voyages and publication of geographical material (another Markham hangout in whose council Church would sit). His library became known to the circle of individuals interested in Latin America, and his knowledge of the subject was enjoyed. A member of the Royal Geographical Society declared, following the presentation of one of Church's papers: "We have all listened to the paper with a feeling of admiration and wonder at the immense amount of knowledge which Colonel Church possesses of the South American continent. There is not a mountain, there is not a river, there is not a plain, etc. etc., of which Colonel Church is not prepared to give us full particulars."[54]

After his death, Church's obituary in the *New York Times* claimed that Church's lecture when he chaired the Geographical Section of the British

Association had been praised in Britain as "one of the most scientific papers ever read before that body."[55] As the project of the Madeira-Mamoré Railway was restarted around the time of Church's death, American academic papers began to make routine references to Church's exploration and writings as the "brilliant studies of Col. Church."[56] New promoters got involved in the puffing up of this new avatar, and naturally started remembering the adventure as one of Yankee intrepidity stifled by British contractors' lack of nerves. The story told in Britain was of course a bit different, yet it is striking that the notice supervised by Markham is silent on the long years spent in tribunals and on the damning evidence that had surfaced on the occasion. A major surveyor, geographer, and anthropologist was dead, and this was all there was to say. This was the generous farewell of one ethno-geographer to his life partner in the traffics of science.

In the few pages she devotes to Church and Markham in her book *Scramble for the Amazon*, environmental historian Hecht speculates that "Church's long association with Markham gave him a legitimacy that was lacking in others hawking upper Amazon real estate."[57] The claim is evidently reminiscent of what was discussed earlier in the context of Burton and the creation of the Anthropological Society—in all cases, legitimacy, received or acquired. This chapter has provided some evidence on how this legitimacy was secured and how it was exploited. It has in particular emphasized the mechanisms whereby finance and science became associated with one another and the contribution made by the technologies of science to hide the connection. Rather than being despised and excluded, Church was elevated and rewarded as a representative of science. His evident talents were immediately put to task, monitored, and controlled, and he was finally rewarded toward the later part of his life for his goodwill and scholarly contributions. After all, had not the Constabulary Force Report of 1839 explained that financial crime was not "characteristic of a barbarous age and of a people subject to blind passion" and that in fact, such criminality could be "deemed an improvement"? In a utilitarian world, even sin is useful.[58]

We may interpret this process, in fact, by using the insights provided by Spencer himself in his "Morals of Trade," in which the great Victorian moralist who had been a subeditor at the *Economist* discussed ways to contain the "merchant who overtrades, the bank-director who countenances an exaggerated report, and the railway-director who repudiates his guarantee." One manner, he suggested, might be to make sure that society would look down on such individuals "as of the same genus as the pickpocket" and treat them "with like disdain." But Spencer doubted that "such higher tone of public opinion" would have any likeliness of being shortly reached. The reason was that the traction of money, which secured status and standing, could not be counteracted in a material age. "When the chief desideratum is industrial growth," Spencer argued, "honor is most conspicuously given to that which generally indicates the aiding of industrial growth. The English nation at present displays what we may call the commercial diathesis [a medical term referring to a hereditary or constitutional predisposition to a disease or disorder]; and the undue admiration for wealth appears to be its concomitant."

Spencer suggested instead catching profit at its own game. The "great task of the age" he argued, could be achieved through the organization "by a stern criticism of the means through which success has been achieved; and by according honor to the higher and less selfish modes of activity." The trajectory of Church, from swindle to science, from cooking the books to auditing them, looks like a perfect implementation of Spencer's program. And this helps to understand the reluctance of the judges at the House of Lords to talk of Church's *mala fides*. For if bad faith was the principle of human action, how could Church be reprogrammed? And if science was to be used to construct an improved, more honorable human, then science itself had to be, as the next chapter will explain, of good faith.

In the end, from the pompous vantage point that imagines science as anonymous, Harvard University's instinct to disband Church's collection was the right one. By dispersing the Church collection across its shelves, it would have managed to blur irremediably the path that existed between anthropology and finance in the shape of a seemingly odd collection that kept together books on Indians, railways, and the business of bondholding. Without the wise and generous help of Brown University, it would not have been possible to establish a conclusive match for the words scribbled from one vulture investor to the next in a book held by Church, and proving the existence of a connection between Church and the vultures. Without this tenuous jungle bridge still holding, could we ever have gone back to the hidden corners of the financial machinations and land hawking? By making sure that his information system would survive him, Church, perhaps by an eleventh-hour attempt at immortality or perhaps also, why not, because the Yankee felt the urge to finally shed the mask he had been carrying for so long and play a last practical joke on the English world, made us the gift of a collection whose absurd combination in itself invites a novel and more thorough understanding of the logic and intentions at work in information accumulation and Victorian scientific knowledge.

CHAPTER SIX

# Acts of Speculation

*B*ona fide or bonâ fide—the Latin expression also occasionally written in
the nominative case as *bona fides*—is found in many Victorian artifacts,
and its ubiquity strikes the eclectic reader. Technically, the formulation refers to someone acting in an honest and sincere way, or to the thing made or
acted in an honest and sincere way. The expression *bona fide*, found through
its negative *mala fide* in the House of Lords' opinion on Church, is used,
literally, all over the place by the Select Committee on Loans to Foreign
States to designate a quality the contractors of the bogus loans precisely
lacked. The fictitious dealings made solely for the purpose of producing a
premium and whetting the appetite of ignorant investors, as the story went,
were described with this very expression. For instance, Henry James asked
the London Stock Exchange chairman Samuel Herman de Zoete whether
the "active dealings" made before the securities were distributed to their
purchasers were "*bonâ fide* dealings, or done for the purpose of launching
the loans."[1] But the select committee was hardly the only venue where the
expression was used: in one letter published in the *Times*, it is used to designate the traveler whose reports were candid. Prospective readers of the
papers left by Anthropological leader John Beddoe after his death were expected to be, according to the will, *bona fide* individuals. Likewise, Hyde
Clarke's intentions, when he began publishing articles questioning the
credit of the Anthropological Society, were self-proclaimed as "*bona fide*,
[having] for their object the best interests of the Society and its Fellows."[2]

Going from there, we see the adjective used identically for actors, their
actions, and the result of these actions. In the select committee, want of
*bona fide* was thought to have characterized those loans made by those contractors who were not *bona fide*. The final report of the select committee contains more than one digression about the proper test of a *bona fide*

124

government bond. In the same report, it was said that everything that had approached the loans ended up lacking *bona fide*: the prices of the loans, the transactions or bargains made in them, and the "City articles" that reported them. In the late nineteenth century, the phrase had become so much of a hackneyed adjective that Isabel Burton could use it spiritedly when she wrote in *Life of Richard Burton* about a "*bona fide* desertion." Sometimes Victorians used it just for the pleasure of proffering the expression. It was evidently in vogue as a way to say "real," such as when Haslewood mentioned the "*bonâ-fide* town of Angostura," which one could find on the map of Venezuela. A similar usage is suggested in Colonel Lane Fox's presentation of the widespread conjecture that modern "savages" provided a modern-day variant of "primitive" civilization: "The existing races," he writes in *Primitive Warfare*, "in their respective stages of progression, may be taken as the *bona fide* representatives of the races of antiquity." There evidently, *bona fide* is used as synonymous for "legitimate," but with the slightly more rarified connotation that this phrase has in the mouth of a scientist.[3] This latter meaning has kept lingering on, forgetful of its origins. I have spotted it in the opening pages of Talal Asad's introduction to *Anthropology and the Colonial Encounter*, where he writes about social anthropology and the other "bona fide social sciences."[4]

There is much to say about the expression *bona fide*, for it does capture in a nutshell much of the spirit of nineteenth-century Britain. In a time that did not take personal and commercial libel lightly, *bona fide* was presumed until proven unwarranted, and the City article editor of the *Times*, commenting cautiously on the Paraguayan loan of 1872, stated that "no doubt [was to] be entertained of the *bonâ fide* character of the scheme" but reminded investors that the promoters, Messrs. Robinson and Fleming, were merely "agents paid by a commission."[5] Given dangers of a commercial libel suit, there was little else to do but take statements for what they gave themselves, monitoring the credit the makers of those *bona fide* statements deserved, and seeing to it later on whether *bona fide* held to its promise on the basis of further evidence. As discussed by stock exchange chairman de Zoete in his interview before the select committee, the prices posted before allotments could be either *bona fide* or made with the intention to deceive, and at the minute when the market opened, the only thing that one could state with confidence was that such prices could be "one or the other; very often they are *bonâ fide*; very often they are for the purpose of launching the loans." And when pushed further by Henry James as to whether he thought this was "a proper state of things to exist," de Zoete responded: "I do not know that I can answer that."[6]

If the modern world has its rational *Homo economicus*, the Victorian world had its respectable *Homo bonafides*. When a company promoter or a banker or a scientist was asked about his company, his bank, or his discovery, he provided some details but rarely all the information. Rather, after some disclosure, he put forward that he was *bona fide* and left it at that. This is why the rising price of the foreign loans was such an important part of their puffing, just like the rising membership of the Anthropological and Ethnological societies. The reported price or the announced membership were the only things one could get to know. Only when prices went down, when membership collapsed, could the questioning intensify because at that point *bona fide* was put to the test. Such was the reason for the ubiquitous use of the expression, which was brandished every time a question was raised: did the questioner have sufficient grounds to raise issues about the *bona fides* of the person he questioned? In other words, focus on the technologies of trustworthiness reveals some aspects of the similarities between finance and science. The rest of this chapter unpacks this parallel, which will go deeper and deeper as we proceed.

## TRUTH AS TRUST IN SCIENCE AND FINANCE

Looking at the existing literature on finance and science, one comes across ideas similar to those expressed above, albeit in a slightly more literary way. Among students writing about the financial history of Victorian Britain, for instance, it has become more and more usual of late to talk about "gentlemanly capitalism." The expression was initially popularized by empire historians Peter J. Cain and Anthony G. Hopkins in their work *British Imperialism*, first published in 1993, where they portrayed imperialism as a policy governed by financial elites in the City and coined the phrases *gentlemanly capitalism* and *gentlemanly order*.[7] The subsequent success of the concept of gentlemanly capitalism owed much to the deliberate looseness of the definition Cain and Hopkins had provided.[8] In their account, gentlemanliness embedded two aspects at least, one strictly descriptive, related to wealth and position and essentially meaning elite, and the other normative, having more to do with a certain aristocratic style that would have been particularly suited to responsible financial endeavors. Illustrating this, they write about the gentlemanly status financiers received only after "appropriate tests of social acceptability had been [successfully] passed" and go on to discuss those "gentlemanly qualities and the social bonding that went with it" that would have been "well-suited to the requirements of merchant banking."[9]

This social definition and the loose elaboration Cain and Hopkins deliberately provided of the associated social technologies have evolved over
time into something slightly different through their use and alteration by
business and cultural historians. Today, gentlemanly capitalism has become
a reference to some kind of professional honesty that is supposed to have
been representative of the social ethos of the Victorian City. This gentlemanly financial sociability has to be understood as something in opposition to, say, the more brutal or outright abusive (plebeian?) style, which
prejudice and the British press in the nineteenth century ascribed to U.S.
finance and railways. Today, especially in the most bigoted variants of business history or financial journalism, it is not infrequent to find reference to
gentlemanly capitalism associated with a positive prejudice for, or at least
nostalgic reference to, a certain type of long-gone "civilized" finance. The
best-selling book Philip Augar authored in 2000 provides an illustration. It
tells the heartbreaking story of an old order made of sustained social and
family structures and a code of honor that stressed probity and loyalty and
was liquidated under the pressures of Margaret Thatcher's Big Bang and the
competition from "harder working Americans."[10]

While the concept of gentlemanly capitalism was being extracted from
the dark bottom of the financial history mine, another group of researchers,
coming from the different perspective of the social and cultural history of
science, came to grips with what in fact was the same or at least a very similar concept of gentlemanliness. Steve Shapin's *Social History of Truth: Civility and Science in 17th Century England* was published in 1994 at about
the same time as Cain and Hopkins's *British Imperialism* and provided a
sort of counterpart to it. Shapin discusses the manner in which truth is recognized as trustworthy. Indeed, science has intermediaries just like there
are intermediaries in finance, raising the problem of trustworthiness. Shapin's answer to the problem of trustworthiness is that belief in one another's
word is established on the basis of the moral estimate of the person upon
whom one depends for information. In the seventeenth century, he argues,
gentlemen were typically seen as trustworthy sources. *Ergo*, gentlemen
were born scientists. In a nutshell, assessments of the credibility of certain
scientific statements were governed by social rating.[11]

Illustrating this with perfection in the historical context of my study,
we have the following quote by Hyde Clarke, drawn from an article "On
the Geographical Distribution of Intellectual Qualities in England," which
he published in the *Journal of the Statistical Society of London* at about
the same time as the merger of the Ethnological and Anthropological societies was taking place. Underscoring the vitality of the old idea that social

standing and intellectual achievements had much to do with one another,
Hyde Clarke reported data that established the correlation between social
and intellectual superiority. This article belonged to a field of reflections
pioneered by Ethnological Society member Francis Galton's famous *Hereditary Genius*. From evidence of a relation between urbanization and the
production of "great men" Hyde Clarke found that on the whole, men of
intellectual distinction trace their origins to the upper classes. "It may be
that such a man is the son of a poor man," Clarke continued, "or of one in
an inferior trade, but the greater men are ascertained to spring from gentlemanly families, or from families formerly in easy circumstances. They may
not all be noble in the technical term of the heralds, but in each country of
antiquity and modern times, they will embrace the new nobles, the nob-
less of the robe and the privileged citizens." The whole argument was then
wrapped up in the language of Darwin and Galton and given as explanation
for cultural nurturing and the accumulation of human capital in which he
found the upper classes excelled.[12]

   Likewise, just as the gentlemanliness of finance is part of the modern
language used to discuss the Victorian City, the sociological perspective on
the gentlemanliness of truth pioneered by Shapin has become part of the
current *vade mecum* in studies of Victorian science. To take one recent
example, Jim Endersby's *Imperial Nature* discusses Victorian learned sociability and its scientific ethos and aspirations in the context of a study of
Sir Joseph Dalton Hooker, son of Sir William Jackson Hooker, who succeeded his father as director of the Royal Botanical Gardens at Kew. This
leads Endersby, naturally, to explore gentlemanly science from the vantage
point of trust: "How was one to know," Endersby writes, "a true gentleman
from a persuasive fraud?" His answer, like that of Shapin, draws on status
and discusses the extent to which scholars were born or made. Studies in
this vein argue that, in the competition for scientific credit, the preexistence
of a social capital (to repeat, a high social rating) provided an advantage. In
turn, references to social capital and social capital accumulation enable us
to explain the persistence of the gentleman scholar as a dominant figure in
British science in the Victorian era and beyond.[13]

   Seen from the vantage point of the relation between trust and truth, the
parallel between science and finance becomes evident. Like the financier,
who originates a financial instrument, works to have its value recognized
and certified, and then benefits from its distribution to a wider public, scientists are very much confronted with this problem of originating, certi-
fying, and distributing. Indeed, scientists make observations and discover
laws, which they then need to get accepted as worthy claims—certified—by

competent peers before distributing such observations or laws to the profession and the wider public. Thus scientists and financiers alike were subjected to the problem of trust, and it is not altogether surprising that they should have found similar and even overlapping solutions.

## TRUST AND ITS TECHNOLOGIES

All too evidently, reference to gentlemanliness obfuscates the reality behind the term. When it comes to the Victorian era, *cognoscenti* ought to have difficulties suppressing a smile at the use of this expression for a country that produced a considerable number of famous swindlers, as researchers are becoming increasingly aware of.[14] Charles Dickens and Anthony Trollope, who knew the men and women firsthand, would have laughed. We may overlook this if we are guided by the gentlemanly compass, but as soon as we land in the large pamphlet literature documenting the most remarkable characters or in the endless litigation trail they left after them, we are bound to conclude that gentlemanliness was just the other side of the coin. Consider vulture investor Richard Thornton, one of the wealthiest Englishmen of his time of whom journalist David Morier Evans has left a suggestive portrait, or John Diston Powles, another prodigal and vulture and a man acquainted with Disraeli, whose astonishing techniques for stripping minority investors of their money and rights are notorious. Or consider later and more well known characters such as Barney Barnato, who liked to fist-fight with brokers in the floor of the Johannesburg Stock Exchange, or Cecil Rhodes, who needs no introduction ever since Hannah Arendt dealt with him. The foreign finance of Britain was handled by a long cast of rogues. These, arguably, were precisely the non-gentlemanly lot, but then what should we say of people such as Lubbock who, as we will see later in the book, manipulated the voting rights of the Corporation of Foreign Bondholders, which he chaired? Because gentlemanliness was a property of the social narrative and editing and not of the character narrated or edited, reference to gentlemanliness is flimsy as a starting point. The phrase *gentlemanly capitalism* resonates more like a splendid example of what foreigners call the British sense of humor when illustrated by portraits of actual Victorian white-collar criminals.[15]

Likewise, and although the matter has not yet received the full attention it deserves, cases of academic indelicacy abound for the period under scrutiny. And they caused serious disruption. To limit ourselves to disputes over plagiarism and the ownership of discoveries involving important characters of the present cast, we have John Lubbock's accusation against

Sir Charles Lyell, which in turn exposed the fact that Lubbock had generously drawn from Danish pre-historians; the dispute, previously mentioned, between Speke and Burton on the ownership of the sources of the Nile discovery, which I suggested played a key role in the launching of the Anthropological Society; and finally, the dispute between anthropologist Owen Pike and Thomas Nicholas in which the latter was accused by the former of plagiarism. The case ended up in Chancery Court and was followed with immense interest from the scientific public and beyond. The usual take is to deplore an age "when conventions for referencing were not as formalized . . . as they subsequently became," a reflection that, again, has counterparts in discussion of lax company accounting standards and their subsequent improvements.[16]

It would be misleading, therefore, to entertain the view that in Victorian finance or science or both, a disembodied cultural norm of gentlemanliness would have ever imposed itself. In practice, deviations were frequent. In fact, if we are to believe Trollope, deviations were the norm—"the way we live now." Reiteration and exhortations to adhere to gentlemanliness at best attest to the failure of gentlemanliness as a way of being, and reference to gentlemanliness is in dire need of elaboration. *Bona fide* as a concept may be more promising and fruitful as a starting point than that of gentlemanliness, although of course they partly overlap, since it is understood that a gentleman is by definition *bona fide*. The use of "gentleman" was confined to certain settings and traditions, such as at the House of Commons, where the routine way to address one another was to refer to the "Honourable Gentleman" whose honorability of course was presumed. But as the example shows, "gentleman" only referred to persons, whereas *bona fide* referred to the character of things. Objects, processes, and things inanimate, once declared *bona fide,* were, in the spirit of positive philosopher John Langshaw Austin's *How to Do Things with Words,* endowed with "performative power, the power to make other things happen." This was a key facet of the *bona fide* individual that, like King Midas who transformed everything he touched into gold, his schemes were presumed to be *bona fide* as well. When Pitt Rivers was conjecturing that "modern savages" were, for any practical research question, equivalent to disappeared "primitive peoples" (one of the widespread hypotheses of both contemporary ethnology and anthropology), he therefore called the "primitives" "*bona fide* representatives" of modern "savages," rather than spelling out the hypothesis. In fact, the language he used was the Victorian way to lay out a scientific hypothesis, which was as much a social as an intellectual pledge.

Thus *bona fide* was, for the bearer and the claims he carried, tantamount to a ticket of temporary admission. As illustration, we have George Earl Church, who was introduced by his friend Clements R. Markham as a *bona fide* geographer and anthropologist. After the introduction, Church could expect his Bolivian companies to be presumed *bona fide* as well. This explains why, as we saw in the previous chapter, Church needed to call at the Geographical Society and joined as a member at the same time he was putting together his Bolivian debt machination. More generally, the constant reports of the voyages by explorers made at the Royal Geographical Society through the reading of their letters ensured that discoveries were attributed to their claimants. It is, in other words, at 15 Whitehall Place (the address of the Geographical Society) that the sources of the Nile were discovered, and Burton created the Anthropological Society in order for him to own another site of *bona fide* performance where *his own* discoveries would be made.[17]

This is the context in which I suggest recasting the social uses of *bona fide* in both finance and science. Not as a social trait but as a technology. And this technology was used to manage individual reputations and that of the society and societies to which those individuals belonged. By its ability to validate a discovery, and record reputations, the Geographical Society provided ownership technologies. What I am arguing here is that the foundations of this system were laid in the 1850s and 1870s and are really background for the later imperial use of geography described more recently by Helen Tilley in her *Africa as a Living Laboratory*, as characteristic of the 1870s and beyond.

In other words, *bona fide* was not a mere cultural ethos clothed with an aesthetic quality. If being *bona fide* was the thing to do in the age of Victoria, it was because of all those *bona fide* outcomes that would result from it, and from which the *bona fide* individual would benefit. From this we see that the heart of the matter was validation. Because the character of the individual proffering a tentative truth was the criterion to assess the degree to which this truth should be received, social technologies to demonstrate character did just as well as the inspection of the facts. They could substitute for them. Ultimately, all that was needed was a social control technology to administer the character of the individual. A demand for truth called for a supply, and truth was thus produced and distributed. This explains the encounter between truth and the organization of society. The control of social behavior had a natural affinity with the production of truth because it provided a ready-made technology for validation and certification of trustworthiness.

This was the mechanism that previous historians have come across when they studied the production of credit in science and finance. Truth in these cases thus came to rely on preexisting features such as belonging to the elite of British society. The member of such a group ought to conform to a number of social norms or else lose membership. The production of truth would thus be achieved through the granting of social or club-like privileges in return for subjection to social control. From this quite naturally follows the reluctance to recognize women as scientists since they could not be, by definition, *bona fide* gentlemen. In other words, because the technologies of truth were embedded in the social fabric, this very social fabric hampered the ability of certain individuals to be trustworthy. From whence the gender prejudice observed in science.

In the language of economists, trust given could be measured by the difference between the benefit from deviation from the norm (the ungentlemanly way) and the cost of gentlemanliness foregone. The optimistic theory, ostensibly, is that if you are so rich that you do not need to rob, you are more likely to be honest.[18] Gentlemanly capitalism would be the money market fund, whose management would go not by the stock market bell but by the four-o'clock tea. The more realist theory, however, is that this calculability of things untrustworthy explains why there were swindlers and plagiarists. Dishonest trading on the *bona fide* status opened valuable combinations, and this endless *in petto* discussion over transgression, not the imagined adherence to the ethos of gentlemanliness, which we must leave happily to previous historians, is what defined the way they lived then. Illustrating this, we have the infamous Honoré Pi de Cosprons, the self-styled Duke of Roussillon, one of the first vice presidents of the Anthropological Society, who understood that being a vice president of a learned society promoted the credibility of his claim to nobility, just like his claim to nobility promoted his credibility as scientist.

Things become even more interesting when we realize that nothing at all made the gentry sole beneficiary of this technology of trust. For if distinction meant control, then control could create other forms of distinction. To some extent, the technology might be separated from the sociological support to which it came so naturally. The logic at work in the economics of gentlemanliness could be generalized and mechanized, and after having given birth to men and things *bona fide*, it could give birth to institutions that would develop new ways to tie men and things *bona fide* to each other. The man and his statements could not exist in a vacuum. But the learned society was a social universe that provided means, to paraphrase Cain and Hopkins, for "appropriate tests of social acceptability." In the arsenal of the technologies available to constrain individuals, associations—by no means a recent invention—were nonetheless the latest and most fashionable avatar. This explains the parallel takeoff in both the supply and demand of societies as reflected by the Anthropological Society bubble or the dramatic contemporary expansion of membership in the Royal Geographical Society.

The Anthropological Society was for the study of "man in all his leading aspects, physical, mental and historical," but there were many other societies for many other purposes. A few years later, an important article by Blanchard Jerrold, editor of *Lloyd's Illustrated London Newspaper* and author of the successful series *The Best of All Good Company*, reported humorously on the phenomenon. The rise of societies produced curious associations for curiouser purposes, which Jerrold described with irony as "The Association for the Total Suppression of White Hats," "The Anti-Flower-in-the-Button-Hole League," "The Society of the Abolition of Green-Tea Drinking," "The Association for the Restriction of Glove-Fastening to One Button," and "The Local Option Snuff Confederation." He added, "And why not?"[19]

Evidently, the contemporaries of Victoria found much value in joining these social groups. I argue that it was because there, to various extents, the *bona fide* of their statements could be subjected to travails that would have them ascertained and weighed. This was conducive of value, explaining the taste for the splashing out of what Jerrold called the "door-plates."[20] In sum, what exploded at this time was the supply and demand of door-plates, a partial solution to the problem of *bona fide*. Every single society, every single acronym, incorporated an ascertaining and validating mechanism, some loose, some coercive, with the degree of coercion ideally indicating the quality of the faith deserved (some believed it at least). And thus James Hunt, in the list of fellows of the Anthropological Society that was put out in 1866 is: "Hunt, James, Esq., Ph.D., F.S.A., F.R.S.L., Honorary Foreign Secretary of the Royal Society of Literature in Great Britain, Foreign Associate of the Anthropological Society of Paris, Corr. Mem. of the Imperial Leopoldino-Carolina Academia Naturae Curiosorum, Upper Hesse Society for Natural and Medical Science, Honorary Fellow of the Ethnological Society of London. PRESIDENT [of the Anthropological Society of London]."[21]

This new formula of the value-enhancing society expanded dramatically and thus to an extent began to compete with the old gentlemanly style. This is what has been described previously as the process of professionalization, conventionally opposed, perhaps not so correctly, to earlier gentlemanliness.[22] The time would come when plate producers would sue illegitimate uses of the acronym. A few decades later, a Scottish Society of

Accountants with a royal charter that permitted its members to use the "C.A." initials for "chartered accountant" brought an action against a "Corporation of Accountants" whose statutes prescribed that members would be allowed to designate themselves C.A. too. The problem was that along with the C.A. label went a number of expectations by the public regarding the "education and conduct and efficiency" of the bearer. In deciding against the copycats, Lord Young insisted that the point was not to deprive the defendants of the right to use a great variety of other designations, but he asked whether the Royal Society would not interdict "a person not a member from using the letters F.R.S."[23] And thus the doorplates were brands, pure and simple.

Thus the contrast between professionalization and gentlemanliness is not as relevant as is the parallel because, rather than substituting for the older technology, the new setup built upon, extended, and generalized it. The survival of the gentleman banker and the gentleman scientist in this context is neither surprising nor worthy of a long digression. In the rush for labels, the gentry, as a class, had a lead and were eager to splash the best of the new plates just like the new plates were eager to splash them. This explains why they were never slow to get their hands on new acronyms and why the societies raced for them. The practice of selling nobility for company directorships has been described before, beginning with Trollope and continuing with financial historians.[24] The nobility devoted itself to this business with the same passion as the middle class and accumulated the new plates with the same determination. Characteristic of this mechanism was the display of the Earl of Derby, fifteenth of the name, son of the prime minister, prominent Conservative politician, and on the Anthropological Society's list of 1866, where he is given as "permanent member:" "Stanley, The Right Honourable the Lord, M.P., F.R.S., D.C.L., F.R.G.S., M.R.A.S."[25] Only the gentry stood to lose more from association with the wrong kind of plate given its position as the social incumbent, giving significance to Buxton's sudden awakening to the "ridicule" of being part of the Anthropological Society.

## TRIAL BY ORDEAL: SPECTATORS, ACTORS, AND THE STAGE DIRECTORS

The outlook of truth was affected by the rise of the societies. Perhaps most prominently, the *bona fide* terminology brought to the fore a legal connotation that was not audible in the gentlemanly variant. As an essentially legal expression, it conjured up a set of legal artifacts: courts, testimonies, trials, verification. The appropriate tests of social acceptability were being replaced by procedures, and the concern that societies had to raise their own profile reinforced the taste for procedure. Sites of truth were thus created. These were scenes where the act of *bona fide* would be performed in a trial-like setting. And just as the *bona fide* deeds were indistinguishable from their *bona fide* authors, the sites of truths ended up enmeshed with the truths they staged. When the Select Committee on Loans to Foreign States was finished with the lack of *bona fide* of the bogus securities, it turned its attention to the Committee of the Stock Exchange, which had signed them off. And Samuel Herman de Zoete, the chairman of this committee, was compelled to acknowledge to select committee panel member Kirkman Daniel Hodgson that the stock exchange committee "had a responsibility . . . in looking to the *bona fides* of the loan."[26]

The theatrical character of the entire subject is a key point in this story. Some writers have toyed with the metaphor of the marketplace of science, a cousin of John Stuart Mill's contemporary marketplace of ideas. It is tempting and convenient because it calls attention to the sites where science was proffered and heard.[27] But the metaphor of supply and demand that reference to a marketplace invites is perhaps less relevant than the question of display and its economics. Indeed, the distribution of science hardly took the form of an auction. Rather, science, like finance, was performed, and like any performance had its actors, spectators, technicians, and associated producers. In this theater, the performers were the producers, and it is no coincidence that the chief inspirer of the Cannibals, Richard F. Burton, was a man who relished the possibilities of the disguise. Just as he dressed as a practicing Muslim to enter Mecca, he could dress as a scientist to enter social arenas. Producer-players such as Burton formed the cast, hired the stage, and called the public in. The society, or club, defined the microcosm where the roles were distributed. A company had shareholders and a board of directors, and the directors of the scientific show were the learned societies' decision-makers. Likewise, a learned society had fellows and a council, and the council was the clique that ran the show. In the list of members published by the Anthropological Society, the overlap between those fellows who "have contributed papers" and those who "belong to the Council" is substantial and almost complete if we try to match paper-givers not only with current members of the Council but also with past and future ones. In other words, the scene was owned by its performers.

This was the universal model of the Victorian play: a society with limited membership holding public meetings usually announced in the press and to which members would have access; a council or committee or board calling the meeting; and a report in the press or in the society's publication

or in a pamphlet providing a transcription of the trial. Anyone who has
spent time poring over the minutes of Victorian meetings cannot help but
be struck by the similarity of the rituals displayed whether they be meetings of bondholders or shareholders or charitable societies or learned societies. The general template was essentially as follows: scenes were created
to decide publicly on character, typically that of the committee, and if the
character of the committee were vindicated, then the entire deeds of the
society were reputed *bona fide.* A typical instance was the general meeting,
held in some public place. In the case of foreign governments' bondholders
or the shareholders of some private company, it was almost invariably held
at London Tavern, "famous for its turtle, dinners, wines, charity meetings
and auctions."[28] In the case of learned societies, meetings were typically
held at the society's premises if they were large enough to accommodate the
crowd or in larger halls when a special event occurred.

Trust was the subject of the plot—Victorian society's single, obsessive
plot. A finding had been made, some event had taken place, a rumor had
circulated, a particular piece of information had been revealed that required action, and a course was to be adopted. *Bona fide* was assumed at the
beginning and would have to be vindicated in the end. Consistent with this
was the cathartic ritual at the closing of the meeting, of the unanimous vote
of thanks to the paper giver, to the author of the report, to the chairman and
to the leaders of the meeting—a pronouncement that credit had been recognized and reaffirmed. Whatever cracks in trust had appeared were thus
slathered with heavy layers of the *bona fide* paint. In some cases, however,
the subject of the trial was unmade in a resounding public meeting that had
the appearance of an ambush. For the Victorian public, the way the "Railway King" George Hudson fell when he ran into a carefully planned ambus-
cade during a February 1849 shareholder meeting for the York, Newcastle,
and Berwick Railway provided the model scenario.[29]

Mirroring this evolution, or rather as cause and consequence of this evolution, a class of individuals asserted itself, devoted to the performance of
social productions. Importantly, as technicians, they were usually specialists of not just one stage but of the Victorian stages in general. In the article
by Blanchard Jerrold alluded to before, and which contained one of the few
recognitions of the relevance of this group, the class of agents who specialized in stage setting are called the "manufacturers of public opinion,"
and hailed, with some irony, as having begun to make of "press-agentry a
fine art." In Jerrold's description, this new, vibrant industry had its "heads
of trade," whom he describes as both actors and stage directors, men he
thought excelled at moving seamlessly between the various societies. This

is how they could put public opinion in motion. Suggestively, he calls them "men of many door-plates," and we recognize here certain characters of our cast, from Hyde Clarke to John Lubbock—multi-door-platedness here less a Victorian sociability than weaponry. And sure enough, Jerrold explains, the member of several learned societies, the society promoter, organizer, and reorganizer, or the administrative reformer, ought to have the qualities of an *actor* so that he may move seamlessly between alternating scenes. His "demeanour at the board of the British Beadles' Recreation Society," Jerrold writes, "would be fatal to him when sitting at the elbow of the chairman of the Impecunious Bishop's Aid Society. A pleasant, discreetly jovial bearing is desirable in the former capacity; and an aspect of sorrow, silvered somewhat with rays of hope, *proportioned to the subscription list*, in the other" [my italics].[30]

## THE ART OF PACKING

It is conventional in anthropological history to contrast the royal performances of power in early modern Europe with the later evolution where these dramatic displays were replaced, or rather supplemented, by transcription, "officializing" procedures, and increased reliance on the services of the scribes, who counted, recorded, and classified. One important artifact of the mid-nineteenth century, which I have come across repeatedly, is striking for the perfect incarnation it provides of the transition from performance to record—the minutes of the society's meetings. Each one was essentially a script, both transcription and play, and the scriptwriter's indications could serve, quite literally, to transcribe the events in the codified language of theater. Befitting Britain's oceangoing prowess, societies often organized their flagship publication in the fashion of a ship's log with minutes from the meetings as entries. In the case of the Anthropological Society, the *Journal of the Anthropological Society of London* was organized by meeting and recorded, in an almost immutable order, the date of the meeting; the chairman of the new meeting; the decision to approve the minutes of the previous meeting; the list of fellows, honorary fellows, corresponding members, and local secretaries voted in; the presents and correspondence received, such as books, periodicals, or Consul Hutchinson's gift of the eight "Argentine" skulls discovered, the presenter explained, when the contractors of the Centro-Argentine Railway had cut through a graveyard;[31] and finally the summary of a paper transcribed from the vantage point of the discussion that had followed it (and thus the paper itself was nothing but the transcription of the performance).[32]

Of course, the relation between the transcript and the events was prob-
lematic. Here, historians of science should take heed from financial historians who had expressed their concerns early on.[33] Transcripts, as *bona
fide* recording, are suspect. Because the meeting had been public and the
minutes published, anyone could in principle judge the conformity between
them, not to mention the fact that the authorities who vouched for the
document pledged their own credit. But in more than one instance, I have
found the typical handwritten transcripts of the Victorian meeting to be
much more concise when they existed at all than the printed, public variant. On the face of it, this would seem to defy conventional presumptions
about the relation between primary (handwritten) and secondary (printed)
sources, which hold that the latter tend to be more condensed than the
former because of editing and the constraints of publication. This arises
because the secretary of the society, after having participated in the meeting, acted it a second time, so to speak, in greater detail and with the intention to display, taking advantage of the silences and ambiguities of the
actual meeting, testing the limits of dishonesty. There was editing in that
the printed document embroidered, expanded on, and developed the manuscript. And thus the script, which is the part we still have, was evidently
itself part of the art of performing.

In the minutes of the already mentioned meeting of Spanish bondholders
kept as part of the Church Collection, one bondholder rises toward the end
of the meeting and, apparently outraged, pronounces the act that has just
been played a "packed meeting."[34] Packing of course was the name of the
game because those very individuals who ran the societies were also those
who called the meeting and staged the act. As narrated by others, although
not exactly in these words, the aftermath of the merger that gave rise to
the Anthropological Institute produced a succession of packing and counter-
packing events that pitted the Ethnological and Anthropological cliques
against each other and must be understood as part of the performance game.[35]

In the minutes of a Mexican bondholders meeting, we find the following
statements: "After the Secretary (Mr. Godfrey) had read the advertisement
calling the meeting, the Chairman rose, and said that the committee had
thought it expedient to assemble the bondholders in consequence of Mr. William Parish Robertson having made a proposition to them which was considered of importance." After one Mr. Lodgen wondered aloud about the
authority of the committee, "Mr. Powles said ( . . . ) Mr. Capel thought
( . . . ) Mr. D. Mocatta proposed . . ." These voices were like the instruments of an orchestra, conducted retrospectively by the transcriber, because
in fact all these men were part of the same bondholders' clique. In the end,

unanimous thanks were voted to the chairman.[36] Thus it is that despite the occasional occurrence of a somewhat dramatized opposition, the outlook for the test of character was almost invariably favorable to the incumbent. A better characterization of these minutes is to describe them as part of a process of manufacturing—packing as puffing and vice versa.

To produce the desired outcomes, the manufacturers could use an array of tricks. The executive committee set the date and venue, chose the subject under discussion, and controlled the evolution of the discussion. It could also plant in the audience members who would make the right comment or provide the right defense at the right time. In the middle of the *Athenaeum* controversy, one of the letters Hyde Clarke wrote to the journal contained a postscript about a General Assembly meeting convened by Hunt and the Anthropological Council to discuss the subject of his exclusion that read: "On the 26th I receive a letter, dated the 22nd, calling a meeting on the 2nd of September, for my expulsion. Who will be in town except the clique concerned? Why this hurry?"[37]

This importance of stage-setting can be illustrated from the following transcription of the little melodrama played out at St. Martin's Place, headquarters of the Cannibals, on June 16, 1868, when the Council of the Anthropological Society met to discuss the merger proposal.[38] It shows James Hunt, leader of the Anthropological Society, coming to grips with a problem of *bona fide* as he is torn between his evident desire not to cut a deal with the Ethnological Society and the need to appear open and congenial.

The synopsis reads like this: At the beginning of Act I, "Dr. Hunt, President, in the Chair," submits a report respecting the "proposed union" of the Anthropological and Ethnological societies, laying before the joint council the proposal from the joint committees to set up an amalgamated "Society for the promotion of the science of man." In response, there follows a first resolution (D. I. Heath and B. Seemann) on the "flourishing" of organizations in several capitals of Europe under the name of "Anthropological Society," this being a sufficient reason to prevent the change of name. The motion produces fifteen votes in its favor and four against. A subsequent motion (B. Pim and R. King) states that the new name is "not better" than "Anthropological" but is willing to "leave the selection of the name . . . to the vote of a combined meeting of both Societies." This one secures even more support with "only one voter against." The meeting is suspended, and there is a static shot on the empty room while Hunt confers with the Ethnological delegates outside.

In the next scene, the Council reenters the room, and Hunt is in the chair again, announcing the rejection of the offer by the Ethnological delegates.

He tenders his resignation from the presidency of the society, and this is followed by resignation of the society's director E. W. Brabrook, also a delegate in the Anthropological Society's committee for amalgamation talks. Resignations are accepted, and the scene ends with a dramatic motion by L. Owen Pike to thank the society's negotiating committee for their "best efforts in endeavouring to carry out the proposed union." But there must be a happy ending, and Act II, which takes place three days later, provides it. It starts with the unanimous election of J. Barnard Davis, F.R.S., to the presidency of the society. He takes the chair only to have a new resolution proposed by Pim and seconded by W. S. W. Vaux "that the resignation of Dr. Hunt, as President of this Society, be not accepted, his services being of such importance to the Society, that they cannot be dispensed with" (all but one voting in favor). Another resolution, proposed by R. S. Charnock and J. W. C. Cox, stated the same about E. W. Brabrook and passed unanimously. Following this, a glorified executive is restored to power, and Hunt resumes the chair to preside once again over the destinies of the society. What is at stake here is not a scientific truth as such but something even more important—the conditions under which truths can be truths. Namely, it is the whole trustworthiness of the learned society itself as an instrument of truth production and thus valuation.

The intentions of the act are easily recognizable. Hunt's authority over the society was complete, and it is of course unlikely that his stepping down from the presidency resulted from his disavowal by his Cannibals. Rather, Hunt-of-the-*bona-fide* could not object to Thomas Henry Huxley's suggestion to discuss a possible amalgamation. Since they both publicly protested of their *bona fide* interest for the science of man, they were compelled to talk and reach some form of agreement. Had Hunt bluntly refused the amalgamation, he would have had to face accusations against his character. On the other hand, since there was probably no intention at all on Hunt's side to carry out the merger, the scenography was a ritual that was really used to absolve Hunt—turning down the proposal, forcing him into a symbolic exit, and then calling him back in an act of vindication of his character. In fact, at the beginning of Act II, he is still physically in the chair, the script says, and never left it.

Thus it is: men in chairs, men rising from their chairs and sitting down again, men voting, doors opened and doors closed, discussion shown and discussion obfuscated, suspense, and the surprising turn of events all summarized by a messenger-participant as in the Greek tragedy and becoming the driver of subsequent dramatic action. The truth of *bona fide*—the truth of trust—is established through "loud cheers" and "hear, hear," through motions, demotions, and promotions. *Bona fide* was a face borrowed and a face lent, and the deft combination of individual and societal character made it possible to circumnavigate social impossibilities and provide enormous sovereignty to the managers of the society acting a narrative. In the language of American philosopher William James (brother of writer Henry James, a contemporary of Victorian Britain): "The truth of an idea is not a stagnant property inherent in it. Truth *happens* to an idea. It *becomes* true, is *made* true by events. Its verity *is* in fact an event, a process, the process namely of its verifying itself, its *verification*. Its validity is the process of its *validation*."[39]

This chapter has cast light on the continuity that existed between the sketches that were played in the Victorian festival—now a bondholders' meeting, now a joint-stock company meeting, now a learned society meeting, now a charitable society meeting. The ubiquitous problem of trust glued these scenes together. In a world of travels, expansion, and distant truths, the inchoate material originating in the most unlikely places could not easily be turned into marketable artifact. For this to happen, an institution and a ritual were required. The pronouncements of the *bona fide* individual in his society's "court" were to be received as true until contradicted, and this system conferred to fellows the ownership of the resulting transient truths on which they could base a variety of profitable social, political, or economic trades. All this was a matter of composition and performance. In one of his controversies with the Anthropologists, Huxley would put it in so many words: he defined himself as a man who has made "grave and public statements, on a matter concerning which he is entitled to be heard."[40] Or, to take up again the example of George Earl Church: the *bona fide* of his railway and navigation promotions, gradients in mountainous areas, and the ferocity of cannibals needed a trial and a process to be transformed into a financial product. The natives, the rocks, the tributaries of the Amazon, had to be structured before they would hit the floor of the exchange in the shape of a 6 percent Bolivian loan, and this was achieved within a social institution. This explains the success of the learned society with enterprising elements of the London Stock Exchange. A new class of creditless individuals joined societies and benefited from the credit of the societies they joined.

The social institution at work, I have argued, was essentially a scene: each time, an act was played out, be it at London Tavern with Mexican bondholders or at 4 St. Martin's Place. The devoted performers, not unlike a theater company, had that cliquish quality noted by others beginning with contemporaries such as Hyde Clarke, who threw the word *clique* as a man from the Stone Age threw his flint weapon, using it against the Anthropological Society in the columns of the *Athenaeum*. Reference to acting (now the Great Act of Probity, now the Great Act of Science) facilitates understanding how easy it was for the men of many doorplates to alternate between scientist and promoter, swindler and swindled. Societies provided semi-public forums that went by the same scenography, and the ticket was the bond, or share, or fellowship. The deftest stage directors knew exactly how to pack a reunion with the right number of supporters and plant the right comment at the right time. They were the specialists in the secrets of Victorian Britain, and they would organize any setting for a fee. This is an important insight, because it will help explain, at a later stage of my narrative, why some characters stepped over the borders of finance and science. The point is that in one universe, such borders did not exist. This is how the many-doorplated man was destined to cross fences. Half actor, half *dalit*, it is no surprise that the man of many doorplates might not have his place in that hall of honor of Victorian decency, the *Dictionary of National Biography*, now misunderstood as a source of biography.

In the end, it is as simple as that. If, as others have suggested, trust and its modalities were essentially social phenomena, then we should be able to see the continuity that existed between the various scenes. Trustworthiness was a whole-purpose instrument that could be used for science, finance, and many other things. This also means, quite importantly, that there was an intimate link between things scientific and things noncompetitive. If trustworthiness rested on restricted entry, then both elitism and knowledge had something in common. Perhaps in this inherently non-democratic nature of science, which came with a vengeance during the scientifically vibrant 1860s, is found the explanation for the gender and race backlash that occurred precisely at that juncture. Indeed, as the rest of the book will argue, it was then that the new imperialism of the last quarter of the nineteenth century would be invented.

While I can see why scholars would have trouble accepting this because they often like to think of themselves as inspired by democratic virtues of which they would also be the watchmen, the fact remains that in the reactionary counterinsurgency of the 1860s, the learned society played the role of the armory. Given this, the last word goes to Benjamin Disraeli, one of the Victorian politicians who had most profoundly understood the ontological nature of the pursuit of truth in the world he lived in, and who, when he introduced the Reform Bill in 1867 (by which the franchise was to be substantially increased), declared about British society something that might have applied very well to Victorian science: "We do not, however, live, and I trust it will never be the fate of this country to live, under a democracy." There was more to this than a formula meant to assuage his Tory constituents and Liberal dissident "Adullmites" as to the lack of revolutionary purpose in his motives.[41] Not all *bona fide* persons are equal, and in this inequality, we find a reason for the survival of the myth of the delinquent state and recurrent oblivion of the Victorian swindler and his later incarnations.

## CHAPTER SEVEN

# Wanderlust: The Upbringing
# of a Victorian Racist

In their accounts of the "Negro and Jamaica" lecture, previous scholars have recorded their legitimate horror at Bedford Pim's racist performance and scoffed at the "anthropologist." But never have they seriously elaborated on Pim's pedigree. Reflecting a mythology that has grown gradually from a failure to explore Pim's actual background, one otherwise well-informed article by a bona fide scholar would have him "hastily admitted" to membership in the Anthropological Society at about the same time the lecture took place. This statement tells a lot about existing prejudices on the subject. It implies that Pim was somehow an illegitimate anthropologist, since his credentials were summarily examined, since this anthropologist was the author of *Negro and Jamaica*, and since the society that admitted him was merely a Cannibal club. As we shall see, however, the claim that Pim was hastily admitted to the Anthropological Society is profoundly incorrect.[1]

There is no dispute that Pim was a pure racist—as pure a racist as Victorian society could show, and it had many splendid specimens from which to draw. In fact, much of the rhetoric in *Negro and Jamaica* may be traced to the classic of the genre, the vast emporium of racist figures of speech of Thomas Carlyle's "Occasional Discourse on the Negro Question" which dealt with what Carlyle saw as the laziness of the freed slaves and the confusion their newfound but illusory equality with their white "born lords" had created. As historian Catherine Hall has emphasized, Carlyle's rhetoric operated through a feminization of the black men in the sugar islands as "pretty, supple, affectionate and amenable," reflecting, she explains, both the Victorian racist's contempt for women and his pervasive fears of black sexuality.[2]

Bearing witness to Pim's belonging to this tradition is the apocryphal description he would make of his stopover for coaling in Jamaica during an 1859 trip that was to introduce him to anthropology, in the coauthored book with fellow anthropologist Berthold Seemann published in 1869. In *Dottings on the Roadside*, Pim recounts the remark ascribed to one of his officers that "it is good news that we are soon to leave Jamaica; for what with a stifling heat at Port Royal during this time of the year, and the overbearing insolence of negroes, almost any place would be a change for the better." Elsewhere, Pim tells the story of a Jamaican who had beaten a dog and would not hear reason, repeating to the admonishing white: "I man like you. I'se good as you." Finally, fitting Hall's hypothesis regarding sexual anxiety, there is this intriguing anecdote in which Pim claims to have been invited to a party in Port Royal that also included black women described as upscale prostitutes. His courting efforts—an attempt at being "civilized"— are repeatedly snubbed, and one woman, losing patience at his attempts to make conversation with her, finally declares, "I come here to dance, Sar. I no come to tark." These recollections were intended to impart a sense of fatalism to the Jamaican Rebellion. "Good Bye Jamaica," Pim concludes with the perfect foresight of hindsight as his ship is steaming away.[3]

No doubt, therefore, that the anthropological surgeon in charge of the autopsy of Bedford Pim's anthropology will find much pus. But if pus he finds, then shouldn't pus be the legitimate object of his study? Not only was Pim a pure racist, but he was also a pure representative of the anthropology that developed in those years, and his pedigree illustrates to perfection several of the themes that I have articulated thus far. In fact, as this chapter and the next will show, Pim was not an eleventh-hour anthropologist opportunistically sworn in by a *mala fide* Anthropological Society just in time to deliver his "Negro and Jamaica" lecture. Rather, his road can be traced back to the 1840s, to the extraordinary publicity campaigns of the Geographical Society, and eventually, as the next chapter will emphasize (for Pim is too important a fellow to be dealt with in but one chapter), to the Anthropological secession of 1863.

If only for the fact that Pim's pedigree has been misread before, it is worth securing a better understanding of Pim's character. But there is more to him as well. Of all the split personalities that Victorian society produced, Pim is an extraordinary specimen. In the *Dictionary of National Biography*, C. Henry Coote declared him to have been "a true-hearted sailor of the old school—brave, generous, and unselfish."[4] On the other hand, Robert Louis Stevenson, the author of the *Strange Case of Dr. Jekyll and Mr. Hyde*, devoted a long, sarcastic article to Pim, which he intended to be the first in a series of portraits of "City men." This was not, it should be remarked, the only occasion in which Stevenson dabbled in anthropological matters.

His famous *Body Snatcher* was inspired by the William Burke and William Hare body-snatching murders and by the role the Anthropological Society's patron saint Robert Knox played in them.

There is, in summary, a mystery to Pim, illustrated by the conflicting superlatives that have been applied to him over time, whether the "true-hearted sailor," the "hastily sworn in" racist or, to use Stevenson's expression, the "salt-water financier."[5] In the conflation of Dr. Jekyll and Mr. Hyde lies the secret of a personality whose significance has been paradoxically obfuscated by his central role in the debates of anthropology in the 1860s. It is time for a restatement.

## FROM ESKIMO TO MISKITO

In 1845 young Bedford Pim embarked as mate aboard *HMS Herald* under Captain Henry Kellett. *HMS Herald* was a survey ship involved in a circumnavigation of the globe, and after cruising Latin America and the Pacific, it was ordered in May 1848 to search for Franklin's naval expedition. From Panama where it was stationed, it headed north to the Kotzebue Sound, the Bering Strait, and Western Eskimo Land.

Of importance to the unfolding of the anthropological tale is the fact that on this same boat was another young man, Berthold Carl Seemann, a German-born botanist and the protégé of William Jackson Hooker, the director of Kew Gardens who had recommended Seemann as an embedded researcher. Both Pim and Seemann were in their twenties when they visited together the places that would provide the focus of their commercial, financial, and scientific enterprises twenty years later. The importance of the experience is attested by a book that Seemann would devote later to the voyages of the *Herald*, where he spent many pages narrating Pim's Arctic exploits and emphasizing their early friendship. The book was dedicated to William Jackson Hooker for his "generous encouragement and ready assistance."[6]

In October 1849 Pim was transferred for a time to *HMS Plover*, where he distinguished himself by launching a sledge party to Mikhailovsky (now St. Michael) in order to try and corroborate from Russian sources the persistent rumors among natives of "white men living and travelling in the interior of Alaska." Pim was made lieutenant on October 2, 1850, and simultaneously transferred to *HMS Resolute*, where he would serve until 1854.[7] After his promotion and transfer, Pim was found behind the proposal, in 1851, to explore the Arctic coast of Siberia as part of efforts to promote an open polar sea. The idea was supported by British whaling captains and endorsed by the Royal Geographical Society.[8] The Admiralty was hostile (diplomatic complications with Russia were potentially enormous), yet Sir Roderick Murchison splashed in the *Times* the Royal Geographical Society's request that the Admiralty should support Pim's voyage by putting him on full pay during the mission. Once again, the Geographical Society exploited the sorrow of Lady Franklin. She had put her trust and last hope, along with a grant, with a young and gallant lieutenant.[9] The extortion succeeded, and a £500 grant from Prime Minister Lord John Russell was secured. Pim, however, would never go beyond Saint Petersburg. Russian authorities proved harder to persuade than Russell. They were lukewarm at the thought of a semi-official British agent animated by the ideas of an open Arctic collecting data in their Siberian backyard.

Our already-accomplished explorer joined the Royal Geographical Society in 1852 at the age of twenty-six. The Franklin searches caused more parties to be dispatched and meet trouble. This was the origin for Pim of further Arctic voyages, which according to his own account "made him a man." In 1853 Pim was instructed to locate, and provide help to, Robert McClure's *HMS Investigator*, which was caught by the winter. The journey on ice, on a "light dog sledge," was "accomplished in twenty-eight days and on 6 April safely reached the vessel only just in time to relieve the sick and enfeebled crew." The experience put Pim in contact with indigenous people, whose directions he had to transcribe in his own system of geographical indications. Pim was further distinguished for having been the first to travel a portion of the Northwest Passage.

To this brilliant geographical pedigree Pim added a few glorious wounds later collected during the Crimean War and in Chinese waters. He was made a commander on April 19, 1858. Between Royal Navy assignments, he would be seen in the Royal Geographical Society's meetings. In April 1859 for instance, Bedford Pim could show his newest plates—"Commander, F.R.G.S., R.N."—as he gave a lecture on the Isthmus of Suez, his latest assignment. With this, he was an authority on canal cutting and the crossing of isthmuses.

Such was the background in which, a few days later, on April 27, 1859, Pim was placed at the head of the *Gorgon*, an old paddle steamer of 235 feet (71.6 m) length and dispatched to Central America. The trip acquainted Pim with Mosquito (or Miskito) Indians and led to an encounter that was to send Pim into the maze of anthropology.[10] Pim did not go straight there, however. The *Gorgon* began by cruising for some long, boring months along the muddy waters of the English Channel and points north until it received its orders and began paddling to the Caribbean. The orders were to secure the Atlantic coast of Central America. Specifically, the *Gorgon* headed to Greytown in Nicaragua (San Juan de Nicaragua) on the Caribbean coast known also as the Mosquito Coast—the former Spanish Main that Columbus had sailed along in 1502.

The context was one of imperial rivalry in Central America between Britain and the United States. A British plenipotentiary, Consul General Charles Lennox Wyke, had been sent to Central America in early 1859 to secure treaties with Nicaragua and Honduras. High on the agenda was the issue of indigenous Indians thus far under British protection. The situation was complicated by the buccaneering tendencies of the local creole and British settlers who, fearing that they would be left to themselves by the retreat of British foreign policy in the area, showed restlessness. They were prepared to hire filibusters to their rescue, such as the private militia of Tennessee adventurer William Walker. In fact, some reports claimed that the inhabitants of the Bay Islands settlements off the coast of Honduras had already called Walker in and a landing was imminent. Given the situation, the Admiralty may have thought that a young man known for his "ability, enthusiasm, endurance, perseverance" as Henry Kellett, Pim's commanding officer aboard both *HMS Herald* and *HMS Resolute* had it, was the right person to send out as senior officer of the Central American Station. It was not long before the Admiralty regretted the decision.[11]

## KING GEORGE AUGUSTUS
## FREDERIC READS ANTHROPOLOGY

A high point of *Dottings on the Roadside* is Pim's rendering of his visit to the king of Miskito, George Augustus Frederic (the name is sometimes given as George Augustus Frederick). The name might have been more befitting a prodigal or projector in the London Stock Exchange than a native ruler, but then this was a Shakespeare-reading monarch raised in Jamaica and crowned in Belize. Careful, as anthropologists are, of leaving an accurate image of the king's features, Pim left a detailed description of the king of Miskito:

> We now had time to take a good look at George Augustus Frederic, the King of Mosquito. He was about five feet seven inches in height, well built, but slight, and of pure Indian blood. His complexion was swarthy—darker than that of a Spaniard, but still fairer than the generality of his countrymen, probably because he was not subjected to the life

of constant exposure and hardship which is their common lot; his face
was flat, like that of a Chinese, cheek-bones high and rather prominent,
the nose small and thin—a distinguishing feature of the Mosquitians,
the other tribes on the coast not being characterized by this marked pe-
culiarity, but, on the contrary, having noses similar to those of other Indians, nay, in some instances, even prominent. His hair was very black,
cut rather short, and parted on one side; it was very fine, and straight,
without the slightest appearance of a curl or even waviness. Having neither whiskers nor moustache, nor in fact the least vestige of a hair on his
face, and with the delicately-shaped hand and foot of his race, he gave
one the idea of being very young; he was not quite thirty, but looked
scarcely twenty.[12]

To complete the portrait, Pim produced a woodcut based on a later photograph. The quality of the original picture was bad enough, but, Pim insisted, "half a loaf is better than no bread." (See figure 5.) The reason for the
painstaking detail in Pim's ethnographic description of the king was that
Pim sought to refute the work of another contemporary anthropologist. In
1855 the American anthropologist Ephraim George Squier (who would join
Hunt's Anthropological Society of London), writing under the pseudonym
Bard, had published in New York an ethnological novel called *Waikna; or,
Adventures on the Mosquito Shore.* "Waikna" was an indigenous term that
meant "man" and which Indians used to designate themselves. The book
was openly racist. It was intended to inculcate to the American public the
idea that Miskito Indians descended from survivors of the wreckage of slave
ships and were thus not Indians but Africans. Pim's description of the king,
with its emphasis on "pure Indian blood," "high cheek-bones," "thin" nose
(as Pim insisted, thinner than that of other tribes on which Miskito Indians
ruled) and hair "without the slightest appearance of a curl" engaged Squier.[13]
    In fact, in the most entertaining passage of the description, Pim claims
that he spotted Squier's book in the king's library, just beside Sir Walter
Scott and William Shakespeare, which enables him to have the king himself dispatching Squier—in Spanish—by stating that "in Nicaragua" Squier
was known as "un alegre menteroso" ("a playful liar," Pim translates for
the reader). To Pim's question of why should "any one in so responsible a
position, the representative of a great country, could be so foolish as to write
what must sooner or later be proved false," the king answered that Pim's
remark showed his wisdom but "does not hold good as regards a Yankee."
The king added, "When I was a boy, I looked upon the Yankee as, next to the
Englishmen, the most honest and truthful people in the world, and I used

FIGURE 5. The anthropology reading native king. "At the end of the jetty, stood the King, dressed in a white jacket, waistcoat and trousers, and felt hat, just as he is depicted in the accompanying woodcut which is drawn from a photograph of his Majesty, taken by the doctor of the *R.M.S. Solent*, under great disadvantages, and with a most imperfect camera; but as there is no likeness of any King of Mosquito extant, I publish this, on the principle that half a loaf is better than no bread and because at all events, it will give some idea of Mosquitian royalty" (Seemann and Pim, *Dottings*, p. 266). *Note: R.M.S. Solent* was the boat aboard which Pim returned to the Shore in 1863.

to read at Jamaica and Belize anything relating to the progress of their nations with delight; But now I scarcely know how to express my contempt for them." And the king, decidedly a fine connoisseur of cultures, would have concluded with "scratch a Russian and you find a Cosack." In a subsequent philological musing, Pim remarked that the king must have *really* meant "scratch a Yankee and you will find the Red Indian."

Pim's speculations on Miskito Indians have had an enduring legacy in anthropological writing quite apart from his legacy as the archetypical racist of the Anthropological Society of London. As I write this, hundreds of papers in anthropological journals include references to the Miskito Indians, and many of those with a historical background provide a reference or an allusion to Pim.[14] For instance, anthropological historian Michael D. Olien has in effect relied on the putative evidence from Pim's account to

deconstruct the politics of Squier's "insidious" characterization of the Miskito Indians as a black population, which, he laments, has "plagued historical scholarship ever since." Olien criticizes Squier's view of Miskito's
blackness in favor of evidence from more serious writers "who had actually
met the young king" and who "all describe him as an Indian." Chronologi-
cally the first of those authorities, Olien explains, was Bedford Pim.[15]


## THE MOSQUITO COAST AND THE BRITISH
## WORLD SYSTEM: A HISTORICAL SKETCH

In order to understand the manner in which Pim interacted with the Mosquito Coast, and how this eventually led him to enter into anthropology, it
is now useful to broaden the perspective for a while and provide a concise
narrative of the politics of the area. Historian Robert A. Naylor has provided what is to date the most thorough study of the place of Miskito Indians and the Mosquito Coast in the history of British informal empire. Unlike Belize (or British Honduras) on the northern frontier of Honduras or
Jamaica, the Mosquito Coast was never formally a British possession. To
the Mosquito Coast, the beautiful words historian Richard White has devoted to another locus of informal imperialism might apply very well: "This
is an imperialism that weakens at its periphery. At the center are hands on
the lever of power, but the cables have, in a sense been badly frayed, or even
cut. It's a world system in which minor agents, allied, and even subjects at
the periphery often guide the course of Empires. This is an odd imperialism
and a complicated world system."[16]

Something similar emerges from Naylor's characterization of British dominion in the Mosquito Coast, which he summarizes as "penny-ante imperialism." A key element of the imperial agency in the area was the securing
of influence through a "King of Mosquito," a British invention. British interests had created this king in the mid-seventeenth century when, under
the reign of Charles I, they shipped to England a young native, later known
as "Oldman," and after an extended sojourn returned him as an English-
speaking broker between Indians and British. Over time, security-minded
interlopers along the Mosquito Coast found value in having a political and
cultural broker to whom they increasingly turned in their various traffics,
eventually passing him as king to British authorities in Jamaica or Belize.[17]
This is how the old alliance between British mercantile interests and the
Miskito people was born, one result being that the English-loving Indians
ended up with names such as Kitt, Morgan, Labrin, Patrick, and Frank, while
their grandees were named George and even Wellington.[18]

While this government-by-proxy originated mostly with private entrepreneurs, authorities in London tolerated and even encouraged the situation, which afforded them influence over and beyond international treaties.
In this gray zone prospered a host of traffics that used connections, money,
and bluff to push their advantage on the back of a passive British imperialism that could own successes and disown mishaps. During the seventeenth
and eighteenth centuries, British interests maintained an alliance with the
king of Miskito through gifts and other encouragements, driving a thorn
into the side of the Spanish Empire. British interests were generally successful at turning natives against Spanish ships and settlements. In particular,
one subgroup of natives, designated as "Sambo-Mosquitos" or "Sambos,"
comes in various accounts as living in the portion of the Mosquito Coast
north of Cape Gracia a Dios (the Atlantic coast of today's Honduras). These
"Sambos" often banded with British buccaneers in expeditions against
Spaniards and shared the plunder. The word *sambo* came from the Spanish
*zambo*, a denomination of the Spanish American caste system that referred
to a mix of Indian and African blacks. In Mexico they were known as *lobos*
(wolves) and in the east of the Caribbean, *garífunas*. The ways these hierarchies were recast by British and American anthropology's obsession with
racial stratification of the Mosquito Coast in the mid-nineteenth century
would be worthy of a full exploration beyond the scope of this book.

A high point of the Anglo-Mosquito alliance was a raid by Mosquito
"King" George and Mosquito-Sambo Chief "General" Robinson, which drove
Spaniards from the coast and destroyed their infrastructures in 1800, some
years after Britain had been forced to demolish its own fortifications along
the same coast by the Anglo-Spanish Convention of 1786. As noted by Naylor, cultivation of the Indians' anti-Spanish attitude permitted the Mosquito
Coast to maintain its role as one corner of a valuable woodcutting triangle
whose other corners were Belize and Jamaica.

Reference to Sambos raises a crucial wording matter that will become
important later. Along with the "scientific" use of the word *sambo* that
we find in the anthropological discourse of the 1860s, where it designates a
certain population living in a given area of the Mosquito Coast, there was
also the more general, colloquial use of the term in both British and American English. In that usage, it refers to black jesters in circuses and shows. It
provided an "amicably" deprecating variant to the word *negro*: the "sambo"
colored "negro" just like the Indian blood colored the individual of African
and Indian ascent. Searching the British press for announcements of spectacles featuring the word *sambos* in the first half of the nineteenth century

reveals a tendency to associate sambos and Jamaica's plantation system or the West Indies with one another.[19]

Thus the word became increasingly used to designate black Caribbeans, though nearly invariably with a joke. Suggestively, the word comes in previously mentioned articles discussing the appearance of "coloured philanthropist" William Craft at Hunt's infamous speech at the British Association in 1863. A self-styled nonconformist reporting for the *Caledonian Mercury* entitled his article "Sambo and the *Savans*." Indeed, Craft's line that "If God had not given [Africans] thick skulls, their brains would probably have become very much like those of many scientific men of the present day," could pass for a joke belonging to the time-honored tradition of the self-deprecating sambo jester.[20] As shall appear later, extirpating the risk of an amalgamation between the Jamaican jester and the pure Indian would become a chief task of British anthropologists.

With the independence of Latin America in the early nineteenth century, the Mosquito Coast became the focus of a new kind of traffic in the shape of the Gregor McGregor "swindle" of 1822. McGregor claimed he had received from George Frederic Augustus I, King of Miskito, a territory that he called Poyais along the Atlantic shore of Honduras. This territory, which the king was ceding to McGregor, had no sovereignty, despite McGregor's attempt at creating a different impression. It corresponded more or less to land that had been inhabited by Paya (or Pawyer or Poyais) Indians, who had been decimated and driven away from the coast by raids from British-armed Sambo-Mosquito warriors.[21] In 1822 McGregor began distributing the securities of Poyais on the London Stock Exchange. At the same time, he sold land grants to would-be settlers, exchanged their pounds against "Poyais dollars" and sailed them off to the Mosquito Coast, where storms and malaria took care of most of them.[22]

McGregor is often portrayed as the king of con men, yet it is important to emphasize that his scheme rested on a more real territory than conventional accounts of his "Land That Never Was" recognize.[23] After all, McGregor's machination was nothing but the reinvention of the old informal alliance between Britain and the king of Miskito, this time narrated in the language of the London Stock Exchange. In fact, as proof of the compact's materiality, a secondary market for Poyais securities and land grants developed in the years after the "swindle" was exposed. Participants to this market—vulture investors—sought to scavenge the land rights and make money with them, recognizing that they had some value. A series of emigration companies were then set up to take advantage of the possibilities

thus opened, and in effect the sale of land by the king or his relatives continued. Naturally, these companies, such as the British Central American Land
Company, were in turn led to push their interests in the area and sought to
lobby whomever they could within the British executive in order to secure
a modicum of official endorsement. In this convoluted context, the local
semi-official British authorities created their own Miskito Council in the
early 1840s only to have it disallowed by the Colonial Office. Subsequently,
a British agent and consul-general took custody of the young Miskito heir
George Augustus Frederic. Officially, the consul-general was an advisor, although as noted by Naylor, "precisely who [he] was supposed to 'advise' in
the absence of an Indian ruler was not discussed."[24]

The true nature of the arrangement becomes even more evident when
it is realized that Bluefield (or Blewfield) emerged as the headquarters of
this Mosquitian Kingdom, although the area was not one where Miskito
Indians predominated initially, nor was it even clearly Miskito territory.
Rather, this hamlet was the capital for a number of creole settlements. According to Naylor, the local creole population was made of English-speaking
"mulattoes" sharing the Caribbean culture. They descended from both slaves
and their owners. For instance, at the time of the McGregor swindle, there
were twenty-nine Hodgson families, who descended from Robert Hodgson's
African slave plantations that had been active in the eighteenth century.
To this group, a variety of fugitives and the survivors of failed colonization attempts had been added along with a later influx of former Jamaican
slaves. The creole population was eager to rely on British authority. When
the British protectorate was resumed in the 1840s, the new Council of State
combined two British agents with five creoles. The Miskito protectorate at
Bluefield may be described as an Anglo-Creole rule that used the Indian king
as totem.[25]

Such an arrangement required permanent adjustments. Under the guise
of the Anglo-Indian alliance, local interests had to work out their own solutions, which would be enforced to the extent that they suited London, Belize, and Kingston, and this required proper calculations as to what London,
Belize, and Kingston were up to. Policy reversals were frequent, officials in
London deciding now to repudiate the local decisions, now to disallow the
fiction of a British territory as opportunity dictated. For example, at one
point British agents canceled a number of outstanding land grants, while at
another they decided to exclude slave-owners established on the Mosquito
Coast from the award granted to former proprietors after the Slavery Abolition Act of 1833.

The rising power of the United States was to further complicate the politics of the area. With its westward push in the late 1840s and the acquisition of Texas and California, the former British colonies became a two-ocean power, and successive American administrations yearned for a Central American route between the Atlantic and Pacific. At the same time, the gold rush sucked in a flow of migrants who looked for the most convenient routes across the isthmus. It is in this context that the arrival of Squier as U.S. *chargé d'affaires* in Central America in 1849 took place, and it explains Squier's insistence on debunking the Miskito narrative. The United States was playing the role that the Spaniards had played before, and the Miskito Indian alliance was used by Britain against them in the same way they had used it against Spain before.

Indeed, the Indian pretext was the primary target of Squier's ethnological novel *Waikna*, as his personal correspondence establishes clearly. He described the young Miskito king (the one Pim would meet later) as "a black boy, or what an American would be apt to call, a 'young darkey.'" According to his correspondence, he was deliberately intent to turn support of "Queen Victoria's august ally of Mosquito into contempt."[26] As they occupied themselves driving away and murdering their own native Indian population, U.S. authorities were leery of providing any form of recognition to a native. "We have never acknowledged, and never can acknowledge, the existence of any claim of sovereignty in the Mosquito King or any other Indian in America," U.S. Secretary of State John M. Clayton wrote to Squier in 1850.[27]

British policy makers, on the other hand, were under enormous pressure to clarify their relations with isthmus countries, and after a decade of bargaining, a series of treaties emerged in the late 1850s. They disposed of the British protectorate in return for the United States recognizing British ownership of Belize. Such was the context for the sending of British Consul General Charles Lennox Wyke to Honduras and Nicaragua with instructions to negotiate agreements with those countries. The resulting treaties, signed respectively in 1859 and 1860, have often been described as a retreat of British interests, including by Naylor, who essentially stops his narrative there as the end point of the history of penny-ante imperialism in the area. Yet upon closer scrutiny, the arrangements that resulted from Wyke's efforts look more like a reinvention of the old formula of informal rule. For instance, on paper, the treaty with Honduras, known as the Wyke-Cruz Treaty, recognized the sovereignty of this country over the northern section of the Mosquito Coast and over the British settlements in the Bay Islands. Likewise, the treaty with Nicaragua, known as the Treaty of Managua, terminated the British protectorate and reduced the Miskito king to a tribal chief while recognizing Nicaraguan sovereignty over the free port of Greytown. In return, however, both Nicaragua and Honduras recognized Indian rights, agreeing that natives would not be "disturbed in the possession of any lands or other property which they may hold or occupy." Honduras and Nicaragua were also made to pay an annual stipend to the Indians, to create a formal reservation zone, and to refrain from interfering with Indian affairs. Given that the two republics were notoriously impecunious, opportunities would sooner or later arise to use a default on payment to the natives as a mean to intervene in local politics. And thus it is that indigenes were once again the vehicle enabling Britain to secure leverage in the area, notwithstanding the fact that, with Indians being given ownership of their lands, those Indians that Britain construed as leaders were made very much able to sell land rights to Britons, whose interests would then be protected under the heading of "Indian rights."[28]

## A RAILWAY THROUGH MOSQUITO

In 1850, a decade before Pim's visit, the big economic question of international interest regarding the isthmus was the construction of a transit infrastructure. There was much diplomatic and financial activity which peaked that year, as transit projects were designed and puffed. Bearing witness to the frantic agitation, there was among many others a pamphlet published by Marmaduke Sampson in the London *Westminster Review* (and immediately after that as a brochure in New York): "Central America and the Transit between the Oceans." The essay reviewed the various routes and political problems to which it found financial solutions. Anglo-American cooperation was much needed, he emphasized. Sampson was to become the editor of the *Times'* City article and as already mentioned, was to fall in the early 1870s on charges of corruption and evidence of lavish lifestyle.[29]

The same year saw the signing of the Clayton-Bulwer Treaty between Britain and the United States, a treaty that sought to defuse rising tensions by instituting international cooperation—"neutrality of transit routes" was the preferred phrase—and ensuring an open Central America. The treaty's introductory article pledged that neither signatory would "occupy, or fortify, or colonize, or assume or exercise any dominion." The treaty considered a canal option through Nicaragua, which would enable boats to steam up the San Juan River from Greytown on the Atlantic to Lake Nicaragua, after which a relatively short canal would be built connecting the lake to the

Pacific Ocean. But the canal failed to materialize and arguments kept crop-
ping up. American diplomats insisted the treaty meant that Britain would
lessen its colonial influence in the region, when their British counterparts
interpreted it as the termination only of *direct* influence. The open conflict
between powers was thus superseded by a cold war intermediated by agents,
allied, and puppets.

When Pim reached the Mosquito Coast aboard *HMS Gorgon* in late
1859, three main roads across the isthmus were now being actively considered, promoted, or completed.[30] The one completed route was the Panama Railroad Company, an American railway that began construction in
1850 and crossed the isthmus at Panama. It was fully opened to income-
generating traffic in 1855. That year, *Herapath's Railway Magazine* trium-
phantly announced, "A Continent has been pierced, and two Oceans united.
A new highway to commerce has been opened, and when the current of trade
will flow through its channel, the world will wonder that this long talked-of
union of Atlantic and Pacific has not been consummated earlier."[31] Having
no competitors, the Panama Railroad charged monopoly prices. The revenue per mile of the road was ten times the European average. This whet-
ted the appetite of competitors, especially because, while the Isthmus of
Panama was the shortest crossing of the land, it was also a lengthy southern
detour for a voyage from one coast of the United States to the other or for
shipments between Europe and the West Coast of the United States.

Slightly superior from that respect was the route opened in 1851 by the
Accessory Transit Company of American entrepreneurs Cornelius Vanderbilt
and Joseph L. White. The full route pended the completion of canal work by
their American Atlantic & Pacific Ship Canal Company, which had acquired
exclusive rights from the government in Nicaragua in 1849. The Accessory
Transit Company provided a New York-to-San Francisco steamship service
that went through Greytown and transported passengers by boat up the San
Juan River and across Lake Nicaragua. From there, a stagecoach took passengers overland to San Juan del Sur on the Pacific Coast, where another
steamer took them to San Francisco. This was a relatively cheap transit, but
in the early 1850s it was described as haphazard. Travelers were exposed to
the stings of mosquitoes, and in 1856, the seizure of power in Nicaragua by
Tennessee adventurer William Walker and the showdown with the Accessory
Transit Company (from which Walker tried to extort protection) closed the
road. In 1857 a raid sponsored by Cornelius Vanderbilt removed Walker, but
rather than reopening the road, Vanderbilt sold out the transit rights to the
company that ran the Panama route, which rushed to do nothing. However,

the attraction of a river-and-land transit persisted and in 1862 the Central American Transit Company, another New York-based company promoted by shipbuilding tycoon William H. Webb, relaunched the project.[32]

The last route to be considered was still further north, through Honduras. This was the route relentlessly promoted by Squier ever since he had become a *chargé d'affaires* for the U.S. government. He secured a concession from Honduran authorities for a road that went from Puerto Caballos on the Atlantic Coast to the Gulf of Fonseca on the Pacific. In 1854 Squier drew up a remarkable map, shown in figure 6, of "Honduras and San Salvador, Central America, showing the line of the proposed Honduras, Inter-oceanic Railway, territories of Indian Tribes and mines etc." The map covered Central America from Guatemala south through Honduras to Nicaragua and showed the location of the Poyais, Seco, Thuacos, Cookra, Wolwa, Rama, and Carib Indians, along with the projected interoceanic railway. The map also displayed ancient ruins and mule paths. This document, perhaps the most detailed contemporary anthropological document available, is revealing of Squier's politics—it has no indication whatsoever of the Mosquito Coast and the Miskito Indians. But then as far as Squier knew, Miskito Indians did not exist, since they were "just" blacks.[33]

As described by historian Charles Stansifer, Squier, having failed to raise the money in New York, went to Europe and, no longer concerned about America's welfare, attempted to muster the help of capitalists from the Old World. British policy makers responded to his call. In 1856 they made sure that a group of British financiers led by William Brown would purchase Squier's rights. This resulted in the incorporation of the Honduras Inter-Oceanic Railway Company in March 1857. Brown was a Liverpool merchant banker and head of the Anglo-American House of Brown, Shipley & Co., but he was perhaps more importantly a Liberal politician (then the ruling party). Because the purchase supported the policy of the Secretary of State for Foreign Affairs, Lord Clarendon, Brown's associates in the original promotion therefore included individuals such as Robert Wigram Crawford, Liberal M.P. and a British East India merchant, and Thomas Matthias Weguelin, another Liberal M.P., a banker, and the sitting governor of the Bank of England. Weguelin was an enthusiastic colonizer. He would later start a colony in Gran Chaco, Argentina, and in November 1871 his son would be killed there in a skirmish with natives.[34]

The objects of the Honduras Inter-Oceanic Railway Company, with a nominal capital of £2.05 million, were vast. They extended from the running of a railway to the management of a telegraph line, from the ownership of hotels and ports to the founding of towns and villages, not to mention

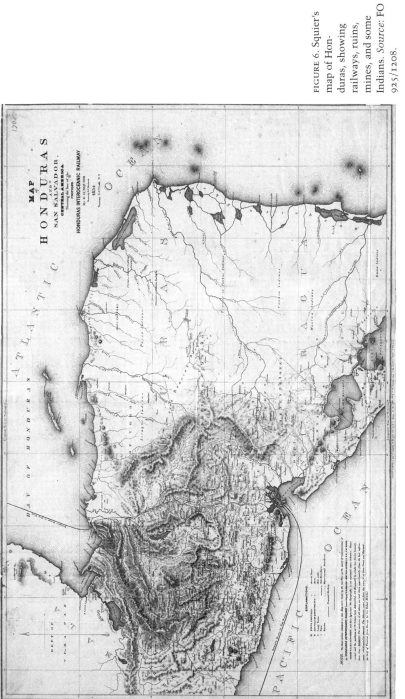

FIGURE 6. Squier's map of Honduras, showing railways, ruins, mines, and some Indians. *Source:* FO 925/1208.

the exploitation of Honduran land grants. The gentlemen immediately set to work and dispatched engineers, but reports deemed Squier's costing optimistic. Adjustments were made, and in October 1859, the company's articles of association were adjusted and the project downsized. The construction of a simple road was envisioned, and the capital was increased by one million pounds sterling. In order to reward the original promoters for their forbearance, it was stated that the company could swap stocks for land owned by the company—territory being as good as cash in an age of land grabbing. As usual, promotion of the project once again was made through the agency of the learned society. The Royal Geographical Society displayed an exhibition of "photographic and other views of the proposed route for the Honduras Inter-Oceanic Railway" on the same day the discussion of a paper on Africa by Dr. Livingstone, F.R.G.S., took place.[35] The political support enjoyed by the project was also advertised to reassure investors, and an attempt was made to recruit French imperial interests.[36] Honduran railway projects continued to be promoted by Liberal politicians with John Russell's long rule as foreign secretary through two successive Liberal cabinets (June 1859–November 1865): such was the highly politicized context for Pim's arrival in Miskito-land.

## ENTER PIM

The *Gorgon* arrived on the Mosquito Coast in the middle of the delicate negotiations taking place between Wyke and the government of Nicaragua. Upon arrival, Pim went almost straight to the "King Musquito," as one of the creative renderings of the redactor of the *Gorgon* ship's log had it. According to the apocryphal interview he himself transcribed ten years later, Pim then asked the king: "What do you say to giving me a concession for your portion of the line?" To which, in Pim's telling, the king answered, "If you really wish it, draw up the document you think necessary and I will gladly sign it, not only to show my friendship to you personally, but also to prove my anxiety not to lose an opportunity of doing anything which may chance to advance the interests of England." He also predicted that Pim would "break his heart over it," for he "little know[s] the disappointment in store." In addition to being a visionary, this native king also had a spontaneous understanding of the capital market as evidenced by his statement that he had "no faith in speculation" and perhaps as well of his being "warlike and the rest." As he warned Pim: "The Nicaraguan will only endeavor to make capital out of you; the Yankees will certainly oppose you; While your own countrymen will desert you in the hours of need, just as they have served me in the new treaty about to be concluded."[37]

But Pim was "deeply impressed with the importance of securing to British interest a transit route through Central America," and this was why, instead of taking the king's advice, he "examined minutely, and studied in detail, the topography of the country, through which [he] hoped to establish an inter-oceanic communication, so extremely desirable both in a political as well as a commercial sense." In so doing he "became convinced that the construction of a railroad from Monkey Point to St. Miguellitto [sic] on the Lake Nicaragua, and the establishment of light draught steamers to run across that Lake, and by a shallow canal into the Pacific, was a feasible project."[38] This is how Pim was drawn to produce his own variant of the interoceanic crossing and as a result his own map of railways across the territory of native Indians, the latter now spatially bound in a Mosquito Reservation (see figure 7).

It was to this effect that on December 20, 1859, "Commander Bedford Pim of the *Gorgon*," on Navy assignment, future leader of the Anthropological Society of London, acquired, as the contract in his own handwriting states, "the lands lying to the Southward and Westward of Monkey Point and included between the said Monkey Point and the point of land marked as Little Monkey Point" and which he later called Pim's Bay (a name still used today). A freehand drawing was added to the document showing the shore and the limits of Pim's land, which extended 500 English feet inland from the low water mark.[39] A second contract recognized Pim's agency in calling the king's attention to "the advantages which would result to our Country of Mosquito by the construction of a Rail Road from Monkey Point to the Lake Nicaragua as a certain, sure and rapid mean of transit from the Atlantic to the Pacific Ocean" and granted the concession to Pim free of cost in exchange for 5 percent of the net profits shown on the balance sheet of the railway company to be created.[40] Lastly, a third contract between the king and William Cope Devereux, the treasurer of the *Gorgon*, granted Devereux two cays off Pim's Bay (Free Cay and Rocky Cay). It may be asked whether this was a reward for Devereux's discretion or assistance in enabling Pim to draw the money required for the arrangement. We cannot be sure, but we note that Devereux's book, *A Cruise in the "Gorgon,"* published 1869, was dedicated to "Captain Bedford C. Pim, R.N., in testimony of regard and gratitude."[41]

In the record of transactions kept by the British consul ("Mosquito Territory Land Grant Record"), the Pim's Bay land grant bore number 57, the railway land grant bore number 59, and the title for Devereux's cays was number 60. Another contract numbered 58 must have been signed simultaneously but is lost. Evidently, a lot of land transfers were taking place. As

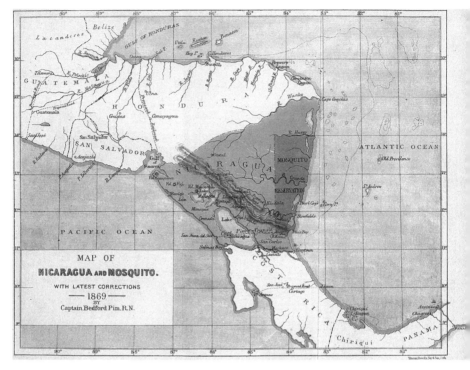

FIGURE 7. Pim's map of Central America, showing Miskito Reservation and competing isthmus crossings. *Note*: On the bottom right (Panama) the Aspinwall transit is represented. In the upper left part, leading to the Gulf of Fonseca, is shown Squier's Honduras transit ("Proposed Railroad"). In the middle of the picture (cutting through Lake Nicaragua from Pim's Bay to San Juan del Sur) is "Capt. Pim's Transit." Note also the Javali mine (ran by Berthold Seemann) indicated in the mountainous range, going across Nicaragua. *Source*: Bedford Pim and Berthold Seemann, *Dottings*, "To face page 1."

the end of the formal British protectorate approached and property rights became unclear, there was a run to transfer land ownership. The run may have been orchestrated by the local clique that sponsored the arrangement with Pim or perhaps even groomed him for the occasion. Indeed, the land contracts showed the witnesses of Pim's transactions. Names included members of the local creole elite, such as J. H. Hooker, from a local family of former slave owners; the Reverend Gustavus Feurig, head of the local Moravian Church; and William Rahn, local agent of the Royal Mail and Steam Packet Company. These individuals knew full well whatever rights they owned would have a shaky legal basis at best and might not be recognized by Nicaragua, which was to provide the law of the land in the future. This must be the reason why the contract took the shape of a "quit-rent," or permanent lease rather than an outright sale from George Augustus Frederic to Pim. The king was in effect leasing the land forever to Pim, his heirs and assigns, for an annual quit-rent of twenty-six shillings (and the Cays to Devereux for one shilling). Evidently, as the quit-rent contracts recognized, there had been a cash payment when the contract was signed, so for all practical purposes this was a plain sale (prudently, no amount was recorded). Yet the legal form protected the purchaser better than a sale. To the extent that, in the future, the only thing that would still be protected by Britain was the rights of Miskito Indians, the rights of the British would be stronger if the land they acquired remained formally with the king of Miskito, explaining why it would become paramount to Pim that, unlike what was happening in the map that Squier had drawn, Miskito Indians should have a "material" existence.[42]

Yet the Miskito-land part of the railway was but one part of the equation. For his road to be considered seriously, Pim needed at least the pretense of some form of endorsement and concession from Nicaraguan authorities. Another issue was Pim's uncertainty about the boundaries of Miskito territory, even though he understood they were part of the negotiation with Nicaragua.[43] Thus, after spending a Christmas week with his crew in his newly acquired Pim's Bay, he sailed south to Greytown. On January 13, after sending a letter to Consul General Charles Lennox Wyke in which he informed him of his purchase, Pim "left the ship to proceed up the River San Juan de Nicaragua for the purpose of communicating with Mr. Wyke consul" (on January 16, according to the *Gorgon*'s ship log). Pim reached Managua on January 22 and tried to meet with Wyke there. The exchange of letters with Wyke shows the British plenipotentiary anxious to see Pim off to Greytown, focusing on his job of securing the coast against the intrusions of filibusters instead of interfering with the diplomatic process.[44]

## PIM AGAINST THE ADMIRALTY

The *New York Herald*, a newspaper that at diverse points in time employed Karl Marx and Henry Morton Stanley (of "Doctor Livingstone I presume" fame), was built on the belief of its founder, James Gordon Bennett Sr., that a newspaper's job was "not to instruct but to startle."[45] On February 13–14, 1860, it ran the story that one "Captain B. Pim, commanding her British Majesty's steamer *Gorgon*" had purchased "from the Mosquito King a large tract of land" near Monkey Point and proposed to use it as departing point for an interoceanic railway. The newspaper claimed Pim had just returned from Managua the day before, and it reported the background:

the completion a few days earlier of the Treaty of Managua guidelines to end the British protectorate of Miskito; the tension that lingered between British ambassador Wyke and his U.S. counterpart, Alexander Dimitry; the interests of Cornelius Vanderbilt's American Atlantic & Pacific Ship Canal Company; and finally the activities of Colonel George F. Cauty, one of the mercenaries Vanderbilt had employed a few years earlier to get rid of filibuster William Walker, which the newspaper explained would have had some connection with the visit of Pim "to Granada and Managua."[46] Pim, who boasted support of the Nicaraguan Parliament in Managua, would have stated that there was "great harmony" between Mr. Dimitry and Mr. Wycke [sic] and alleged that Vanderbilt was "a party" to the project.[47]

The article reached the British commander-in-chief of the North America and West Indies Station, Vice Admiral Houston Stewart, aboard HMS Indus in Bermuda in early March, and as far as Stewart was concerned the newspaper was up to its self-assigned goal to "startle." He immediately reported to the Lords of the Admiralty, predicting they would be "as much surprised" as he had been upon learning "what a prominent part Commander Pim of the 'Gorgon' who is now stationed on the Coast of Central America for the protection of British interests is represented as taking." Pending verifications, Stewart was to dispatch immediately to Jamaica HMS Icarus, under the command of a man under whose orders Bedford Pim had served, Commodore Henry Kellett. Kellett had orders to find Pim and "relieve the Gorgon and recall her to Port Royal [in Jamaica] for employment elsewhere."[48] Simultaneously, Stewart wrote to Pim a note asking him whether the New York Herald was correct in reporting that he had abandoned his ship to proceed to Managua or perhaps Granada, and what could be the "circumstances of a public or pressing nature" that had led "the senior officer on the Coast" to leave his ship.

The letter concluded with language that sheds an interesting light on the nature of informal empire:

I have also to request that you will state what foundations there is to the assertion that you have become proprietor of a track of land near Monkey Point. I cannot of course prevent officers from acquiring Property where they please, but situated as you are, charged with the protection of important public interests, such a proceeding could only tend to place you in a false position, and possibly to embarrass the Government at a critical moment. As a matter of precaution therefore I have directed Commodore Kellett to relieve the Gorgon and to recall her to Port Royal.[49]

On April 2, 1860, while aboard the *Gorgon* at Great Corn Island off Blue-fields, Pim composed his answer to his superiors, stressing his sheer concern for the public interest. The addressee was Rear Admiral Sir Alexander Milne, an officer who had just relieved Stewart from the North Atlantic Command and a man who was about to play a critical part in British Atlantic diplomacy during the Civil War. Pim explained that his sudden expedition to Managua had been triggered by the reports "from newspapers and other sources that Mr. Wyke, Her Britannic Majesty's Minister to Central America had failed to negotiate satisfactorily with the Government of Nicaragua and was on the eve of leaving Managua." It had been his "imperative duty," therefore, to "put his Excellency in full possession" of his "admirably calculated" project, which would provide a "safe and expeditious mean of reaching our great Australian Colonies, China etc. to say nothing of British Columbia and San Francisco."[50]

Pim also alleged that he wanted Wyke informed of the "rapid filling up of Greytown Harbour, which must soon cease to have a Port for even small craft and therefore utterly ruin the transit route via River San Juan." Regarding his purchase of land from the king of Miskito, he considered it "proper precaution as very lately an American had acquired a portion of Mosquito Territory, which may ultimately give much trouble." If their lordships wanted evidence of Pim's *bona fides*, they might want to reflect upon the risks he had thus taken, "exposed day and night in the open Canal to the vicissitudes of so notorious a climate." Why risk his life, if it had not been from the commander's "strong sense of duty"?[51]

On April 16 Milne asked Pim, now based in Jamaica, for further clarifications as to the "particulars of the land purchased" and "what steps, if any, [had] been taken to form a settlement there," to which Pim responded with another long letter in which he again stubbornly vaunted the merits of his project, "which must *ultimately* be a grand success: for even with all the difficulties of Rapids etc. on the San Juan route when Greytown was a harbor, it *more* than competed with the Panama route; and the known facilities of the present project would therefore it was considered certainly monopolize by far the greater part of the Northern traffic."[52]

Milne thus reported to the Lords of the Admiralty that he now understood that the purchase of Monkey Point and the formation of a settlement had been made "apparently on speculation." This was an "objectionable proceeding" given the "special position" Pim held at the time of the purchase. He was deferring to the decision of their lordships and had not yet "expressed his opinion to Commander Pim" but had decided to go to Nicaragua and visit the consul in Greytown to investigate the subject.[53] Milne went to Colón with the *Emerald* in early May 1860 and made his own report on the Panama Railroad, which contrary to Pim's allegations he found amicable to British interests.[54]

The pending question for the Admiralty was to decide what to do with Pim. As has been emphasized by Naylor, competition between the Colonial Office, the Foreign Office, and the Navy was not unheard of, and there was a risk that in disobeying the Navy Pim had really played to the hand of some other bureau. After all, had not Pim in the past sought to explore Siberia on a grant from the current foreign secretary and prime minister John Russell? The Foreign Office was thus approached tongue in cheek, but John Russell reported being just as dismayed as the Lords of the Admiralty. The prime minister's opinion, the second secretary of the Admiralty reported to Milne in late May 1860, was that "the proceedings of Commander Pim should be disavowed and disapproved" (and evidently so, since Pim's efforts ran head-on against those of the House of Brown, which was attempting to proceed with a Honduras route at the very same time).[55] In summary, the Admiralty was "unable to account for an officer in Commander Pim's position having failed to perceive that his proceedings and conduct on the occasion were liable to be misinterpreted as the acts of an officer in Her Majesty's Service." Pim was to be informed that he had "incurred the marked displeasure of their Lordships and has proved himself to be unfit for the duties with which he [had] been entrusted." This was written in late May as the *Gorgon* was moored in Bermuda.[56]

With ratification of treaties pending, it was urgent to make sure Pim would not create further havoc, and he was therefore removed from the scene of the crime and sent back to Britain, reaching Portsmouth in July 1860, where the *Gorgon* was refitted. It is probable that the Lords of the Admiralty asked for Pim's oral clarifications, at least from the way they subsequently hurried him and the *Gorgon* off to the coasts of Brazil and Africa on the usual anti-slavery assignment—spot slave ships, investigate, report.[57] William Cope Devereux's narrative in *A Cruise in the "Gorgon"* gives colorful evidence of Pim's continuing interest in railways. While the *Gorgon* was at anchor in the Bay of Rio, Pim paid a visit to the newly extended Don Pedro II Railway Company and enjoyed the trip, said Cope Devereux, while being "clad in Arab coat, fez, and umbrella."[58]

Meanwhile, the diplomatic situation clarified. In December 1860, the substance of the treaties with Nicaragua and Honduras was officially communicated to British commanders, and the termination of the Mosquito Protectorate was announced. British officers were now expected to focus on the "intrusions of Filibusters." In January 1861, Consul Green was reporting to Milne in Halifax on the smooth transfer of Greytown to Nicaraguan authority.[59] Pim was now harmless and could be discharged. While in Cape Town in late March 1861, the *Gorgon* met *HMS Fury*, whose command Pim took with instructions to bring her back to England.[60] The officer in charge of filling the ship's log may have been struck by the straight line that the *Fury* took back home and drew a little map to this effect. On June 19, 1861, the *Fury* reached Portsmouth, and Pim was "compulsorily retired" from the Navy, his pay capped. In *Dottings on the Roadside*, Pim concludes: "after this journey, circumstances prevented the active prosecution of my plans until my return to England from foreign service."[61]

Such are the episodes behind the *Dictionary of National Biography*'s cryptic statement about Pim that he "was somewhat harshly censured by the Lords of the Admiralty in May 1860."[62] As for Pim, he narrated the facts as a tale of the political incompetence and of cowardly British policy makers. His scheme was bound to succeed eventually, he repeated, but he would have "soon found that it was not expedient, politically, to endorse" his plans. The reason he could never "enlist the sympathy, far less the moral support, of the English Government" was because this government "was rather inclined to throw cold water than otherwise on any attempt to compete with the Panama railroad, having, doubtless, the threatening shade of Mr. Monroe before its eyes."[63]

And thus the mystery of Pim is growing impenetrable. We left him a foe—painted by Stocking and virtually all research on British anthropology as a fake anthropologist and the racist author of "Negro and Jamaica"—but upon closer scrutiny, we found the man who wore a fez to visit a Brazilian railway built by American engineers almost a friend, the source of important information, helping another anthropological historian, Michael D. Olien, to refute Squier's racists theories. Pim and his *Dottings on the Roadside* now become an early source of firsthand anthropological description of the Mosquito Indians, correcting Squier's misrepresentation.

If we wanted further evidence of the absurdities produced by the splendid ignorance of one another's work inside the social sciences, we could always point to the adventures of Pim and the king of Miskito. But then if Pim corrected Squier, and if we take from previous work (such as Barnhart's recent biography of him) that Squier was a completely real American anthropologist and the founder of American anthropology, then are we not bound to conclude that Pim, being no less true than Squier, was, logically, an anthropologist too? Isn't it time to move on and recognize anthropology where we actually find it? As the previous narrative has shown, Pim's anthropology was born at the linchpin of the micro-politics of bureaucratic control and the macro-politics of empire, as demonstrated by the dispute between Pim and the Admiralty. Thus, the roots of Pim's anthropology took hold on a loophole of informal empire, but the plant would never have reached full maturity without the complementary mechanics of the learned society and the London Stock Exchange, as the next chapter will show.

## CHAPTER EIGHT

# Salt-Water Anthropology

The obstacles the Admiralty and Foreign Office had thrown in Bedford Pim's way proved unable to deter his newfound passion for things Miskito. Later, his mother, Sophia Soltau-Pim, summarized in an acrostic the ineffable riches of "MOSQUITO," the U and I of which provided initials for "Under the soil and on the tree / Infinite wealth for industry."[1] Thus the years following Pim's discharge from the British Navy saw him busying himself with getting around what he read as the obstructions of bureaucrats and policy makers in order to avail himself of the opportunity he had carved across Miskito and Nicaragua.

Pim's frantic activity was multipronged. The ship's log of his subsequent traffics would give a map of the sites of emerging resistance to Liberal policy making—first the London Stock Exchange and next, accessible through a revolving door, the Anthropological Society. In tracking Pim's various interventions, this chapter articulates the contours of the modes of anthropological dominion as they developed at that point. As we shall see, Pim's anthropology may be interpreted as an instrument of ownership, which provided the crutches of science to the limping property rights he had on Monkey Point, the bay, and the railway line. More generally, as we shall see, this provides a picture of the manner in which money and anthropological knowledge embraced one another, brought together by matters of foreign ownership.

## MONKEY BUSINESS

Just as Squier's inability to find someone to underwrite his project in New York had led him to cash it out in London, with the help of British policy makers (his archenemies of the day before), Pim's "patriotic notions of progress" did not hold much longer than the time it took for the Navy to discharge him. Almost immediately afterward, he went to Paris, where he would have received encouragement from Napoleon III, an early student of the crossing of Nicaragua himself. He later reported meeting with French financiers such as the powerful Pereire brothers, Emile and Isaac, who were at the time extending their railway empire across Continental Europe. Unfortunately for Pim, the eminent capitalists remained sold on the idea of a canal as opposed to a railway.[2]

In his first attempts at selling his project, Pim had been told that more data was needed. Securing the certification of one civil engineer would give further strength to his scheme. And thus it was that in October 1863 Pim followed his trip to Paris with a second exploration of Nicaragua, and for greater security he took two civil engineers rather than one. Beyond surveying, the goal of the expedition was to secure from Nicaragua the concession he had failed to negotiate during his brief sojourn in Managua in 1860. Arriving in the middle of a war between Honduras and Salvador on the one hand and Nicaragua on the other, Pim managed to meet with Nicaragua's president, General Tomás Martinez. With the "support" of Martinez, the Nicaragua legislature provided in February 1864 a concession completing the Mosquito rights. "With this concession," Pim stated later, "I lost no time in returning to England."[3] After a brief passage through Britain during the summer of 1864, Pim headed again to the isthmus in November 1864 for a third voyage. This time he took with him Charles White, a consul for Nicaragua for officialdom; a mining engineer "of repute" for enhancing the railway with ore extraction projects; and an artist for visual testimony.

Pim now had enough material to start his financial projects. The two most widely advertised promotions that grew out of the contract with the king of Miskito were the Central American Association Ltd. and the Nicaragua Railway Company Ltd.—they were "Pim's own," so to speak. The Central American Association was a company with capital of £150,000 launched in March 1866 and dedicated to prospecting and acquiring minerals, negotiating loans to Central American governments, and managing and valorizing concessions from those governments. The Nicaragua Railway Company Ltd., launched in November 1866 with capital of £1,000,000, would build a road across the isthmus.[4]

The financial logic of the Central American Association was to serve as an instrument to coordinate a heteroclite collection of Central American land-rights holders and absentee landowners, enabling them to get rid of their dubious claims. After acquiring on the cheap such a variety of ill-secured land grants, the association would use political and legal harassment to raise their value in the market. In the language of one advertisement, the association's secretary would provide "active and energetic" interventions.[5] The association also served the purpose of "securitizing" investment schemes that came its way. The various mining and land rights that ended up in its hands could be structured and launched on the market (See figure 8.) The Javali gold mine, floated by the association in 1868 and run by Berthold Seemann, Anthropological Council member and Pim's friend and coauthor, provides an illustration.[6]

Like other contemporary vulture funds, the Central American Association relied on a regulation of the London Stock Exchange, which required that a country with unpaid debts could not issue fresh loans to the market. The rule had initially been introduced with government debts in mind. But since many private claims had been issued with the support of governmental authorities or with some form of state guarantee or partnership, it was usual to play on public-private ambiguity and harass originators of such private loans. The ultimate objective was the collection of a pound of flesh: force other financiers into cooperation with the vulture. In other words, competitors were encouraged to settle with the Central American Association before applying to the London Stock Exchange.[7]

Such was the way the association managed to get involved in Honduras government finance in the late 1860s. Using its advocacy on behalf of people who held old "Poyaisian land securities" as a bargaining chip, the association managed to secure an adjustment from the authorities of Honduras when this country sought to borrow.[8] And it did not matter for the outcome that the Poyaisian securities were a dead letter since the collapse of McGregor's schemes forty years earlier. In this way, Pim the anthropologist had found a role as rabid collector of land concessions. A few years later, as he participated in one of the myriad bondholders' meetings in which he was familiar as the manager of the Central American Association, Pim was heard exhorting bondholders to bring a "gentle" pressure to bear on Nicaragua or Honduras "and insist upon payment, if not in money then in land, in exchange for their bonds."[9] Land, it appears, was the ultimate currency.

Pim conceived the Nicaragua Railway Company as complementing the association. The land at Monkey Point would sell better if there were a railway, and the railway would be more profitable if there were a larger population of colonists. According to Pim's inimitable prose, "the joint enterprises of colonization and transit might then have been made with mutual advantage to travel harmoniously together."[10] However, it turned out

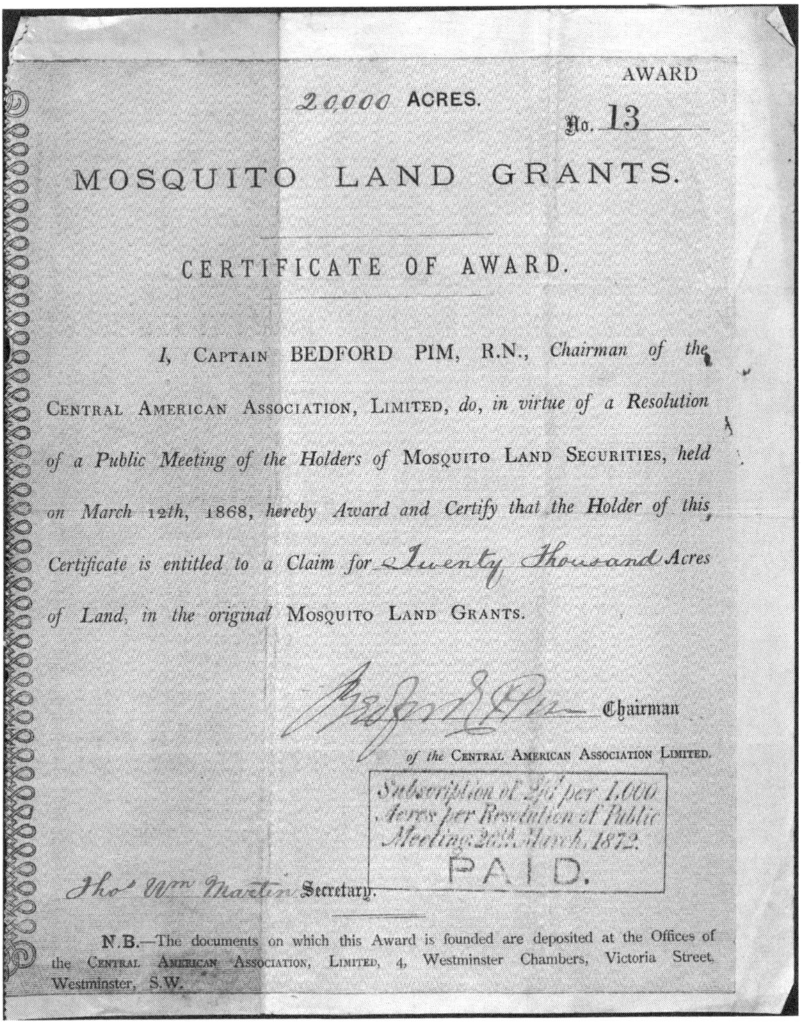

AWARD

20,000 ACRES.

No. 13

MOSQUITO LAND GRANTS.

CERTIFICATE OF AWARD.

*I,* CAPTAIN BEDFORD PIM, R.N., *Chairman of the* CENTRAL AMERICAN ASSOCIATION, LIMITED, *do, in virtue of a Resolution of a Public Meeting of the Holders of* MOSQUITO LAND SECURITIES, *held on March 12th, 1868, hereby Award and Certify that the Holder of this Certificate is entitled to a Claim for* Twenty Thousand *Acres of Land, in the original* MOSQUITO LAND GRANTS.

Bedford Pim Chairman

*of the* CENTRAL AMERICAN ASSOCIATION LIMITED.

Subscription of 2/6 per 1,000 Acres per Resolution of Public Meeting 20th March, 1872. PAID.

Thos Wm Martin Secretary.

N.B.—The documents on which this Award is founded are deposited at the Offices of the CENTRAL AMERICAN ASSOCIATION, LIMITED, 4, Westminster Chambers, Victoria Street, Westminster, S.W.

FIGURE 8. Mosquito land grant certificate from Central American Association Ltd., signed Bedford Pim. *Source*: "Mosquito Land Grants," 1872, from Papers of Charles Stephen Hill, University College, London.

that the Nicaragua Railway Company, having failed "to find enough subscribers," was wound up in July 1868. Yet in April of the same year, Pim had secured an act from the legislature of New Jersey in order to incorporate, with a group of associates, a "Pacific Transportation and Nicaragua Railway Company" with the very same railway rights as the company previously incorporated in London. In *Dottings*, Pim explains that after a trip to New York where he tried to work out an arrangement with W. H. Webb's Central American Transit Company, he returned to London and found a "gentleman" with whom he had originally hoped to organize "the carrying out of his plan" willing to buy him out.[11]

The association and the railway company formed the hub of a ring of projects centered on Central America and promoted by a clique of individuals at the center of which we find, systematically, Bedford Pim flanked by his associate, Berthold Seemann, who thought of "tropical America as the field of colonization of the future."[12] A stable group of individuals surrounded Pim, occupying alternating or jointly various roles, from original promoters to directors to solicitors to stockholders. The interlocking network of promotions included in particular, beyond the Central American Association Ltd. and the Nicaragua Railway Company Ltd., the Chontales Gold and Silver Mining Company Ltd. (1865) and the Javali Mining Company (1868). Beyond even that, the network included the Foreign Lands and Mineral Purchase Company Ltd. (incorporated in 1864) and the Mineral Rights Association Ltd. (1866). Pim himself is found in still more related enterprises focusing on land and colonization. An example was the Emigrant and Colonists' Aid Corporation created in December 1869 for "the purpose of securing the success of the emigrant and the colonist without making him an object of charity, but paying a fair return for the capital employed on his behalf and at the same time carrying out the designs of the benevolent and the philanthropic." The board of the Emigrant and Colonists' Aid Corporation included the prominent middle-of-the-road Conservative politician R. N. Fowler, a banker and a man active in leading charities, who would later feature on the board of the Corporation of Foreign Bondholders.[13]

The list of Pim's associates gives shape to the contours of his network. One was the Confederate explorer Captain Matthew Fontaine Maury, a geologist and meteorologist whose brother-in-law, Lieutenant W. L. Herndon, had explored the valley of the Amazon with Lieutenant Lardner Gibbon. We already came across Maury when we traced the origins of the colonization of the Amazon in the 1860s inspired by the London Stock Exchange. In the middle of the Civil War, Maury had sought refuge in London, bringing his goodwill and know-how to British investors.[14] Another member of the Pim group was John Henry Murchison, the cousin of Sir Roderick Murchison and a mining investment expert who was later associated with Richard Burton in an African commercial enterprise.[15] There was one George Henry Money, who would later be on the board of the Daira Sanieh, the foreign-controlled company that administered the estates the khedive of Egypt had given as security for further borrowing after 1870. With the Egyptian default in 1875, the Daira Sanieh became a lever for the Western takeover of Egypt. There was George C. Bompas of the law firm Bischoff, Coxe and Bompas, who was employed as solicitor of several of the companies in the group and would later be employed as lawyer in the George Earl Church debt contract with Bolivia.[16] Not incidentally, Bompas was a fellow of the Royal Geographical Society, of the Geological Society, and of the Paleontological Society. There was also Henry Sewell, the stockbroker of several of the companies in the ring. He may have been the same Henry Sewell who joined the Anthropological Society in March 1866 as local secretary in Real del Monte in Mexico, site of the Real del Monte Mining Company, which appears from time to time in the press in relation to other mining promotions in the group. There was John Field, another stockbroker whose name comes up in the context of the Nicaragua Railway launch. Field was a partner of the London Stock Exchange brokerage Field, Wood, and Haynes and was often seen with John Diston Powles, Haslewood, and Gerstenberg in the inner circles of the bondholding syndicate. He was also a director in Gerstenberg's Ecuador Land Company. There was George M. Bowen, a partner of the firm Thomas Manning & Co. in Liverpool, who was a British vice consul in Nicaragua in the early 1850s and by this time had become a consul *for* Nicaragua and as such was involved in negotiating land grants with bondholders. There was Arnold Rogers, a dental surgeon, expert in cranial issues, vice-president of the Council of the Odontological Society, and a member of the Geological Society of London since 1859. And there was George Salmon, the lawyer who in 1864 had assisted Pim's acquisition from his former paymaster on the *Gorgon*, William Cope Devereux, of the rights to the cays Devereux had himself secured from the king of Miskito as found in the previous chapter.[17]

As can be seen, the network included a mix of stockbrokers, lawyers, and scientists, and such connections were an essential element of the puffing up of financial schemes. They were seen in the many social-promotional events that were put together by Pim and his associates. For instance, in 1867 a dinner was organized to honor General Tomás Martinez, the former president of Nicaragua who had come to "discuss the land rights" with British investors. Pim chaired the dinner, many of his partners in business attended, and he was further joined by two companions from the Anthropological Society, Charles Carter Blake and of course Berthold Seemann. According to the *London Standard*, glasses were raised to "the health of the Chairman," to "Nicaragua Commerce and Transit" and "To the Ladies."[18]

## A CAREER IN ANTHROPOLOGY

After being discharged from the Navy, Pim was still very much connected to scientific societies, and they threw on him the kind of official light he needed to go on with his projects. In fact, Pim's discharge marked the take-off of his academic career. If his expeditions lacked the shine of the Navy, they could nonetheless be advantageously portrayed as so many scientific endeavors. For instance, the second voyage to Nicaragua would be branded not as a financial survey but as adding "extensively to the topographical knowledge of the interior of Nicaragua." Furthermore, just as any honest Victorian explorer would, Pim set to work on his Great Railway Epic, *Gate of the Pacific*, published in 1863, in which he narrated his 1860 encounter with the possibilities of Miskito-land. Promotional articles followed. Pim read a paper on the subject of the isthmus at the 1863 meetings of the British Association for the Advancement of Science in Newcastle-upon-Tyne, pushing the idea of an interoceanic railway. The paper was included in Section E (Geography/Ethnology) and was thus delivered under the same auspices as Hunt's "Negro's Place in Nature."[19]

Pim's investment in anthropology can be dated from the next year. Just after meeting with the then president of Nicaragua General Tomás Martinez and securing land rights from the legislature in Nicaragua, he arrived in London in June 1864. According to the records of the Anthropological Society, one of the first things he did upon arrival was join on June 14 the Anthropological Society as local secretary in Nicaragua. He had been preceded there, and perhaps canvassed for, by his old fellow explorer Berthold Seemann, who was one of the society's first converts, having joined in December 1863.[20]

In September 1864 Pim participated again in the meetings of the British Association for the Advancement of Science held in Bath, where minds were fixated on the showdown between Burton and Speke and the death of the latter. Pim, still a geographer, had an article on the "Volcanic Phenomena and Mineral and Thermal Waters of Nicaragua." On September 14, however, he participated in the general assembly of the British Association—the so-called general committee meeting—to do a favor to his new friends from the Anthropological Society: A planted supporter of the Anthropological Society, he sponsored a failed motion that the British Association's Section E should in the future include the word "anthropology." This request was to become in subsequent years, as many authors have emphasized, a major bone of contention between the Anthropological Society and the British Association.

As suggested, Pim had most probably been tipped for doing this, and it was part of a well-rehearsed act. The report made by the Anthropological Society's Honorary Secretary Charles Carter Blake referred to Pim as his "friend" and the "eminent Arctic voyager."[21] It was astute of course to use a geographical hero for the purpose of securing the recognition of anthropology in the geographically dominated Section E. Pim was one in a group of men promoted by Sir Roderick Murchison's Royal Geographical Society and from whom Burton's secessionist Anthropological Society drew some of its most characteristic members. From that point on, very much like Burton, Pim remained involved with his initial Geographical club, but the bulk of his scientific investment was more and more with the Anthropological Society.[22]

This puts in context Pim's role in the aftermath of the Jamaican Rebellion. As said, the politics of Nicaragua, or more specifically the Mosquito Coast, had long been connected to Jamaica. Kingston was an important coaling spot for steamers cruising in the Caribbean, and the governor of Jamaica had substantial authority over the region. If a paper on the topic had to be given (pending the return of W. T. Pritchard a local secretary for Mexico who would be appointed "Special Commissioner" of the Anthropological Society to report on the "anthropological causes of the recent insurrection of negroes in that Island") the choice of Captain Pim was thus a natural. One only needed to put together the favorite themes on the irrecoverable backwardness of "Negroes" and add the varnish of a plausible knowledge of the area. And thus Pim's lecture was announced by Hunt, in an "extra meeting . . . on the causes of the Negro insurrection in Jamaica" to be held in Saint James's Hall on February 1, 1866.[23]

The event also heralded Pim's further elevation in the Anthropological Society. In January 1867, as Pim was again on his way to Nicaragua, he joined the society's Council. As we have already seen, his participation in the *Athenaeum* controversy attested to his place in the inner circle of Cannibals. After the death of James Hunt in 1869, he moved further up by joining in 1870 the even more prestigious circle of vice-presidents of the society and as such chaired and moderated many meetings. Unsurprisingly therefore, he would participate in the first Council of the Anthropological Institute of 1871 and while he stepped down afterward, he remained a member of the institute up to the time of his death in 1886. Judging from his jawbone, therefore, there could not be in fact a *purer* specimen of the successful race that made British anthropology in the 1860s. The suggestion that he would somehow be an accidental anthropologist (and "Negro and Jamaica" somehow accidental anthropology) is, while comforting, baseless.

## LEVERAGING MISKITOLOGY

Science helped Pim's business through the regular certification role identified in previous chapters. The ultimate purpose was to lay railway track
and promote migrations. And thus Pim sang a song about a future that
would "bring Mosquito and Nicaragua together, so as to form a united State,
and then to connect their interests still more firmly by a road, laid down
for the most part by immigrants, who, on proper encouragement, would
have made the intervening country between the oceans their home."[24] As
observed previously in similar cases, this required publicity. The launching
of the corporations that would connect the interest of Mosquito and Nicaragua was accompanied by the release of a supporting literature advertised
on scientific circuits.

The civil engineer for the railway company, John Collinson, C.E., one of
the two Pim had taken to Nicaragua in 1863, was turned into a student of
geography and anthropology. He joined the Anthropological Society and was
asked to put out a pamphlet containing an update of Pim's early figures for
a railway across the isthmus.[25] His *Descriptive Account of Captain Bedford
Pim's Project for an International Atlantic and Pacific Junction Railway
across Nicaragua* compared the various routes, underscoring the greater cost-
efficiency of Pim's route at the same time as it made ethnological remarks.[26]
Because the origination of long-distance projects required the certification
of many different truths, one had better be conversant in many different
sciences.

The data Collinson provided was an example of these optimistic accounts of which company promoters were so fond. This applied with peculiar strength to the question of diseases. Collinson's pamphlet vaunted
the "general healthiness of the country," which was "above the tropical
average" as "proved by the age the natives attain . . . Indeed in almost every
city and town people point out with pride their very old men, a custom
dating from long before the conquest."[27] The theme of the healthiness of
the place would continue to pop up in Pim's writings. In *Dottings*, he again
described the territory as a "healthy" garden of Eden with "bright, clear
weather," and he would put in the mouth of the ship's doctor the assertion
that he had been "much struck by the appearance of the residents; most of
them have been more than years in the country; in fact the English consul
is the only sickly-looking person in the place. Yellow fever is unknown."[28]
Such assertions were disputable at best given a number of factors: the actual prevalence of fevers, which had plagued previous attempts at European

settlement; Pim's own justification for his behavior when he told the Admiralty that "most assuredly [he] would not have undertaken such a journey, exposed day and night . . . to the vicissitudes of so notorious a climate except under a strong sense of duty"; and the death of Berthold Seemann, taken by the disease in Nicaragua in 1871.[29] In fact, by embarking two engineers with him in 1863, had not Pim anticipated the risk of losing one? Yet making statements about the healthfulness of a region in the context of a learned society meeting conferred some weight to them. Those who knew better could be co-opted in one way or another.

The timing of Pim's scientific and promotional activities underscores the depth of the entanglements between science and business. In the fall of 1867, at the time when Pim was experiencing trouble with the distribution of his Nicaragua Railway securities, Collinson was involved in two sessions on Miskito Indians and Mosquito territory, one at the Anthropological Society on November 5 and the other at the Royal Geographical Society on November 25.[30] The presentation at the Geographical Society dealt with technical features of the railway by providing relevant computations about gradients, but it also provided the draft of a tourist guide-length list of Mosquito words. At the Anthropological Society, the chairman was that day Berthold Seemann, who introduced Collinson's talk by recalling that Pim had established "some years since" the "practicability of constructing a railway across the Isthmus of Nicaragua, to connect Atlantic and Pacific Oceans. The scheme was at first regarded as incapable of being realized but several eminent men [such as Collinson, author of the paper!] had been sent out with the view of placing the practicability of the project beyond doubt."[31]

Collinson's list of native words was referred to as a "Vocabularies of Mosquito Dialects." Richard Stephen Charnock and Charles Carter Blake, two prominent Cannibals (the first had sat in the Council during the first three years of the society and subsequently became a vice president, while the second had been the society's first honorary secretary before sitting in the Council), discussed the vocabularies in a subsequent paper. One variant ("On the Mosquito and Wulwa Dialects") belonged to the group of "martyr" articles that were withdrawn in protest by their anthropological authors during the 1869 meetings of the British Association for the Advancement of Science when the higher authorities of the British Association once again refused to grant right of citizenship to anthropology. As the reader will recall, the showdown was to break James Hunt's heart and cause his sudden death.[32]

The relationship between science and business is also discernible in the way Pim's business concerns oriented work at the Anthropological Society. C. Carter Blake's voyage around the world in 1867–68 began by calling in Nicaragua "on the borders of the Mosquito country." Anthropological reports stated Blake was spending time at the Chontales ruins in the territory of Pim and Seemann's Chontales Gold and Silver Mining companies. A few weeks later, Blake's correspondence was read at the Anthropological Society. Pim told the audience he had seen the archaeological remains but did not have the opportunity of examining them himself. This important task would fall upon "Dr. Carter Blake, to whom the Council of the Central American Association had directed that every facility should be offered for his investigations, and for adding to the scientific knowledge of this interesting district."[33] Then Pim warned society members, "I am sure you will agree with me that if the aborigines are not thoroughly handled it will not be [Blake's] fault. My only fear is that, in his zeal and affection for anthropology, he may be tempted to send us skulls and skeletons fresher than we could quite approve of."[34] The minutes are silent as to whether loud cheers greeted the jest.

## WHAT ANTHROPOLOGY CAN DO

To understand better what Pim tried to do after he completed his survey of the Mosquito Coast and how anthropology turned out to provide solutions to one of his more acute problems, it is simpler to start from the context of the coast, described in the previous chapter. This will help to identify the alliances on which Pim relied and cast some light on the reasons that led him to reach for anthropology. Although Pim's egomaniacal tendencies somewhat conceal the existence of constituents behind the impression that they included only himself, his conscience, and Queen Victoria, it is possible to identify Pim's associates on the Mosquito Coast.[35] As we saw, among the witnesses to his contract with the king we find J. H. Hooker, a prominent member of the local creole elite and member of a family of former slave owners who considered themselves inadequately indemnified following abolition. In several places in *Dottings*, Pim gives indications of his acquaintance with this group.[36] The poor Hookers, he explains, were "unfortunate claimants." Having "neither wealth nor influence" to lobby effectively in London, they would have had to "whistle for their money." Pim for his part declared himself "interested in the slave question."[37] Elsewhere, Pim dwelled on how the British government had unjustly ignored the requests from a number of local slave owners in Corn Island to be indemnified for their emancipated slaves (Corn Island was a creole plantation settlement off the coast of Nicaragua that had been under formal British jurisdiction until the Treaty of Managua transferred authority to Nicaragua).[38]

Another informative document is the minutes of a meeting organized in Bluefields in May 1867 (which Pim printed, as he would have done for the minutes of a bondholders' meetings at London Tavern). He called this gathering "the first public meeting ever held by other than white men in that country," stating that there were "about seventy inhabitants" providing the embryo of a Miskito representative "self-government." The occasion was a protest against some decisions by the local chief justice, backed by the remnants of British administration including the sickly looking Consul Green. In the group of Pim supporters, we see no Indians but recognize names from former slave owners, such as the Hookers and the Hodgson clan of Bluefield.[39]

The creole population at Bluefields and along the coast included all shades of black, and the ruling class tended to include whiter elements as they kept amalgamating mahogany traders. These individuals, who were important economic actors in the area, worried that the Treaty of Managua would expose them to the hostility of Nicaragua, which would interfere with their affairs. This was the group that had most definitely benefited from the myth of the Anglo-Miskito compact. As Naylor summarized it, in the past "this hodgepodge of private parties resorted to the traditional practice of documenting past British ties with the Shore and emphasizing the current British responsibilities for the welfare of the Mosquito nation . . . They were careful to 'educate' newly arrived superintendents to Belize about British responsibilities for the welfare of the native kingdom and they persuaded them to play a direct role in Mosquito affairs."[40] From the available evidence on Pim's eagerness to defend the creole elite, it seems natural to assume that they probably introduced Pim to the matter and, as they had done with earlier representatives of Britain's informal rule, educated him accordingly.

By involving Pim, these local interests made sure to interest him *materially* in the Miskito myth. Making him the owner of shaky property rights was probably the best way to align Pim's interests with theirs. Indeed, Pim's Mosquito rights were not worth much. They consisted of a leased beach and islets and promise for land surrounding an imagined railway transferred in the middle of a treaty that was precisely about disenfranchising such rights. It was unclear what court would recognize the curious deeds signed by George Augustus Frederic and now kept at Tulane Library in Louisiana.

Victorians such as Pim, who were not nearly as frivolous about property as they were about many other subjects, knew this full well. In the end, the rights Pim had purchased with his quit-rent were just as solid as the narrative on which they rested. One owned land because one had purchased it or received it from the king of Miskito and *because the king of Miskito had a special relation with the British monarchy.* This was how the ownership title was played out—not so much on a land registry despite the efforts of the British consul to develop such an instrument but through the *representations* that right-holders would make to British representatives.

In other words, now that he owned land in the kingdom of Miskito, Pim faced precisely the same problem generations of previous interlopers had contended with—to what extent could he have his own property rights enforced? The modernity of Pim's story was that he sought to achieve this through science. Yet it was not Pim's Central American Association's use of the road show in learned societies that was original. Such techniques were used routinely for virtually every single promotion, as we saw for instance in the case of Gerstenberg, Bollaert, and John Field's Ecuador Land Company. Pim innovated by using the learned society not only as a place to own truths about climate, geography, and the life of indigenous populations, but also as a technology to in fact *create* a form of property from scratch, something that amounted to quasi-property rights.

Faced with the problem of the absence of a court that would recognize his land titles, Pim imported Miskito issues in a special court called the Anthropological Society of London. There, Pim would be able to create interest, provoke discussion, generate debate, and invite controversy. Discourse and acting at the learned society would substitute for enforceable legal rights by transcribing the old interlopers' narrative into the language of a budding science. The old discourse about a compact that existed between Miskito and British, between the king and the queen, was now rebranded as an anthropological description of the fine features of the Indian king and brought to the higher sphere of the learned society. This explains why people such as Pim ended up "acquiring" territories in such regions; their connections with the institutions of science put them at an advantage over other buyers regarding the exploitation of weak foreign rights.

And thus it is that, as he puffed his companies on the London Stock Exchange during the second half of the 1860s, Bedford Pim postured invariably as a man of science *and* a defender of Indian rights. This advocacy led him to declare in one meeting after the other that he had been "appointed" to "protect and defend the interests of the inhabitants of the Mosquito Reservation,"[41] and retained the keenest and "liveliest interest in the progress of the country and the welfare of its people [the Miskito Indians]."[42] As Pim actually claimed, it was quite fortunate that an Anthropological Society existed to take care of both the Miskito king and the Nicaraguan *tapirus bairdii*, both endangered mammals against whom, he claimed, the bureaus in London had issued "search warrants."[43] As for the means used to these humanitarian and environmental ends, a sailor should be pragmatic. He was quoted by the *London Standard* stating his determination to "protect and defend the interests of the inhabitants of the Mosquito reservation in such a manner as shall seem best."[44]

## TRANSCRIPTIONS AND APPROPRIATIONS

As argued, Pim received the Miskito fetish from the interlopers of the coast, which they were determined to exploit as long as they could. He then reworked it and put it in the language of science, and from this developed an array of threads, each of them feeble if considered separately but, taken together, hopefully as effective as the spider web is for catching a fly, enabling Pim to "cling to the hope that [he] may yet see the seed [he] had sown bearing fruit abundantly."[45]

The capitalist logic of the stock exchange thus informed and shaped the anthropological discourse that could be heard at St. Martin's Place. The starting point was that the lease had been made by a Jamaica-educated Indian ruler. To enhance the value of the rights that proceeded from this ruler, it was therefore necessary to construct this ruler into a *bona fide* seller and thus endow him with the right kind of character. Anthropology provided just this. This explains why, in his drive to transcribe his capitalist concerns in the language of science, Pim invented a new anthropology of Miskito, which came in direct conflict with Squier's claim to the "blackness of Miskito."

As argued in the previous chapter, Squier had sought to challenge British rights to the Mosquito Coast by assaulting in *Waikna* the king of Miskito, a debased "black penny": "Physically," he had claimed, "the Mosquitos have a large predominance of Negro blood; and their habits and superstitions are African rather than American."[46] Likewise, Squier emphasized that pre-Columbian artifacts observed in the Chontales area included sophisticated petroglyphs, so that the ruins revealed an earlier, higher civilization from which Miskito must have decayed if any link had existed with it in the first place. This cast doubts on the Miskito Indians' "right of first occupant."[47]

Discussions at the Anthropological Society before Pim was involved

show the official take of the Cannibals on the subject. In May 1863, one
of the very first meetings of the society had Bollaert emphasizing both the
deleterious effect of the mixing up of "races" and the difference that existed between Indians and "Zamboes." After stating that the "principal species" as "regarded colour in particular" were "1) White; 2) Brown; 3) Red;
4) Black; etc. from which proceed endless varieties, by commixture," Bollaert claimed that the "fusion, or rather *confusion*, of the White, Indian, and
Negro elements . . . is unfavourable to a strong, healthy, and prolific progeny, which produces numberless *varieties* of Mulattoes and Zamboes [i.e.,
Sambos-Miskitoes]." Another Cannibal, Carter Blake, concurred, referring
to the authority of Robert Knox on the demoralizing effect of the intermixture of races and asserting that "the Zamboes appear to have reached the
lowest depth of moral degradation."[48]

This view opened the possibility of rescuing the character of Miskito.
When Squier employed himself to paint Mosquito dark, his archenemy in
anthropology and railway Bedford Pim would do exactly the opposite and
was careful to construct the Indian as pure. This explains the importance
of the already discussed ethnographic portrait of the king and the inclusion
in *Dottings* of the woodcut mentioned in the previous chapter (although
the handsome three-piece white suit is probably the only truly discernible
element, perhaps the most important one, too). Pim's mercurial concerns inform his conspicuous attempts at distinguishing between the black "sambo-
miskito" element and the Indian element of the population on the Mosquito
Coast, as well as his concern with establishing a continuity in the history
of Mosquito ownership of the Coast, as this would reinforce the idea of a
*bona fide* race. Indeed, we already quoted Collinson mentioning that "people
point out with pride their very old men, a custom dating from long before the
conquest."[49] Thus spoke Bedford Pim:

The sambo is the result of a large admixture of the negro family amongst
the aborigines of the Mosquito shore, the offspring of a number of male
blacks wrecked from a slaver very many years ago. This dark element is
now self-supporting, having been much stimulated by a large influx of
escaped slaves. All trace of the Indian share in this family has now disappeared, and the woolly hair, thick lips, and flat nose, of the pure African,
prevail in all pristine vigor.

Some writers have mistaken this hybrid people for the aboriginal, or
at all events the predominant race, and describing their vices and customs, have occasioned a low estimate to be formed of Mosquitos. One

American writer in particular published a book called *"Waikna"* which he doubtless wrote for "strategical" and diplomatic purposes—Samuel A. Bard (E.G. Squier) former *chargé d'affaires* from the United States to Central America. The real possessors of the country are pure Indians, whose king must be of pure blood, and a direct descendant of those Caciques who have ruled the land from time immemorial.[50]

After this piece of bravura followed the description of Miskito Indians as, well, "typically Indian"—the "swarthy" complexion, the "coarse black hair," and above all the nose, "sharp, thin and small."[51] Thus in Pim's anthropology, the pure-bloodedness of the king was also extended to the rest of his people, and when Pim met with a company of Indians from different tribes, he would note that the Miskito "appeared to [him] to be lord of all he surveyed, for he domineered over his companions" and was about the "most intelligent native [Pim] ever met," and *therefore* obviously not a mixed blood.[52] In other words the barrier, which Pim found existed in people's mind between pure and mixed races he erected still higher, making sure to put his client on the right side. In Pim's anthropology, the important issue was the notion that the Indians were the *real possessors* of the country, an ownership transcribed as the pure-bloodedness of the king, a "direct *descendant* of those Caciques who have ruled the land from time immemorial." Underscoring the continuity between Pim and earlier schemes brewed on the Coast, we may note that MacGregor had claimed to be a former cacique, too. Sign of the times: when MacGregor felt he should *be* a cacique, Pim felt he should *write* about them in his anthropology.

The conclusions for Miskito science followed logically. Miskito were natives who lived up to their word. This comes in the middle of an astonishing argument about Miskito as trade union–resistant people (investors' concerns whether the company would find appropriate workers within these "strongly built" Indians cannot have been far away). Political agitators, Pim explained, had better not approach the shores of Mosquito, because "if any adventurous trader in trade-unions were to appear amongst them and propose a strike, I very much question if he would escape for life; he would be probably clubbed to death, a just punishment according to their lights for suggesting meanness and absence of good faith in carrying out a bargain."[53] Thus were these good Indians, ones who would honor the contract they had signed with Pim—gentlemanly Indians, so to speak. This was because they were a people who, Pim emphasized during a discussion at the Anthropological Society, were "teaching, somewhat in the order of Mr. Disraeli, industry, liberty, religion."[54] Beyond the shadow of reasonable doubt, Miskito were Tory, empire-loving Indians.

As the background of Pim when he made his "Negro and Jamaica" presentation becomes clear, one hopes the mystery surrounding him dissipates. He was not quite the hastily recruited fellow that previous researchers have imagined. If one agrees with my hypothesis that the Anthropological Society should be understood as a student and heir of the promotional tactics of the Royal Geographical Society, Pim the commercial trafficker was a pure anthropologist, a Franklin hero. He was the right man, in sum, for the task at hand.

At its core however, this chapter has been only partly concerned with Pim himself. Rather, it has provided a discussion of the role of anthropological science in the micro-technologies of globalization. While Pim continued a century-old tradition of British trafficking with the Coast, he also improved on all his predecessors by involving science and the learned society in the process. It was his contribution to turn a learned society into an instrument of ownership. The technology evolved by Pim cannot be understood without reference to the power of science, which was by now an autonomous force. Evidently, previous "swindles" such as McGregor's provide background to Pim's own invention of a colony, to which a railway would now be appended (for this was the mid-nineteenth century). But more importantly perhaps, Pim innovated by organizing, on the heels of the American anthropologist Squier, a Mosquito Question that he managed to transform into a subject of study for the Anthropological Society of London.

The history of Pim thus makes patent the fundamental interested-ness of science. And indeed, *Negro and Jamaica* contained a preface that made explicit reference to Pim's broader commercial interests in the area. Perhaps with a cold sense of humor, Pim disclosed this as a conflict of interest, which actually *enhanced* the trustworthiness of his claims. Being the owner of land in Central America, he explained, he would have benefited from an exodus of whites out of Jamaica toward his colony. Why, says Pim in substance, he does control a million and a half of acres of land in Central America, which could readily host white refugees from Jamaica if the Negroes got their way in the colony—"but the plain duty which every Englishman owes to the land of his birth made me put aside these and similar considerations."[55] A man of science and trustworthiness, he had felt compelled to speak the truth and lend support to Governor Eyre's "excellent policies."[56]

The resulting expertise and command of the tools of science was also a powerful instrument to remain an insider. Being conversant with the language and techniques of anthropological science helped Pim maintain a position, substantial and symbolic, that provided him with material benefits and trading tools. The command of science gave legitimacy to interventions, which ultimately provided information to the trader. This evidently contributed to Pim making his infamous comeback as special commissioner for the debt of the government of neighboring Honduras in 1872 (an episode narrated in a future chapter).

Most importantly, the scientific forum permitted radical reinventions under the guise of scientific debate, hypotheses provision and their refutation. More decisively perhaps than the ownership of truth, the learned society enabled ownership of these transient truths known as hypotheses. This explains the dizzying plasticity of contemporary discourse, a consequence of the dizzying plasticity of business needs. For instance, the latter part of *Dottings* articulated a project that would seem incomprehensible to anyone who would take the content of the *Negro and Jamaica* essay for evidence of a commitment to racism. Indeed, at this point in the aftermath of the Civil War, Pim's new grand idea was to *welcome* black people in Central America who would, he stated, prefer to leave the United States because racism in the United States was extreme to the point that "anyone with even a tinge of 'colour' or the semblance of wool is open to be called a nigger." Former slaves, relabeled the "industrious coloured population of the Southern States," were becoming apt candidates for colonizing Central America. Since President Abraham Lincoln called for a Liberia in Central America, Pim the shopkeeper eagerly displayed his wares.[57]

The stock exchange modality was again in full swing. Racism was "mispricing" and the mispricing of the North American "black" by the American population provided the anthropologist who happened to own land grants in Miskito territory with a profitable arbitrage opportunity. Pim, who had studied the characters of peoples and discussed them fortnightly at the Anthropological Society, found "hope" for those blacks "who have a white intermixture however slight in their veins."[58] He could see how, on the Mosquito Coast, the African-American migrants "would be hailed as country-men warmly welcomed and finally be placed in a position to do some good for themselves." Suddenly, the intermixture of races was no longer a handicap, and Miskito Indians were identified with the black Miskito-Sambo. Pim also expected that in the United States, white Americans would "not but feel that the departure of the irrepressible nigger [is] good riddance," notwithstanding the fact that in Miskito-land, the "climate is healthy (yellow fever is unknown) and is indeed in every respect superior to that in the vicinity of Mississippi."

So writes the captain who could never refrain from expressing his ideas in the guise of a prospectus. What a happy world there was in store, where everybody would be satisfied and more blacks profitably mass murdered under the pretext of colonization and the knowledge of races. Can one think of a better illustration of the "inter-mixture" of anthropology and trafficking, or of a better case study of the use of the science of man as an instrument to create or appropriate wealth, to harness the knowledge to design a more "effective" world—for a fee? Perhaps now we can better see the common thread in the different narratives of Pim the Navy captain, the author of "Negro and Jamaica," the anthropologist, the founder of Miskito science, the entrepreneur in colonization, the promoter of railways and mines, and the vulture investor. The unifying theme is not race, evidently, or else mahogany is race. From exploration to anthropology, from the promotion of a railway to the defense of Indian rights, and from the ultra-racism of "Negro and Jamaica" to his indictment of American racism and concerns for the huddled masses of an "industrious coloured population" fleeing Mississippi, Pim's adventures provide an interesting window on the versatility of anthropological discourse in Victorian times. A careful examination of Pim's deeds and words, of Pim the non-accidental anthropologist, essentially blows to pieces previous attempts at constructing racism as something with precise contours. Given the destructive effects that a more detailed portrait of Pim has on the earlier literature's darlings, one can understand the previous instinctive urge to exclude Pim from the world of anthropologists.

Thus Pim's character provides a good example of the problems associated with any attempt at telling the scientific wheat from the white-collar criminality chaff. Such was Victorian anthropology—probably much less tied to any specific cultural pattern than has been repeated ad nauseam, but plastic, mercantile, insane, powerful, dangerous, perhaps, and above all opportunistic as the true stock market speculation must be. To return to our opening question about the real nature of Pim asked at the beginning of the previous chapter, we may conclude that it was an absurd quest to begin with. Just as it is baseless to disentangle what was genuine scientific interest from what was lucre in the anthropology of the 1860s, it is baseless to attempt telling, in a character that fascinated R. L. Stevenson, the Dr. Jekyll from Mr. Hyde. As the gallant captain would have had it: "Sailors, always were and always will be a restless race, or, indeed they would not be sailors."[59]

The legacy of Pim's grand fantasy to compete against the Panama route endures. A few years ago, a plan was hatched to use Pim's Bay (Monkey

Point) as the site of a projected $350 million seaport, to be financed by Iran and Venezuela. Later, American and South Korean investors expressed interest in developing a container terminal there. Still more recently, Chinese telecommunications billionaire Wang Jing signed with Nicaraguan President Daniel Ortega a contract that grants a fifty-year concession to Wang's HK Nicaragua Canal Development Investment Co. The project includes a canal, an oil pipeline, two deepwater ports, two airports, and of course an interoceanic railroad. The games of empire are endless, and like neurosis, repetitive. Interested readers can use a tool such as GoogleEarth, search for "Monkey Point" and survey the small groupings of houses at Greytown and Bluefields on the Atlantic coast of Nicaragua. They will find that as of today, there are hardly more of them than at the time of Pim's voyages. In this political interstice still live a few Miskito Indians, the subject of many academic papers and enduring controversies.

## CHAPTER NINE

# The Violence of Science

Among Hyde Clarke's appearances in the press during the second half of 1868 discussed in chapter 1, the reader might remember Clarke's letter to the *London Standard* calling for contributions to a "Testimonial Fund" established to honor Dr. Charles Tilstone Beke for his role in the Abyssinian hostage crisis.[1] Beke was a pioneer in then-fashionable biblical archaeology, a science that consisted in rummaging Middle Eastern deserts in search of spots mentioned in the Scriptures. He had spent a good part of his life crisscrossing the region with the purpose of reconstructing the book of Genesis from geological data and making his own contribution to the debate on the sources of the Nile. His knowledge of the area is what got him involved in the British expedition to Abyssinia, in the middle of which he published a noted book, *The British Captives in Abyssinia*.[2]

The British expedition to Abyssinia resulted from a complicated affair that had started in early 1864 with the revelation that Abyssinian ruler Emperor Tewodros II had thrown two Protestant missionaries and a British consul, Captain Charles Duncan Cameron, into chains. The crisis simmered for a few years as a succession of missions sent to liberate the hostages failed. It morphed into a full-blown political controversy that eventually dented the political capital and reputation of John Russell, who was blamed for mishandling the crisis, first as Liberal secretary of state for foreign affairs and then as prime minister. After the fall of Russell, it was eventually resolved in 1868 by a military campaign undertaken by the new Tory government led by Benjamin Disraeli. The campaign was hailed as a triumph because it resulted in the rescue of all the hostages, the slaying of a few hundred natives, and the death of Tewodros, reportedly, from suicide.[3]

British pro-consul Napier had been provided an army of dark men from

India so that he could fight darker Abyssinians, and with this army he
trained a menagerie numbering by one count some 25,000 camels, elephants,
horses, donkeys, sheep, and cattle. The British press organized a nail-biting
narrative made even more exciting by the lack of a cable, which introduced
hazard in the way news would arrive (the typical delay could be as long as
three months). Photographers had been dispatched. Suggestive of the public
fascination with the episode, publishers rushed to circulate eyewitness accounts of the campaign. The following excerpt from Captain Henry Hozier
of the 3rd Dragoon Guards, at one time Napier's assistant military secretary, illustrates the peplum quality of these accounts in its description of
the triumphant march of the British force on its way to the fortress of Magdala: "It was a fine sight to see the long line of red, Royal Engineers . . . ,
Sappers, 33rd and 115th Regiments, the 4th King's Own in their grey ka-
kee, the Beloochees in their dark green, the Royal Artillery in blue, and the
mountain batteries on mules, winding up the steep and picturesque path
that led to the Fahla saddle; . . . Sword and helmet sparkled in the morning
sun, the banners were unfurled, the breeze was just enough to display their
gay colours and the proud names woven thereon, and all nature seemed to
contribute to the splendour of the pageant."[4]

The reasons for the Abyssinia campaign of 1868 have intrigued students
of British history. No doubt those ugly-faced Africans needed to be taught
a lesson (the campaign inspired *Punch*'s infamous "Now then, King Theodore! How about those prisoners?" cartoon, a commonplace testimony to
Victorian racism). (See figure 9.) No doubt there was the urgent need to
show that Britain had the stamina to meet Russia's push beyond the Cauca-
sus and the threat this posed to India. On the other hand, there was something of a profound reversal at work in the pledging of £9 million for just
one absurd campaign, a dilapidation that horrified Liberal politicians. This
ushered in a reversal of the habits of parsimony that ruled mid-nineteenth
century British "Whig" imperialism. Indeed, the Abyssinia campaign coincided with Disraeli's first term as prime minister, the last effort of his ascension to the "top of the greasy pole," as he liked to put it. It also became
a building block of his subsequent promotion of empire and articulation
of a new imperial policy.[5] Disraeli's most famous contemporary speeches
reflected a new, aggressive rhetoric that replaced the earlier low profile and
concern about cost minimization. In the one he delivered in July 1868 to
greet the triumphant legions of Napier, who was made Lord Napier of Magdala, Disraeli fully exploited the glitz of the subject: The expedition had
carried "the artillery of Europe" on the backs of the "elephants of Asia,"

THE ABYSSINIAN QUESTION.

Britannia. "NOW, THEN, KING THEODORE! HOW ABOUT THOSE PRISONERS?"

FIGURE 9. From *Punch* (by John Tenniel, of *Alice* fame): "Britannia. 'Now, then, King Theodore! How about those prisoners?'" *Source*: *Punch*, August 10, 1867. British science was deeply involved in the events that led to the slaughtering of "King Theodore" and his army. Key hostages were members of the Anthropological and Ethnological societies, maps were supplied by the Geographical Society, and Clements Markham was advisor to Napier as he led the Anglo-Indian forces to the storming of Theodore's fortress in Magdala.

"over African passes which might have startled the trapper and appalled the hunter of the Alps." The wide screen was evidently part of Disraeli's notion of globalization.[6]

Historian Freda Harcourt, in a classic paper devoted to the emergence of the new imperial spirit at the juncture of the Abyssinia campaign, has suggested looking toward domestic factors. She writes about how Disraeli would have identified and exploited new forces at work in British dispositions toward empire. In her own words, "there is ample evidence that from 1866 to 1868 *compelling influences* bore upon the British polity, that Disraeli recognized them and used them."[7] But she remains elusive as to the origin and nature of these compelling forces. In this chapter, I develop an argument about these influences. I study the contribution made by science in the chain of events leading to Napier's invasion of Abyssinia. I argue that an important aspect of the crisis revolved around a confrontation between on the one hand Prime Minister John Russell and the learned societies on the other hand. Since the negotiation of the Central American treaties discussed in the previous chapter, the foreign policy of Britain had been run by John Russell a key architect of informal empire.[8] But in June 1866, the Liberal coalition that backed Russell collapsed, and his government fell. The episode resulted in the formation of a Conservative ministry led by the Earl of Derby and later by Disraeli. And as Harcourt has implied, Disraeli's genius in exploiting the Abyssinia campaign was to paint it as a sanction of Russell's policies. It was a statement on imperial management and indeed a plain anticipation of Disraeli's famous Crystal Palace speech of 1872, when he would claim that Tories owned the promotion of empire and that Liberals had conspicuously aimed at "disintegrating" the empire of England.

One intriguing aspect of the crisis that has gone unnoticed so far is its connection with anthropology. First, several captives were members of the Anthropological Society of London. Consul Duncan Cameron, who was thrown into jail by Tewodros and started the whole matter, was thirteenth among the first batch of fellows voted in by the Anthropological Society in 1863.[9] Dr. Henry Blanc, who would come to Abyssinia with a mission for the release of captives but ended up detained with the rest, was another Anthropological Society fellow.[10] Second, the coalition of interests who were especially vocal in complaining about the government's lack of decision in Abyssinia included anthropologists. Here we need not look further than Bedford Pim, whose fallout with Russell following the purchase of land from the king of Miskito narrated in the previous two chapters had led him to embark on a career as anthropologist and entrepreneur. Third, when "Johnny"

Russell fell, much to the delight of anthropologists, the new foreign secretary was Lord Stanley, son of the new prime minister, the Earl of Derby. Lord Stanley, we have seen, was also a fellow of the Anthropological Society who had been signed in by Burton.

There were just too many anthropologists around for this to be coincidental. In fact, it is possible to distill the logic of the irruption of anthropology in the public sphere using an expression coined by Hunt at that very time and which Bronislaw Malinowski would reinvent in the 1920s: "practical anthropology." Anthropologists had data and theories, but they had politics too, and the expression "practical anthropology" summarized the politics of anthropological science as Hunt, Burton, and their followers conceived them. Practical anthropology and evolving perceptions as to what constituted a proper foreign policy formed an important part of the new forces, the "compelling influences" that Disraeli undertook to harness. In 1868 Pim and Seemann dedicated their joint book "To the Right Honourable Lord Stanley" using a language that emphasized their praise of the new political course and scorn of earlier ways. They wrote that Lord Stanley's "efforts to restore English prestige abroad and elevating the standard of our foreign policy" commanded the respect of all parties.

This new interpretation of the intellectual background of the rise of Disraeli will enable us to outline the deep significance of Disraeli's political insight. The Abyssinia hostage crisis was the straight outcome of a confrontation between a principal of empire (Russell) and a local agent of empire who happened to enjoy the backing of a learned society such as the Anthropological. When such a showdown occurred, it was unclear who would have the upper hand. In a nutshell, what Disraeli knew is that policy makers had now to contend with the violence of science.

## INFORMAL EMPIRE AND ITS COMPLICATIONS

British consuls typified the conflicting objectives of public duties and private pursuits at the heart of the management of informal empire. Being proficient in foreign languages and residing in foreign countries, they were an essential cog in the information machinery of the Foreign Office, feeding the headquarters with dispatches that were carefully archived and stored and even read. Also, it was notorious that British consuls were underpaid, explaining why some of those who entered the consular service thought of their assignment as a subsidy permitting other activities. Such activities, because of the consuls' mandate to help and promote British commercial interests, generally had a mercantile side, which generally took over as it should. Defending a proposal to appoint an adviser to a suggested consulate in Abeokuta in 1861, with the mandate of finding ways to encourage the cultivation of cotton and development of a cotton trade, Russell told the Commons that "he must endeavor to find a man of experience who would be useful in promoting the interests of British trade."[11] Thus consuls were, to use a modern expression, "incentivized." To put it in yet another way, they were encouraged to engage in all kinds of commercial prospecting even when there existed in principle formal rules against "consul trading." They were great originators of mercantile information and commercial promotions, and were closely associated with the spirit of informal empire they best epitomized. As Hyde Clarke claimed in 1868, one British governor-general once told him, "Mr. Clarke, every thief has got his consul."[12]

When Charles Duncan Cameron was assigned in 1861 to Massawa (or Massowah, according to the transcription of the time), he replaced Consul Walter Chichele Plowden, who as military adviser for the ruler of Abyssinia had died from injuries received in an ambush. Massawa is a Red Sea port on the coast of modern Eritrea, which at the time was an Ottoman protectorate. Plowden had been a consul there during the 1850s, and his territory covered a region whose most extensive part was Abyssinia (modern-day Ethiopia), a Christian kingdom unlike Muslim Massawa. There was hardly any trade to speak of, and the politics of the area were chaotic, but the area's location between India and Egypt had political significance. Not incidentally, a tributary of the Nile, the so-called Blue Nile, begins in Abyssinia, making the region a bordering part of that which came under discussion during the sources of the Nile controversy.

As a true consul in a political post in Britain's informal empire, Plowden had employed himself in travels beyond Massawa. This is how he came to provide detailed and much-quoted dispatches containing political analyses of the rise of Tewodros, where he pleaded for British endorsement of the new leader. The new Abyssinian leader, he wrote, was a "remarkable man" who "under the title of Negoos or King Theodorus, has united the whole Northern Abyssinia under his authority, and has established tolerable tranquility, considering the shortness of his career."[13] Plowden cultivated friendship with Tewodros just as he had with Tewodros's predecessor Ras Ali, and in extensive reports that would have been worthy of the *Anthropological Review* had it existed in the 1850s, he sought to distill the cultural foundations of the political order prevailing in Abyssinia as he understood them. In his mind, the European Middle Ages provided the best comparison, and he likened the difficult position of the Abyssinian ruler to that of Louis XI of France, a king who permanently fought his barons and threw them in a cage after he had defeated them.

Consular dispatches make fascinating reads. Plowden's are intriguing in the way they embroider narrative and social scientific conjecturing and because they reveal the strategies of language and communication in informal empire. For instance, rather than describing the policies he intended to pursue and seeking Foreign Office endorsement, Plowden ascribed to the king of Abyssinia policies he, Plowden, had evidently inspired or encouraged. This avoided a loss of face for everyone in case the direction of advising was censored as impolitic. Thus Plowden construed as the king's own spirited policy reforms such things as political centralization, clemency toward the defeated, abolition of the talion law, and so on, as well as the banishment of Roman Catholic priests, which Tewodros decided after "perusing the history of the Jesuits."[14] In such documents, of course, we can only get a glimpse of Tewodros's own ideas and objectives. It is plausible that he looked with some interest toward those eager Britons who came a long way to declare their readiness to serve him. This was especially so because this ruler was a Christian surrounded by Turks, and this may have created in his opinion natural connections with other Christian rulers. As Tewodros would later write in a letter to Queen Victoria, and if we trust the work of the translators, he had understood his relation with Plowden as one that would make him (Tewodros) "known" to Victoria and would "establish friendship" and of this he was "very glad."[15] Maybe Plowden had not fully explained that, gradually over the 1850s, British attitudes towards the Turks were becoming friendlier in a context of the weakening of the Ottoman Empire, competition with Russia, and projects for the opening of new trade routes.

Plowden's replacement in Massawa, Charles Duncan Cameron, was a career military adviser who had been active in the Crimean War. The reasons for continued British close surveillance of Tewodros included an emerging French interest underscored by their purchase in 1862 of the territory of Obokh (Djibouti) and the desperate quest for cotton production sites to replace American cotton locked out by the blockade of Confederate ports during America's Civil War. Egypt was showing its potential, and regions with like climate attracted considerable attention. The instructions that Secretary of Foreign Affairs Russell gave to Cameron were a chef d'oeuvre of ambiguity in the best tradition of informal empire. Cameron would be based in Massawa, but his instructions explicitly stated that he should use this position to make himself acquainted "with the affairs of Abyssinia." Russell recognized that, since the death of Cameron's predecessor, Her Majesty's government was so "imperfectly informed" that he was "unable to lay down any very precise rules for the guidance." The usual generalities followed. It was important to cultivate good relations but essential to avoid taking sides; one had to watch the agents of other European governments but avoid competing against them; and of course it was important—and not to be forgotten—"to impress upon any native rulers who may directly or indirectly encourage or permit the traffic in slaves the abhorrence in which it is held by this Government."[16] With such instructions in his pocket, Cameron did what he understood he was expected to do. He began, like his predecessor Plowden, like any honest consul, his own shop of micro-imperialism.

Difficulties started when, on October 31, 1862, Cameron sent a long dispatch to Russell that opened with a description of the reception Tewodros had reserved for him at his camp "in a reclining posture, with a double barrel gun and two loaded pistols by his side." It then discussed at length the subject of the embassy that Tewodros wanted to send to Britain but could not because he was surrounded by Muslim rulers in Massawa and in Sudan who would not let his emissaries go, especially because Tewodros was not a model neighbor himself. The letter concluded with an indication that he, Cameron, intended to proceed to the "neighborhood of Bogos" (a region inhabited by a pastoral people in the highlands immediately north of Abyssinia, now part of Eritrea) whose inhabitants "have been long under our special protection, and for whom we formerly interceded with the Egyptian Government on the occasion of certain predatory inroads, which, from intelligence I have received may again be renewed." Cameron's dispatch also contained copies of the letters he was sending around, including the conditions for a peaceful settlement of pending disputes between Abyssinia on the one hand and Massawa and Sudan on the other. Upon reaching Russell, these letters triggered a strong response and, in April 1863, Cameron was specifically instructed that it was not desirable "to meddle in the affairs of Abyssinia." He was ordered back to stay with Tewodros. The message was repeated in strong and stronger language in two subsequent letters in August and September 1863, including a reference to the office's impatience at Cameron's repeated use of the expressions "envoy" and "mission" when he held "no representative character in Abyssinia."

During a period that extended between the initial dispatches in the fall of 1862 and the fall of 1863, the rebellious consul kept plowing ahead, evidently playing on all the margins of uncertainty and ambiguity that existed in the imperial relation. The need for political continuity, alleged or real, and the existence of local support, invented or material, were invoked to provide excuse or motivations for explorations whose eventual test, the agent hoped, would be their success. For instance, when motivating his proactive policies of offering protection to some tribes, Cameron argued he merely continued Plowden's course of action, and since Plowden had not been disavowed, wasn't it compelling evidence that this had been the policy? Though he could not refer to archives because he was in Kadarif, Sudan, at the time of writing, Cameron felt one had to "suppose [this] was known to, and not disapproved by, Her Majesty's Government."

In this instance, his exploitation of the margin of uncertainty that surrounded colonial monitoring was to end badly for Cameron. Tension in Abyssinia escalated as the overstretched consul faced the growing anger of Tewodros, to whom he had promised a lot—as it turned out, a lot more than he could deliver. The smart ruler of Abyssinia, a deft politician we are told, must have understood that Cameron was the weak link in the imperial machinery. Seeing his desire to secure British political support constantly frustrated, Tewodros turned his anger against Cameron, who was thrown in chains in December 1863 after the stakes had been raised by doing the same with two missionaries a few days before. Tewodros may have thought of them as subaltern figures, a way to send Cameron a signal.[17]

When they heard of the news, the offices in London were sorry. But while they were making some efforts to secure Cameron's release, they could not help thinking, and later stated publicly, that Cameron had gotten what he deserved. The spirit in which Russell held the matter is well summarized in an instruction to the mission sent in 1865 to liberate the hostages where it was stated that Consul Cameron was to be "employed hereafter in a different part of the world and will never have occasion to return to Abyssinia."[18] As far as Russell was concerned, this was just another matter in the day-to-day government of informal empire—one more British agent overseas gone loose. And this would have been the end of it had it not been for the brutal attack that one scientist called Beke led on Russell, contributing to Russell's downfall and paving the way for the Abyssinia campaign.

## THE ART OF SCIENTIFIC AMBUSH

The news of Consul Cameron's detention reached London in early 1864 along with that of the two missionaries, Reverend Henry A. Stern and his assistant Mr. Rosenthal, who belonged to a church that promoted Christianity among the Jews. After the news was confirmed by a succession of messages received between April and June 1864, several individuals contacted Russell to offer their good offices. One Edmund C. Plowden, cousin of the former consul and a former major of the Indian army, volunteered his "zeal and activity" for Cameron's release. One Henry Dufton, a cloth merchant, offered his son as messenger on the basis of a recent four-month stint in the Abyssinian court. But of all these expressions of goodwill, none were more determined and persistent than those sent by the geographer, explorer, archaeologist, ethnologist, and self-appointed fixer Charles Tilstone Beke. Beke advertised himself as a scientist with knowledge of the region, but as always, the commercial interest was lurking. Son of a merchant and himself a merchant at the beginning of his career, Beke had always been concerned with promoting commerce with Abyssinia. He had submitted in the late 1840s a proposal to the Board of Trade for the establishment of a textile factory in the "high table land of Abyssinia, behind Massowah," which the Board of Trade politely declined.[19] Precisely at the time Cameron reached Abyssinia, Beke sent a further memorandum on the subject of British commerce within the Red Sea. In one of the reports with which he peppered then-Secretary of State John Russell at the beginning of the hostage crisis, Beke went back to these suggestions and emphasized Abyssinia's enormous potential as a supplier of cotton. He also had the idea of using Abyssinia as the main route for a telegraphic line between India and Britain, a line that would circumvent both Russia and the Ottoman Empire and therefore secure communications between Britain and its main colonial possession.[20]

Russell's administration conspicuously, if politely, declined Beke's proposals, and when they were made public, he explained in Parliament that he had the matter under control and was confident that things would sort themselves out soon. Because the Liberal administration was unwilling to commit resources for a rescue operation and was also reluctant to rely on the advice of the overly enterprising crowd of entrepreneurs and scientists, the situation just dragged along. As the missionary society to which Stern and Rosenthal belonged campaigned against government inaction and Conservatives decided to take advantage of Cameron's continued imprisonment, the matter exploded in Parliament at the end of the spring of 1865 when Beke decided to step up his attacks.

Beke's *The British Captives in Abyssinia* was first published as a pamphlet in 1865, with an enlarged second edition coming two years later on the eve of the campaign itself. It provides a guide to the logic of the attack Beke launched on Russell and the Foreign Affairs undersecretary Austen Henry Layard, himself perhaps not coincidentally a traveler and biblical archaeologist. It contained a violent indictment of Russell's former policies. In a vengeful way, the preface described how Beke had been proffering "advice and assistance in the most friendly spirit to the Foreign Office and to the Board of Trade, though (as [he had] regretted to see) without any good effect" and how provision of this free expertise had in fact been "displeasing" to the "late Administration." Beke now retaliated by poking holes in the late government's policies. The infeasibility of a military campaign, which Russell emphasized publicly, only revealed the flawed "notions entertained by our Government respecting the approaches to Abyssinia in the event of a war"—notions that were "if possible, even more erroneous than those concerning the climate and physical character of that country."

To drive his story home, Beke took issue with specific claims that had been made publicly, and the preface drew the reader's attention to the passages in the book in which he claimed to destroy Russell's assertion, made in the House of Lords, that he did not want to risk an army across the "deadly plain which separates Abyssinia from the sea." The reader would be shown "what that 'deadly plain' is." In fact, he went on, the climate in Abyssinia is "congenial to European constitutions," and thus an expedition was absolutely feasible.[21] As time passed, continued survival of the hostages dented Russell's credibility, and the missionary societies agitating in favor of Stern and Rosenthal ensured that the matter would not die. To the extent that there was still a crisis and not just a few families in London mourning their dead, the executive must have been less informed than its critics. The corollary was that the relation between science and politics needed a reshuffle.

Beke's various episodes describing Europeans being thrown into chains underscored this very point. To take just one, his narrative of the last hours before the jailing of the two missionaries and Consul Cameron detailed how on the morning of October 15, 1863, in a context of deteriorating relations between the king and European residents, one of the missionaries had unfortunately showed up at the king's camp and, in the king's presence, happened to bite his thumb "under alarm and excitement." But this action, Beke-the-ethnologist explained, is in that country "considered a threat of revenge." It was, Beke affirmed, what produced the ensuing anthropological catastrophe.[22] Evidently, the reason Russell needed an advisor with anthropological flair was that he could tell when to bite a thumb—and when not. With magnanimity, Beke concluded that Her Majesty's ministers might be—partly— forgiven. They were responding to the public, and the public, ill-informed on this "dark" and "difficult" subject, was not "alive to its importance" until properly "enlightened." It was to the scientists to generously volunteer this enlightenment, for the disinterested purpose of achieving higher-grade foreign and colonial policy.

## JOHN RUSSELL VERSUS THE ANTHROPOLOGISTS

Beke's successful raid on Russell epitomized the triumph of science over politics. It also underlined Russell's biggest political mistake. He had miscalcu-
lated Beke as just another overly enterprising, self-appointed agitator without
realizing that, behind the plates displayed on the cover page of *British Captives in Abyssinia* (they boasted "C. T. Beke, PH.D., F.S.A., Fellow and Medal-
list of the Royal Geographical Society" as well as author of "Origines Bibli-
cae," "The sources of the Nile" etc.) lay a new political force, the scientific
interest. As Russell was to learn, those societies that were the hotbed of imperial enterprise—the Geographical and the Anthropological—came down
unanimously against him. Illustrating to perfection the dynamics at work,
we have James Hunt's Anthropological Society anniversary address of January 3, 1866, providing an indictment of Russell's misguided policies. After
reminding the audience that Cameron, who had been detained since 1863,
and Blanc, who had just been sent to Abyssinia with a rescue mission, were
both members of the Anthropological Society, Hunt deplored the failure of
Russell's policies, which had left Britons at the "mercy of the King of Abyssinia" when "a little prompt action on the part of our then Secretary of
State [i.e., Russell, now prime minister] might have saved" both the "poor
neglected Fellow of this Society" from his "sufferings" and "the name of
Englishmen from disgrace." As Hunt continued, "I think I do not go beyond
the bounds of the President of this Society when I publicly proclaim and denounce the apathy which has existed in the government of this country with
regard to" Captain Cameron.[23]

The matter takes special significance once it is realized that the escala-
tion of the dispute coincided precisely with the controversy over the Jamaica
Rebellion. And indeed, in Hunt's anniversary speech of 1866, the Abyssinia
affair was put in explicit parallel with the Jamaica crisis to demonstrate the
"evil effects of the ignorance of anthropology both in our statesmen and our
politicians." Russell typified this regrettable incompetence, Hunt claimed.
The British prime minister was "perhaps the man who, more than any other
statesman of our time, [had] shown himself incapable of seeing the facts in
their true light. Educated in the pseudo-philanthropic school of Wilberforce
and other well-intentioned men, he is ignorant of the merest elements of
the science of comparative anthropology." Hunt established the connection
between the anthropological incompetence of the executive in the Abyssinia hostage crisis and its attitude towards Governor Eyre who had been
recalled to London. As Hunt declared, the "merest novice in the study of
race-characteristics ought to know that we English can only successfully

rule either Jamaica, New Zealand, the Cape, China, or India by such men as Governor Eyre."[24]

Hyde Clarke's public endorsement of Beke long after the fall of Russell, with which this chapter opened, can now be reviewed and better understood. By crediting a scholar for a decisive role in the liberation of the hostages, Hyde Clarke was evidently claiming, just as did Hunt, a role for science as a positive force in imperial management. Consider the words he used in his *London Standard* letter:

> While the benevolent rejoice, in the welcome news of the deliverance of European captives from Abyssinian chains, and in the tragical end of their atrocious ruler, as detailed in your paper of this date, and while others exult over the material success of the British and Indian forces employed in the Abyssinian expedition, there is one to whom the gratitude of all classes of the British people is eminently due, and who ought not to be forgotten or left in the cold shade at this time of triumph. It is acknowledged by all who knew the obstacles which were in the way of any successful undertaking for the deliverance of the British captives in Abyssinia, that to Dr. Charles Beke's knowledge, experience, philanthropic zeal, and intimate acquaintance with that country, and to the valuable information which he has from the first afforded to the government, this unparalleled success is in a great measure attributable.[25]

Given the violent conflict between Hyde Clarke and the Anthropological Society at about the same time, the near perfect alignment of Hunt and Hyde Clarke on the subject of the Abyssinia campaign reveals the compelling influences at work in British society, which according to Harcourt, Disraeli "recognized and exploited." They did not consist in some abstract change of mood but in the formation and expansion of a new, voracious power best represented by the rising membership of learned societies devoted to the study of imperial topics—the Anthropological Society, for example. It is this precise, focused, and recognizable pressure Disraeli and the new Conservative government of 1866 had identified and to which they responded with the Abyssinia campaign of 1868. In essence, Disraeli had understood the rise of anthropological science (of which Russell was such a bad student) and saw the value of harnessing it. Disraeli catered to these emerging interests, whose expansionary nature—or imperial character, to use a sullied word—should by now be obvious. He gave them the opportunity to enjoy a bombast against the policy weaknesses of the former Liberal prime minister, to savor the scenic beauty of an exotic military campaign, and to experience the excitement of a manhunt followed by the death of Tewodros. And to be sure, Clements R. Markham, the scientist and future member of the Council of the Anthropological Institute was put at the elbow of Commander Sir Robert Napier as official geographer of the Abyssinia campaign.[26] The Cannibals avidly received this bone of political goodwill as compensation for having been so conspicuously ignored by John Russell. As to "Old Johnny," as *Punch* called him, he would subsequently become famous among historians for his blunders.[27]

There was no going back, and, in a speech before the House of Commons that presented the compliments of the Liberal Party to Her Majesty's victorious forces, Gladstone, the new leader of the opposition, acknowledged that this Abyssinia campaign was a new kind of war. It was an "Expedition" Gladstone declared, whose "general character" was of "no very common order in military history," not one of those "desperate conflicts . . . waged with equal or near equal force between nations or between armies alike possessed of all the resources of modern warfare."[28] Indeed, such had been a war where armies had been ambushed by scientists, a war where retreat had been cut off by a learned society medalist, a war that was not owned by the military, whose prowess would be not in the valor of its warriors but in the sanction it would receive from the learned society. In sum, it was a conflict of which science was stakeholder. The Abyssinia expedition exposed this decidedly new element, and it is striking that in the *duetto* both Disraeli and Gladstone sang at the House of Commons this novelty became the theme of the chorus. But because Disraeli could not publicly state the true causes of this war—and because he had, as great politicians must, a sense of humor—he spoke of the Abyssinia campaign's "purity of purpose" in an "age accused . . . of selfishness and too great regard for material interests," a case where "a great nation . . . vindicated the higher principles of humanity." With this he confused later research, which has even read in the Abyssinia campaign an early case of "humanitarian intervention."[29]

## PRACTICAL ANTHROPOLOGY

Accepting the above elements has important implications for the way the relationship between anthropology and empire is construed. As indicated in the preface to this book, one view considers colonialism as some kind of economic-political-cultural matter that preceded the development of anthropological science and then points at possible conflicts of interest for late-coming anthropology. These conflicts would have arisen more recently when scientists, anxious for data access, put themselves at the elbow of colonizing authorities. An illustration usually given for this comes from anthropologist Bronislaw Malinowski's 1929 classic article "Practical Anthropology," where he advocated the use of anthropological science in colonial management. In Malinowski's words: "Scientific knowledge on all these [colonial] problems is more and more needed by all practical men in the colonies. This knowledge could be supplied by men trained in anthropological methods and possessing the anthropological outlook, provided that they also acquire a direct interest in the practical applications of their work, and a keener sense of present-day realities."[30]

It is conventionally suggested that any discomfort arising from such a reading would come from the display of science touting politics, begging for attention, prostituting itself. Yet there is a deeper and, I find, more perturbing significance running through this quotation, a significance that is heir to the early years of British anthropology in the 1860s. It becomes apparent when we continue reading beyond the previous quote and reach a development that provides a more menacing variant:

> The essentially unwarranted act of the Canadian Government, who abolished the institution of the Potlatch [a gift-giving festival practiced by indigenous peoples of the Pacific Northwest coast of Canada and that had been outlawed by the authorities of the Dominion] has in every respect completely disorganized the life of the Natives, and it has produced most untoward economic consequences. As we know from all parts of the world, a completely detribalized community, if it is not to die out, is extremely difficult to manage. We have here an example of how an unscientific spirit leads to serious practical errors.

Now the mask of the prostitute drops to show the face of a racketeer. No begging here; more like an indictment. This second strategy for securing the attention of colonial authorities differs from the first in the same way a threat differs from a prayer. By emphasizing the errors associated with the abolition of Potlatch, Malinowski as representative of a scientific community was not begging for positions but rather menacing authorities by subjecting them to a negative review.

I am arguing that Malinowski's diatribe against bad colonial policy was nothing but the replica of the ruse anthropologists had used against Russell and Layard at a time when the institutions of empire were still being shaped. In fact—and this may be uncomfortable to admit for holders of conventional

views on Victorian anthropology—when discussing the subject of the political uses of anthropology, Hunt discovered and explored the same path
traveled by Malinowski about sixty years later.[31] Malinowski organized his
thoughts in an article he called "Practical Anthropology." Hunt, pioneering
the technique when he visited the territory, used almost the same words
in the title of his article—"Anthropology a Practical Science," published
in the first issue of the *Popular Magazine of Anthropology*. The article was
written in the midst of the Eyre and Beke controversies, and the extent to
which the very language used by Hunt anticipated Malinowski is striking:

> Had anthropology been studied with that ardour now shown by the so-
> cieties in Paris, London and Spain, some half a century ago, we should
> probably not have witnessed and deplored the horrors of an Indian mu-
> tiny—a New Zealand war—a Jamaican insurrection. A better knowledge
> of Anthropology might have prevented such a juncture as that presented
> by the spectacle of the semi-civilized Abyssinian monarch paralyzing
> the energies of the British Empire and retaining in his ruthless and despi-
> cable grasp such men as Captain Cameron and his companions. Again,
> we hear from Queensland and other colonies of the extermination of ab-
> original populations by the hands of the settlers, ignorant of the natural
> causes of the passing away of savage races. . . . Anthropological sciences,
> like all sciences is passionless on the point, but the better knowledge of
> its deductions and principles would have instilled some feeling of pru-
> dence and pity into the murderers, who seem to revel in the unnatural
> process of extinction.[32]

In this respect, anthropologists were just like ethnologists, and in fact,
this was one of the subjects on which they fully concurred. They may have
had different tastes, but anthropologists and ethnologists saw themselves as
holding the key to a knowledge that was to be used to certify the quality of
colonial policy, and their politics in this respect were similar. Among many
other cases, Sir B. C. Brodie's Ethnological Society's presidential address of
1853 had vigorously discussed and resolved the question of ethnology's uses
and found it to be of great practical importance especially, he added, "in this
country":

> In this utilitarian age there are . . . some who regard Ethnology as offer-
> ing matter for curious speculation but as being in no degree worthy of
> a place among those sciences which admit direct and practical applica-
> tion to the wants of society and ordinary business life. . . . Setting aside,

however, these considerations and admitting that it affords us no assistance in the construction of steam-engines or railways; that it is of no
direct use in agriculture or manufactures; still it may be truly said, that,
even according to his own estimate of things, the most thorough utilitarian who looks beyond the present moment will find there is no science more worthy of cultivation than Ethnology. Is there anything more
important than the duties of a statesman? And can there be any more
mischievous error than that of applying to one variety of the human species a mode of government which is fitted only for another? . . . Surely
much advantage would arise and many mistakes might be avoided, if
those who have the superintendence and direction of the numerous colonies and dependencies of the British crown would condescend to qualify
themselves for the task which they have undertaken by studying the
peculiarities of these various races, and by seeking that information on
these subjects which Ethnology affords.[33]

Likewise, in the last public fight between Huxley and the anthropologists before the amalgamation (on the subject of "Celts and Teutons"), both
parties acknowledged that the "science that deals with the natural history of
man" has had a good deal to do with "practical politics" and the expressions
*political ethnology, anthropology and politics,* or *race in politics* were used
as captions for the various articles that the controversy stimulated.[34] The
strategy of practical anthropology was always to invest a site with ownership
techniques—the scientific discourse—so as to deprive political authorities
of some of their former prerogatives in matters within, or adjacent to, this
site. The technique recalled Trollope's Mr. Alf, the journal editor in *The Way
We Live Now* who had discovered that "eulogy is invariably dull" and then
exploited his discovery. Likewise, the emerging science sought to extract
a surplus from owning the capacity to either compliment or defame. Such
were the processes that accompanied the rise of ethnology in the 1850s and
that Hunt sought to appropriate and perfect when he launched anthropology
with Burton in the 1860s—to be heard by policy makers or subject them to
criticism.

In so doing, Victorian anthropologists were trying to appropriate colonial policy for themselves, and naturally they called into question the
understanding that existed between British policy makers and incumbent
societies such as the Ethnological. Being the first to have used the technique, indeed as a manner to foster their anti-slavery agenda, ethnologists
had already managed to develop links with the executive, as attested by
the presence of Austen Henry Layard as Russell's undersecretary for foreign

affairs or the use of Prichard's manual for ethnological data collection by the British Navy. The anti-establishment quality of the Anthropological Society now finds its social demographic interpretation. The newcomer was not different in its political aspirations, only bound to be more radical than the incumbent, play on existing rifts, and make more noise as it did.

This conclusion may be used to shed some light on the complicated relationships both societies had with religion and religious leaders. As a result of the abolition of slavery in the 1830s and its subsequent implementation as a political program, religious leaders had found themselves, through missionaries, in charge of millions of souls. Technically speaking, the church had claimed both former and current slaves as part of its flock. In managing them, religious institutions ended up associated with the ethnologists. And the link was kept because the religious needed science to tell that they were doing the right thing and science needed religion to have subjects: the link was ecological.

Now, with the expansion of global trade, a new group of travelers, exemplified by Burton, began writing on those subjects and reached new conclusions. The result was a competition between missionaries who turned natives into Christians and anthropologists who turned them into knowledge. The shift became especially perceptible during the 1860s, and with the Jamaican Rebellion and the Abyssinia campaign, a new discourse emerged, one where issues of slavery were reworked as issues of race. This reworking was foundational for anthropology, at least, and this may explain why, following the much-publicized Reade paper against missionaries, Hunt was very willing to lend the *Journal's* space to ample discussion and further articles that *attacked* Reade's paper. He included objections from the social reformer and humanitarian John Colenso, lord bishop of Natal. The point that was under discussion in the ensuing controversy was not at all the Africans, although a lot of saliva and ink was indeed wasted discussing their character. The native was a pretext. Rather, Hunt was attempting a takeover by anthropology of what was called missionary work, of which anthropology would become, through the operation of speech proffered at St. Martin's Place, a referee and a stakeholder—such a powerful referee and stakeholder, in fact, that it would be in position to go as far as declaring missionary work useful or useless. The extent to which anthropology thus *needed* religion cannot be too strongly emphasized.

The linguistic preparations needed for this operation were made in the open and relied on translating common language into scientific discussion. Missionary work was renamed for the occasion a social science, more precisely a special branch of social science, itself a sub-field of anthropology, that was in Burton's words "popularly known as missionary enterprise."[35] The African missions controversy occupied about a fifth to a quarter of the thick 1865 volume of the society's *Journal*, with statements favorable to the missions hardly absent. After this, anthropology would claim to own the subject of missionary enterprise. By opening their pages to discussion of race, anthropologists could acquire a form of property rights over missions and missionaries.[36] In a sense, the anthropologists were getting back to the abolitionists' famous motto, "Am I not a man and a brother?" by asking, "Very well, but then what is a man?"[37]

Whether through a moral or legal doctrine that permitted man's ownership of man or through the technologies of religion or through those of science that embed property rights in rituals and institutions, the constant element in all these processes was the construction of a form of ownership. Science was just the latest, trendiest form. The best example is perhaps the last words of Pim's speech on "Negro and Jamaica," which have escaped previous commentary, concealed as they were behind the loud cheers that greeted Pim. The cheers were those of racist men, but they principally cheered at the sound of the resolutely modern claim, which they all shared, that anthropology rather than the Foreign Office was the legitimate place to discuss empire. So spoke Bedford Pim: "Who then, will discover the true art of governing alien races? I answer, the statesman, who makes anthropological science his study, and the basis of his efforts for improving the condition of mankind."[38] With this statement about statesmen who make anthropological science their bedside reading, Pim underscored a subsequently neglected aspect of the Anthropological project—the political instrumentality of anthropological science and its centrality in the debate over empire that developed precisely at this point. By promoting science, anthropologists wanted to acquire a form of leverage or ownership of imperial and colonial policies. In fact, once we master the self-righteous emotions provoked by the offending language in "Negro and Jamaica," the whole purpose of the lecture becomes evident. It was an attempt at governing public policy, a raid by a new lobby against the incumbent administration and a move undoubtedly inspired by political concerns.

## PIM AND DISRAELI

To complete the demonstration, however, it is useful to show explicitly that a connection other than logical, opportunistic, or ideological existed between the Tory leader and the anthropologists. One piece of evidence supporting this has already been mentioned when we discussed the sociology of the Cannibals in comparison with that of the ethnologists. Van Keuren's partial survey of the political preferences of members of the Anthropological Society's Council underscored their tight connections with Conservatives, while the Ethnological Council comprised an overwhelming majority of Liberals. My finding that the split in learned society membership operated along similar clubbish sociabilities (to repeat, the Ethnologists' predominant connection with the then-predominantly Liberal Athenaeum Club and the Anthropologists' links with the Conservative and other clubs) also suggests that what Van Keuren identified at the level of the councils of both societies was true for regular members too.

But these macro-trends have a micro-history counterpart in the shape of suggestive traces of Disraeli's fingerprints at the intersection between science and finance. Fingerprints are probably all we can hope to find. Before they were made accessible, Disraeli's papers, kept at the Bodleian at Oxford, had been subjected to the unsparing care of Disraeli's lawyer, Sir Philip Rose, who undertook, the research aid explains, to "examine and arrange them."[39] Rose, a lawyer, started his political career by being a fixer for the Conservative Party. He was its electoral agent. Given how electoral seats were won, this meant Rose was responsible for overseeing the purchase of electoral results. In 1859 Rose faced accusations of corruption and was forced to withdraw formally from his role as agent for the Conservative Party only to be replaced by Markham Spofforth, his assistant at the law firm Baxter, Rose, Norton and Co., so that the mill could be kept running. As the inimitable *Dictionary of National Biography* deplores, political agents working for parties were exposed "to the corrupt side of electioneering."[40]

Rose's increasingly close connection with Disraeli (he was immortalized as "Lord Beaconsfield Friend" by a famous caricature in *Vanity Fair* in 1881) is significant for the story here because Rose's law firm was involved in drafting the contractual agreements necessary for launching on the stock exchange many of the international companies and foreign government debt securities of the foreign capital export boom that accelerated after 1866. Although the matter was never raised in the select committee, those law firms advising contractors must have been privy to the machinations the committee set to examine. In the infamous case of Honduras, the name of Rose's law firm comes up on the contractual documents the select committee put together.

On the other hand, the links between Disraeli and the foreign capital market, while never systematically explored for lack of sources, are notorious. More than one thread starting in the stock exchange ends up with Disraeli, even when, or precisely when, his correspondence is mute on the subject.[41] I have already mentioned Disraeli's early career as a puffer of Latin American securities employed by John Diston Powles, who was to become the eternal godfather of the foreign debt vulture industry and whose correspondence with Disraeli in the prime minister's papers stops inexplicably in the mid-1850s when Powles's power became enormous. Yet after this period, as noted by historian Robert Blake, Disraeli still boasted that he had "all America and the Commercial Interest at his back," something that might have happened through his association with Powles.[42] According to historian Michael P. Costeloe, Disraeli had a long legacy of encounters with the South American market, both trading (on inside information?) and investing.[43] During the 1860s this link materialized in the alliance between Disraeli and Barings, the incontrovertible authority of the Latin American market in London. This was the time when Thomas Baring, the family politician and a Disraeli confidant, was acting as chairman of the City Conservative Committee, and together they planned a political siege on the City, considered to be an unassailable Liberal stronghold, with the help of their supporters in the foreign debt market.[44]

The death in September 1867 of Powles, the godfather of the vultures, opened a crisis of succession in the business of bankrupting countries. The scheme promoted by Isidor Gerstenberg was to try and corner the industry through the creation of the Council of Foreign Bondholders, which was launched two weeks before the Liberal return to power in 1868. Almost simultaneously, and a few months before the *Athenaeum* controversy in March 1868, Philip Rose, who "served as a director on several public companies," launched the Foreign and Colonial Government Trust, a very innovative piece of legal-financial engineering.[45] As its name suggests, the new trust specialized in foreign government bonds. Thus while the substance of the communications that must have taken place between Disraeli, Rose, and people in the foreign stock exchange is lost, profiling of the intermediaries who took care of managing the memory gives away most of what is needed. Rose had been in charge in these operations, and their conspicuous absence from Disraeli's papers is in fact additional evidence of their materiality.[46]

Likewise, the connection between Disraeli's eventual triumph in 1874 and the stock exchange is evident. In fact, the return to Parliament of a new generation of Tory financiers at this time was sufficiently striking to Trollope for his decision to have his stock exchange swindler Melmotte become a City M.P. Melmotte is believed to be patterned after Baron Albert Grant, who had been elected as Conservative member for Kidderminster in 1874. But there were many other models to choose from, such as

Charles E. Lewis, who joined Parliament as honorable member for London-derry in 1872 and was another pillar of the Conservative group in the City. Lewis was involved in many bondholder groups, had joined the commit-tee of Rose's Foreign and Colonial Government Trust in 1871, and in 1875 introduced an important bipartisan proposal to reform the bondholding industry.[47]

Yet another characteristic man, one who needs no introduction, was Bedford Pim, who joined Parliament as Conservative M.P. for Gravesend in the general election of 1874. On him, I discovered in the Disraeli papers a golden nugget. The Pim folder in the Disraeli papers is thin, but it does inform on an exchange that started in the last days of 1878 between Dis-raeli and Pim, the latter trying to secure from the former a promotion to the rank of rear admiral or, failing this, the highly coveted honorific title of Knight Commander of the Order of the Bath—the prestigious "K.C.B." doorplate. Initially, Pim complained that, had it not been for "political ani-mus" against him from the Admiralty, he ought to have been promoted to rear admiral in March of that year, and "Your Lordship needs no reminder of my political service for the past fifteen years."[48] But His Lordship did need a reminder. Disraeli had received a memo from the Admiralty stating Pim's naval colleagues were skeptical that "his services at sea have been of such an exceptional character as to call for promotion over the heads of other officers," just as they were skeptical that the K.C.B. alternative made any more sense, given the "circumstances of his services."[49]

Hard-pressed by Liberals who never missed an opportunity to denounce what they described as Disraeli's "immoral administration," Disraeli's an-swer to Pim was thus polite but negative, its substance conveyed to Pim by the prime minister's private secretary, Algernon Turner. This led an enraged Pim to spell the matter out in a subsequent letter: "I cannot concieve [sic] it possible that Your Lordship has yourself read my letter . . . , the tone of the reply being so contrary to my preconcieved [at least his spelling is con-sistent] notion of Your Lordship's leading characteristic, especially towards one whose allegiance to you had been so ungrudgingly given as mine, and where services to you and the Party have been second to none."[50]

What were the "services" Pim alluded to? The "past fifteen years" re-ferred to put us back around 1863–64, at precisely the time when Pim be-gan his association with the Anthropological Society. In 1864 Pim and his business associates launched the Chontales Company, and the board of di-rectors included Tory grandees such as Horatio Nelson, Third Earl Nelson and supporter of the anti-Corn Law movement, as well as Conservative M.P. Cavendish-Bentinck (also a George Augustus Frederick). The first was

most likely a remunerated decoy director. The second was a specialist bondholder, later active in foreign debt instruments in which Philip Rose was involved, such as the Foreign and Colonial Investment Trust and the Council
of Foreign Bondholders. The next year, in a by-election, Pim ran with one
Colonel Dawkins to represent Totnes, Devon. Pim was later reported to
have attacked with "gallantry" the "pocket-borough" of Edward Seymour,
the Liberal first lord of the Admiralty. The expression *pocket borough* referred to an electoral district where open and free competition was impossible because a dominant local landowner could evict electors who did not
vote for the man he wanted. We can only imagine the pleasure Pim must
have taken in challenging the first lord of the Admiralty in an election. As
for "gallantry" a subsequent parliamentary report investigated bribery and
corruption at what turned out to be an infamous election. It established
that Pim had gotten involved at the prompting of Markham Spofforth from
Baxter, Rose, Norton and Co. who assisted Philip Rose on political matters
and had become Rose's successor as chairman of the conservative central
committee, after a scandal erupted in 1859 (but Rose was still informally
in charge). The report established that Pim and Dawkins had put together
about £4,500 in order to wage a bribery contest against the Liberal party,
who did not remain "backward." The resulting display of corruption was
so egregious that Totnes would be one of the first two boroughs disen-
franchised by the Reform Act of 1867 while Spofforth's conduct would be
blamed. One criticism was that he had brought in a candidate of "the commercial spirit."[51]

Not coincidentally, Pim's subsequent elevation to the rank of captain
of the Navy in April 1868 (when he had been compulsorily retired in 1861)
occurred under a Tory administration and looks very much like a political reward, giving yet another perspective to the "feelings of warm admiration" Pim had expressed in 1869 toward Lord Stanley, Disraeli's foreign
affairs secretary and permanent member of the Anthropological Society.[52]
And if Disraeli turned down Pim's request for further promotion in 1878,
it fell upon a Conservative administration to grant him the privilege he so
badly wanted. In 1885, under Prime Minister Salisbury, Pim the rogue commander of the *Gorgon* was finally made rear admiral.[53]

This provides more than hypotheses as to the content of those services
for which Pim would later consider himself insufficiently rewarded, and
they evidently provide a background to the "Negro and Jamaica" episode.
This outlines the conclusion that at the time when the famous controversy
occurred, there was a three-tiered deal to be cut between those overlapping
groups—the foreign debt specialists in the stock exchange, the scientists

who roved the outer boundaries of the British Empire, and the executive. In summary, what was special about the mid-1860s was the organization of a powerful scientific interest, which Disraeli figured out. His romantic talent for the grandiose provided a gift-wrap. Sir Philip Rose truncated the relevant archive. This, plus the human penchant for confessing the wrong kind of sin, delayed the examination of facts, and we must be grateful to Pim for having left, intertwined with his scientific path, a money trail.

This chapter has provided a new perspective on the political significance of the rise of anthropology in the 1860s by painting a picture that is at odds not just with the conventional perspective that thinks in terms of the encounters social scientists had with established colonial powers but also with the view of a colonial modality where scientists happened to come across a colonial reality already formed. Instead, the Beke and Eyre cases show that colonial policy was shaped very early on by learned societies such as the Geographical, Ethnological, and Anthropological. The violence science contributed to the process is attested by the toppling of John Russell and the poor memories historians have kept of "Johnny."

Focusing on the Abyssinian hostage crisis, I have argued that the rise of the Anthropological Society was one episode in the showdown that took place during the 1860s between the agents of British imperialism and the policy makers in London. The rise of the Anthropological Society was one of the many processes through which this modification of the balance of power happened, and it provided its members—agents of empire—with what political scientists call a collective action institution. The Anthropological Society had been conceived by an academic entrepreneur and a British consul. Its international organization, with local secretaries, corresponding members, and fellows residing abroad, paralleled the networks of informal empire and opened an outright competition with the information system of imperial Britain. We are dealing here with a pattern—Cameron's combination of aloofness, ignorance, or deliberate misinterpretation of the orders from the top, interaction with a local ruler, sudden passion for a lost tribe, membership in the Ethnological and Anthropological societies, and eventual ability to reinvent himself as a hero, with the help of Disraeli, evokes a striking parallel with Pim's operations in Nicaragua.

This finding also encourages revisiting the controversy over Governor Edward Eyre of Jamaica and his actions during the Jamaica Rebellion. An element of the dispute, which has been obfuscated by discussion of the civil rights of George William Gordon, the wealthy "mulatto" politician murdered by Eyre, is that Eyre typified the explorer-scholar, and his work on Australian aborigines commanded authority as a treatise in the management of what would later be called subject races. In his book, the later murderer presents us once again with this characteristic Janus-like nature observed in other anthropologists. He deplored that the "progress and prosperity of one race should conduce to the downfall and decay of another" and that "it is still more so [deplorable] to observe the apathy and indifference with which this result is contemplated by mankind in general."[54] Yet Eyre's membership in the Anthropological Society is at best mentioned as an almost curious coincidence aggravating the Cannibals' scorecard when this membership was in fact the heart of the whole matter. The Jamaica Rebellion was also a rebellion of organized agents of empire against the British executive.

The parallel between Cameron and Eyre makes this point in the clearest light. The essences of the two episodes are similar—in both cases, science and imperial governance. This parallel has been concealed by circumstances that caused the subsequent dynamics of the two episodes to diverge. Eyre's response to the Jamaica Rebellion evidently raised the problem of violence and mistreatment of black people and thus conjured up the old humanitarian coalition, which sought to exploit the episode politically. By contrast, the Abyssinia episode did not produce the same political fault lines and thus generated another form of opportunism and exploitation. Tewodros was a Christian ruler, and his behavior was perfect dinnertime evidence of the need for Britain to civilize (and thus the label "humanitarian" that was affixed to the expedition). If one wanted to be a bit more extreme, one could rant at the inability of centuries of Christian faith—let alone recent conversion—to civilize, thus making the use of some degree of force not only inescapable but legitimate.

Indeed, I am struck by the fact, neglected in previous research, that the gentle Ethnologists never organized a campaign to prevent the killing of Tewodros and his army nor to prevent the destruction of the precarious political order he had introduced in his kingdom. For political entrepreneurs of a new Conservative sensitivity, the beauty of the Abyssinia affair is that it opened the possibility that the former owners of the imperial question— progressive humanitarians and clergymen of all stripes, as well as their Liberal patrons—could be dispossessed to the benefit of scientists. In fact, the offensive had come from the scientists themselves. Through the actions of scientists, Russell had been construed as presiding over the actions of an incompetent, timid, and ill-informed executive. This opened a major controversy on efficiency, stamina, and information in imperial management. Russell was identified with the traditional policies of informal empire, which strenuously abstained from commitment.

Such was the situation Disraeli inherited when he came to power in February 1868, replacing the Earl of Derby at a time when the engine of the Abyssinia campaign was already running. Perhaps nobody better than Disraeli could know about the force of science. After all, the first major experience of his political career had been the implosion of Peel's Conservative government on the subject of the repeal of the protectionist Corn Laws in 1846, an event that led to a major reconfiguration of British politics. In this instance, another theoretical idea borrowed from another group of social scientists—the economist's laissez-faire doctrine—had been used as policy motivation and resulted in enormous political shifts. In fact, there is a striking parallel between the way economics had paved the way for free trade in the 1840s and the way anthropology opened the road for empire in the 1860s. Indeed, in one of his letters to Russell, Beke ominously compared himself to Richard Cobden, the man whose Anti-Corn Law League had played a decisive role in the repeal of the Corn Laws; a man, Beke explained, who did not have any formal political function. Ironically, when Cobden's action had eventually caused the fall of Peel's cabinet and put an end to his career, leading to his replacement in 1846 by John Russell, the Abyssinia crisis would do just the same to Russell.

It came to a politically fragile Disraeli to see the value of throwing open the cage where the Cannibals were kept and sending them against the shortcomings of earlier imperial management. The Cannibals were grateful and delighted at the magnificent dawn of a new age, and they roared with gratitude. These were the new forces within British society, which Freda Harcourt may have referred to when she suggested that Disraeli really redesigned imperialism between 1866 and 1868. Can it be a simple coincidence that this happened at the same time the Anthropological bubble peaked? These were the new forces within British society, which Disraeli, in his speech introducing the Reform Bill in 1867, described in his characteristic style as "numbers, thoughts, and feelings." Such new feelings had been expanded since the Reform Bill in 1832, and it was desirable to "admit [them] within the circle of the constitution." The franchise had to be expanded, he explained, because of the "spread of knowledge."[55] These forces were not a new mindset or disposition, but a power and a precise technology. They were illustrated by the rise in British civil society of a new type of outward-oriented learned society, itself driven by economic motives. They consisted of the increasingly voracious appetite of increasingly assertive scientists. These triumphs, of Beke over John Russell and of the rebellious learned society over the Foreign Office, were a figment of the rise of science in an age of mercantile expansion. I conclude that the incorrect portrayal of the

Anthropological Society as a radical minority movement and the lazy portrayal of the conflict between ethnologists as the tender Darwinians and anthropologists as the despicable Cannibals obfuscates the more central and embarrassing scientific origins of empire. Despite the editing by Sir Philip Rose, evidence for this is still found in the Disraeli papers, archived at the time of my visit within a stone's throw of the Pitt Rivers Museum where the nauseating plunder from early imperialism is still proudly displayed.[56]

## CHAPTER TEN

# The Man Who Ate the Cannibals

In this book, we have repeatedly come across Hyde Clarke the Ubiquitous—and the man of so many doorplates. As Blanchard Jerrold explained, the manufacturing of public opinion was about when and where a given plate should be shown. Depending on circumstances and who would ask about his plate, Hyde Clarke's answers could vary. To the clerk sent to his home by the 1881 census, Hyde Clarke declared himself secretary of the Corporation of Foreign Bondholders, and the clerk, unfamiliar with this mysterious organization, wrote down "Secretary of a Corporation."[1] But when Hyde Clarke wrote to the *Athenaeum*, a weekly journal for those who followed the latest progress in the sciences and arts,[2] and complained about the management of the Anthropological Society in a letter printed on page 210 of the August 15, 1868, issue of this journal, he described himself as a "fellow" and spoke about "our Society."[3]

As readers could see, the letter to the *Athenaeum* was addressed to "James Hunt and to the Council of the Anthropological Society." It opened with an allusion to the inefficiency of the society, which had dispatched Hyde Clarke's copy of the *Review* "after a very long delay, explained as arising from some mistake in the office," then complained about "lampoons" on prominent individuals and the Ethnological Society, which could be read on pages "323, 324 and 327," of the society's periodical. The letter then moved on with a diatribe against Hunt and the Council, denouncing the ownership of the *Review* by some "unknown individual or individuals." The letter, which alleged that clarifications had been asked before, ended with threats that, if the Council needed to get its hands "strengthened," then it might be done "by giving publicity to these observations, in the hope of reaching the body of our fellows."[4]

This is how the *Athenaeum* controversy started. Hyde Clarke's letter triggered responses by prominent Cannibals, followed by responses to the responses, then rejoinders by yet other members of the Anthropological Society until December 19, 1868. Counting letters published, letters embodied in letters, and so forth there were about twenty such documents in the *Athenaeum* involving about five different correspondents. This does not count the echoes in other newspapers—the *Pall Mall Gazette* in particular.[5]

According to its main scholar, the *Athenaeum* was, by the time the controversy erupted, a well-established newspaper that had contributed earlier to the reputations of Charles Dickens, Thomas Carlyle, John Ruskin, and Alfred Tennyson, men whom it had alternately praised or criticized. But it had always been partial to John Stuart Mill, who was the only writer to "escape censure in that journal, either early or late, for wasting powers undeniably of the first order." Now Mill had recently been, with Huxley, a leading figure in the high-profile anti-Eyre group, which lobbied in the aftermath of the Jamaica Rebellion for trying Governor Eyre, while Dickens, Carlyle, Ruskin, and Tennyson were all pro-Eyre partisans. It is probably no coincidence, therefore, that the columns of such a newspaper should have been used by Hyde Clarke to assail the credit of one society that had itself gotten closely involved in the Eyre controversy on a side opposite to Mill and the Ethnological Society.

In the world of performance we have described, challenging the credit of the Anthropological Society had value in itself. The insult was intended to have an effect regardless of its veracity, the art being to create a context of plausibility, which appearance the *Athenaeum*, with its reputation for independence, gave. This was fully perceived by the leaders of the Anthropological Society, who went on repeating that Hyde Clarke's behavior was "calculated to injure this Society." The *Athenaeum* reached to the best possible readership, and its tradition of taking something from both sides in major disputes as well as its ostensible policy of fighting superficiality and one-sidedness ensured leveraging whatever criticism would be published there.[6]

The reasons for the outbreak of an open conflict in the anthropological-ethnological cold war have neither been understood nor really explored, with the bulk of previous writers making at best half-hearted efforts to ascertain the extent to which Hyde Clarke's accusations were true or not and generally finding they were, as Stocking did. To understand what happened, it is necessary both to look at the substance of the controversy between ethnologists and anthropologists and to examine closely the methods employed. I approach this through the lenses of the capital market and its techniques, a relevant perspective because the accusations Hyde Clarke leveled against the Anthropological Society were an import from the world of financial speculation and used the language of finance. In that language, a speculator who sells a security of a company or government forward is called a bear and the operation a "short." The surest way to make a bear speculation profitable is to spread bad news just after the security is shorted. The price of the security collapses and the bear speculator can pocket a profit on the day of the delivery. In a like fashion, Hyde Clarke was "bearing" the stock of the Anthropological Society. By observing Hyde Clarke more carefully than has been done before, by measuring him against the right linguistic background, we will indeed discover the same talents at work, not only in the destruction of the Anthropological Society but in the simultaneous blasting of a company in the Levant and, some years later, the bursting of a foreign government debt financial bubble in which Bedford Pim was involved. I am not aware that anyone has noticed this most troubling fact, the substance of this chapter.

## THE UBIQUITY OF HYDE CLARKE

In its obituary for Hyde Clarke, the magazine *Engineer* remarked that of him it "may be well wondered 'How one small head could carry all he knew.'" It also reported the claim (not infrequent about British scientists of the time) that he was "familiar with a hundred languages." But the journal could not say "whether this is true or not, although the statement could be safely accepted in a figurative sense."[7] Just like we found out for Pim, there is a mystery to Hyde Clarke. But unlike Pim, Hyde Clarke's traces are well hidden, making him a more difficult character to pin down—but let's try to follow him.

During the 1830s, this engineer had been a contributor to *Herapath's Railway Magazine*, a leading trade journal. If we go by the afterlife Hyde Clarke now lives among economists, we find him taken seriously as a pioneer of business cycle theory. In the aftermath of the sub-prime crisis, there has been a Hyde Clarke revival in the blogosphere.[8] Allusion is often made to a paper in the *Railway Register* called "Preliminary Inquiry into the Physical Laws Governing the Periods of Famines and Panics," which itself followed conjecture offered in 1838 in an article published in *Herapath* called "On the Mathematical Law of the Cycle." These papers are hailed as pioneering texts in cycle theory. Their read has an intriguing Kabbalistic feel, which may be traced to Hyde Clarke's background as a Freemason. In obscure language, Hyde Clarke claimed to have observed that business cycles were subjected to mathematical laws, although the method used to organize his long-run observations is difficult to figure out. In his "Mathematical Law" for instance, he claims that his interest in cycles had "begun in 1832 and 1833, when [he] saw the dawn of a period of these speculations, which from a close acquaintance of the history of the mania of 1823, 1824, and 1825, [he] was induced to look upon as a recurrence of the same phenomena" although in the two different articles, different wavelengths were suggested (10 years in his early paper but 54 in his later one). "Mathematical Law" also insisted that contributing to a better "knowledge of the functions of the Creator" would enable us to demonstrate the "dependence" of the world "on superior intelligence."[9]

The legacy of these curious papers was ensured by economist Stanley Jevons, who looked into the subject of the relationship between sunspots and business cycles and liked to quote Hyde Clarke, whose ideas as a result found their way into the works of the historian of economic analysis Joseph A. Schumpeter, likewise a great lover of business cycle theory. That in turn ensured the production of Hyde Clarke as a founder of modern theory. Upon inspection, however, Hyde Clarke's "Preliminary Inquiry" reads more like an anticipation of modern "technical trading," an approach previously taken by a number of traders and shunned by economists. It sought to uncover invisible regularities in price movements and purports to inform trading strategies. Its modern avatar is perhaps the algorithms with little counterpart in economic theory, which computer-assisted trading borrows from rocket science.[10]

Hyde Clarke's *Railway Register* was a puffing newspaper. It belonged to the crop of journals created to encourage investors to subscribe to railway shares, and indeed, Hyde Clarke's early works complimented the shareholders of the new railway companies for holding securities that would "turn out very lucrative investments."[11] The bubble peaked in 1845, and railway shares began an unstoppable decline. When this great truth of cycles was revealed to Hyde Clarke, his opinion swung to the other side. He better understood the deep structure of the business cycle, and this led him to much circumspection. The 54-year cycle he felt he had now discovered enabled him to claim in 1847 that the trough of the depression was within reach, "but it was hard to tell whether it would come in 1848, 1849, or 1850."[12] The *Railway Register* was put to rest that same year.

An article by Andrew Odlyzko describes Hyde Clarke as a producer of some early statistics intended to monitor declining railway revenues at the time of the collapse of the railway bubble. This locates him within the statistical machinery, which the "bearing" of the railway mania put in motion and which is said to have provided the hotbed for the birth of modern accounting.[13] The link went from engineering to accountancy, and Hyde Clarke indeed fits the bill of an engineer turned accountant. But his monitoring of railways extended beyond simply tracking numbers. According to railway mania historian John Palmer, a student of the role of the media in the railway bubble, a contemporary edition of the *Railway Directory* in Oxford's Bodleian Library exhibits Hyde Clarke's handwritten annotations. The notes are scribbled on the margins of lists of railway directors. This document has the feel of the list of Ecuador bondholders' representatives found in the Church collection.[14] What were Hyde Clarke's motivations when he tracked down directors?

In 1849, Hyde Clarke was next seen in India on an assignment to report on the Indian telegraph system, and like several other individuals who had been involved in puffing British railways in the 1840s, he started dabbling in Indian railways and communication infrastructures during the next decade, a large part of which he spent in India. He became secretary of the Northern Bengal Railway Company, which had £2 million worth of capital and whose chairman was Sir Rowland Macdonald Stephenson, an early promoter of railways. Clarke was later involved with Indian affairs as honorary agent for the Darjeeling settlement in connection with projects for extending the settlements in Nepal. Historian Dane Kennedy describes Hyde Clarke at this point as an "agent for Indian railway and planter interests."[15]

In 1857, just before the Sepoy Mutiny, Hyde Clarke published a promotional pamphlet, *Colonization, Defence and Railways in Our Indian Empire*, dedicated to the chairman, deputy chairman, and committee of the East India Company. The pamphlet was intended to persuade readers of the benefits brought to the highlands by European colonization—benefits made possible, he emphasized, by the availability of such lines as the Northern Bengal Railway Company. European colonization in this area by a class of "superior settlers"—soldiers, planters, and pensioners—who would oversee Indian laborers would supply a "copious reserve" prepared to pounce on insurgents in the plains at the first sign of trouble.[16] With the outbreak of the Sepoy Mutiny, Hyde Clarke capitalized on the increased concern with security, publishing an article "On the Organization of the Army of India with Especial Reference to the Hill Regions."[17] The sales pitch echoed traditional concerns such as we have already come across, for instance when Hyde Clarke recommended "the establishment of a large land revenue, available for internal improvements, for promoting immigration, and for redeeming the national debt." At the same time, Hyde Clarke was trying to stir the interest of the Council of the Royal Society of the Arts, pushing the idea of the creation of a colonial think tank (or more adequately a "colonial center" in his language) in London, with a special section for India, another for Australia, and yet another for English America. The subjects suggested were "railway extensions, irrigations, canals, European colonization, tea cultivation, fiber products, the iron manufactures and the copper mines."[18] The parallel with infrastructure and settlement projects promoted by other anthropologists or ethnologists such as Church, Haslewood, Gerstenberg or Pim is striking.[19]

## SMYRNA, RAILWAYS AND FREEMASONRY

Railways, again, were to provide the stuff of Hyde Clarke's next assignment. This takes us to his role in the infamous Smyrna-Aidin Imperial Ottoman Railway Company created in 1856. It has not been the subject of any significant monograph, although like so many other cases explored in this book, its story merits careful consideration. Scattered material nonetheless helps reveal a generally unknown part of Hyde Clarke's career, situated after his Indian railway promotions and before his reinvention as the many-doorplated secretary of the Corporation of Foreign Bondholders. This episode also provides a perspective on the way in which he later joined both the Anthropological and the Ethnological societies.[20]

Hyde Clarke was involved in this company as its secretary and representative in Smyrna in the early 1860s in the ambience of frantic speculation in the Middle East that followed the Crimean War, described so aptly by David Landes in *Bankers and Pashas*.[21] Hyde Clarke had been deeply involved in what emerged later to have been disturbing aspects of the early management of the project. The story of the Smyrna-Aidin Imperial Ottoman Railway is a long catalogue of mismanagement and abuse of outside investors by directors further confused by higher and lower politics. This much was established by a report made in 1867 under pressure from shareholders and the Board of Trade. Just as had been the case for the navigation and railway companies created by Church, the project was described in the report as "under-capitalized," and exactly as was the case for Church, a contractor-insider was lurking in the shadows. In convoluted language, and an evident reluctance to say it in so many words, the report of 1867 revealed how the directors of the Smyrna-Aidin line had abused both shareholders who were "kept in ignorance" and the holders of debentures issued in 1861.

Clarke's early role had been to put his writing talent at the service of the directors. One episode in which Clarke had been involved was the 1861 debentures issued when the capital of the company had been depleted without much railway track having been laid. In support of this new call for resources, Hyde Clarke published a pamphlet that ended with the usual stanza of the Great Railway Epic. The refrain included the Ottoman Empire, which, in failing to complete the project, had nonetheless "laid the foundation of a railway system by inaugurating two great works, one restoring the project of Trajan, and connecting the Danube and the Black Sea, and one opening the gates of commerce for the wealth of Anatolia to flow more prosperously through the port of Smyrna."[22] The difficulties of the Smyrna-Aidin line, he concluded, were not unheard of in railway work since "even the London Metropolitan Board of Work," the most trustworthy of all utility projects, had initially found great difficulties in getting its act together. This is the same man who would become the great slayer of Anthropological puffery. In the jargon of the time, the technical name for such a railway line as the Smyrna-Aidin was a "contractor's line," one promoted by producers of equipment who sought to sell the infrastructure rather than complete the line, letting ordinary investors and creditors foot the bill.[23]

It is also from this period that Hyde Clarke's involvement in Freemasonry can be dated, in association with the Scottish Rite. The role of the Freemason lodges and their rivalries in informal empire have been discussed by other authors in the context of Mexico.[24] The situation was similar in the Ottoman Empire, where higher officials and Europeans participated in Masonic lodges. The growth of Freemasonry in the Ottoman Empire followed the Crimean War, which heralded Western penetration in the area. Freemasonry tended to attract supporters of reform and modernization. Examples include Ismaïl Tewfik, the son of the khedive of Egypt, and Grand Vizier Fuad Pasha, who had played an important role in containing French imperial humanitarianism following the 1860 Maronite slaughters by the Druze minority in Syria. Both the Khedive Ismaïl and Sultan Abdülaziz, although not Masons themselves, endorsed Masonic organizations.[25]

Masonic lodges, or "Orients," as some of them were known, mushroomed throughout the region. Hyde Clarke, during a sojourn in Constantinople in 1860, began his Masonic labors promoting structures under the Scottish Rite, where the central authority is known as a Supreme Council. In 1861, claiming to having the authority of the Supreme Council of France, he constituted at Constantinople, for Turkey and Egypt, a Supreme Council of the Ancient and Accepted Scottish Rite. It is probably around this date that he acquired a command of the Turkish language.

Following this initial impetus however, the Supreme Council remained in a dormant condition since by now its founder was in Smyrna, where he created yet another lodge in 1863 amid his duties as secretary for the Smyrna-Aidin Railway Company and his senior membership of the Im-

HYDE CLARKE,
Formerly resident at Smyrna.

FIGURE 10. Hyde Clarke in the 1860s. *Source*: Robert Morris, *Freemasonry in the Holy Land* (New York: Masonic, 1872), 50.

perial Cotton Commission. In 1864 Hyde Clarke, again invoking the authority of the French Supreme Council of the rite, legalized the Supreme Council for Turkey and Egypt and changed the location, or "Orient," of the dormant Istanbul Supreme Council to Smyrna. In plain English, he had claimed to have received from a Freemason authority in France the right to secure official recognition of the Smyrna lodge. Clarke's claim was challenged in 1868 when it appeared that he had not in fact asked for any authorization, leading to a dispute. Hyde Clarke eventually managed to receive support from American adherents to the Scottish Rite by obtaining recognition from a Charleston-based Supreme Council, which controlled the Southern Jurisdiction of the United States of America. The local grand master, one Albert Pike, found in favor of Hyde Clarke, arguing "Fraud is no more to be presumed than crime and we cannot presume it of so reputable a gentleman, scholar and Mason as Illustrious Brother Hyde Clarke."[26] Upon visiting Smyrna in 1868, Robert Morris, an American Freemason, addressed an assembly of brothers and mentioned the "greatly beloved" Brother Hyde Clarke, who had unfortunately returned to England after "doing a good work for Freemasonry in the Levant." Morris said his own mission to identify Masonic resources in the Holy Land had Hyde Clarke's "valued approbation."[27]

Hyde Clarke's connections with Freemasonry may have played a role in helping him become so quickly an important member of the Anthropological Society. As we have seen, the society's triangular logo affirmed a Masonic origin. In 1877, former Anthropological Society Council member Kenneth R. H. Mackenzie, himself a Freemason, published a *Royal Masonic Cyclopaedia* where the entry "Cannibal Club" stated that "there were many Masons members," and listed Captain R. F. Burton, Dr. James Hunt, Dr. R. S. Charnock, Mr. Algernon C. Swinburne, Mr. J. Frederick Collingwood, the author himself, and "many others."[28]

## THE PROFITS OF PANICS

In the narrative offered by historians of anthropology where all the above is absent, Hyde Clarke the scholar is born in the late 1860s, fully armed, an Ethnological Council member and in the more informed accounts, an amalgamation negotiator and fellow of the Anthropological Society.[29] In fact, Hyde Clarke initially joined the Anthropological Society as local secretary abroad, and thus his anthropological career provides a direct counterpart to Pim, who moved from local secretary in Nicaragua to the Anthropological Society as regular fellow once back in the British capital. Likewise, Hyde Clarke became in 1865 the society's contact in Smyrna, from where he sent archaeological material and photographs of remains to London, while he met with pioneers of architectural and ethnographic photography operating in the Ottoman Empire such as Pascal Sebah (who took a rare picture of Hyde Clarke himself in Constantinople) or Alexander Svoboda. Clarke also dabbled in humanitarian relief when the eastern Mediterranean was hit by a cholera epidemic.[30]

In 1867, having moved back to Britain, he was greeted by the Anthropologicals for his achievements in Asia. The printed minutes of the society meeting of May 14, 1867, record that Hunt required a vote in favor of Hyde Clarke who, "while he remained in Smyrna, had discharged the duties of the office in a manner highly satisfactory to the Council" and begged to propose their best thanks to Hyde Clarke, adding that he hoped Hyde Clarke would continue to "give his valuable aid to the Society." The motion was carried unanimously. Hyde Clarke then answered modestly that he had merely "endeavoured to do his duty" and that "the tribute of thanks they had given him" was "unnecessary."[31] Following this, Hyde Clarke began participating in society meetings, and his scientific opinions are recorded in several instances during the year 1867. As further indication of the excellent relations developing with the society, Hyde Clarke soon became Corresponding Secretary for Asia in London, and "from his well-known and able researches, there was no doubt that the appointment would afford great satisfaction to the Society."[32]

By comparison, Hyde Clarke's mingling with the Ethnological Society took place later. He joined it in the first half of 1867. Less than one year after this, in January 1868, relations between Hyde Clarke and the Anthropological Society soured. The trigger appears to have been critical comments he made during the Anthropological General Assembly, leading the Cannibals, in their next Council meeting on February 4, 1868, to resolve "not to continue Dr. Hyde Clarke in his office of Corresponding Secretary for Asia."[33] Almost simultaneously, Hyde Clarke was promoted through the ranks of the Ethnological Society, joining its Council on May 19, 1868. The coincidence of the downfall in one society and elevation in the other is intriguing. Anyway, Hyde Clarke's senior position within the Ethnological Society was acquired just in time to become a credible negotiator in the abortive talks of amalgamation. He sat there along with blue-blooded ethnologist Balfour and veteran Robert des Ruffières, who had joined the Ethnological Society in 1853. The latter, just like Hyde Clarke, was a significant member of the Anthropological Society as well. According to later charges by the anthropologists, Hyde Clarke displayed rabid aggressiveness, such as when he told Anthropological negotiators that their society was like a "toad under the harrow" and had better accept the offer made to them without bargaining. Among other ungentlemanly behavior, he was reported to have quit after "half an hour" an early session of the discussions in June 1868, stating that he "was compelled to leave to . . . 'reorganize the Statistical Society.' "[34] The verb *reorganize* is striking because it was used at exactly the same time as the reorganization of the Scottish Rite Supreme Council in Constantinople and its transfer to Smyrna were being challenged, not to mention the pending reorganization of the Smyrna-Aidin Railway Company of which Hyde Clarke was also a stakeholder: Clarke, the *reorganizer.*

The letter Hyde Clarke subsequently sent to the *Athenaeum* journal and which started the public controversy over the Anthropological Society, was released after amalgamation talks had failed in June 1868 and just

before the British Association for the Advancement of Science meetings
were about to be held in Norwich.[35] Clarke complained about the "lampoons" Hunt had printed against Sir Roderick Murchison and the Ethnologicals. In truth, the "lampoons" Hyde Clarke complained about were banal
language by the standards of the Anthropological Society. There was the
usual lash against Murchison, who had claimed in a letter to the *Pall Mall
Gazette* that he had been supporting the merger of the two societies but had
been, Hunt suggested, really an opponent of rapprochement. There was also
the normal lash at Huxley, who was to become the new president of the
Ethnological Society, "welcome news" according to Hunt because he would
have to give up "self-imposed duties such as the prosecution of Governor
Eyre." Hunt deplored sarcastically that this would transform the Ethnological Society into a "sort of Darwinian Club." And there was the mandatory
lash at "ladies and other friends of Ethnological science."[36] Such language
abuses were improper, Hyde Clarke suggested, since at the same time "our
Society [meaning the Anthropological Society] . . . is in professed amity
with them and that we lately sought amalgamation with them."[37] They
suggested Hunt's claim that he wanted amalgamation might not have been
*bona fide.* And *this* was contemptible.

   But these objections of form were merely rhetorical preparation for a
frontal attack. First, Hyde Clarke explained that the *Anthropological Review* was not owned by the society itself but by an unnamed party. The society settled the purchase of copies from the unnamed party, and as a result
was in debt with the printer.[38] Second, management's lax enforcement of
payments produced a large number of what Hyde Clarke called "non-paying
fellows," a number possibly larger than the paying ones, which he set at
400, without giving sources. Because of this, he claimed, the new fellows
"canvassed for among general practitioners and clergymen in the country
in a manner unexampled in scientific societies [had] left the Society by hundreds." The society was deep in debt, and its members ought to know that
they were liable for this debt, although the language used was confusing as
to how exactly this would be the case. The recommended remedy was the
liquidation of the society's flagship, the *Review,* where the "lampoons" had
been published.

   Taken together, the statements amounted to charging that the society
essentially provided revenues to the undisclosed owner of the *Review* while
leaving the debts with the society and exposing its fellows. Later in the controversy, the technical word to describe allegedly selective payments was
"preferential payments," a serious accusation in a world that did not take

differential treatment of creditors lightly. The accusation had been well-crafted because, as all insiders knew, the proprietor of the *Review* was Hunt himself. The kind of money that flowed in was hardly enough to make him a rich man. The tendency of clubs and learned societies to be permanently cash-strapped was common in Victorian Britain, but the allegations about wrongdoings could force Hunt to come out as the owner of the *Review*. If that happened, the rest of Hyde Clarke's accusations would gain credibility. Hunt was feeble if he remained silent and feebler if he spoke.

Hunt's rejoinder to Hyde Clarke, published in the next issue of the *Athenaeum*, reflected the difficult position in which he found himself. Shrewdly, Hunt decided to be polite and ironical. He thanked Hyde Clarke for his solicitude before poking some fun at Hyde Clarke's numbers. Reviewing membership statistics, Hunt argued that it was difficult to see how fellows could have resigned "in the hundreds." In the meantime, Cannibals initiated a procedure to expel Hyde Clarke "for conduct calculated to injure the Society." A vote from a Special General Assembly meeting was called in, and the meeting was held on September 2, 1868.[39] Hyde Clarke's response to the impending procedure against him put the matter beyond repair, as had been evidently the objective from the beginning. In a letter published in the *Athenaeum* on August 29, 1868, Hyde Clarke came up with a list of fifteen points for which he requested answers and, waving again the threat of the liability of fellows, he warned individual members of the Anthropological Society of London that they incurred the debts of the society. As for himself, he claimed, he feared being still "enrolled under some idle designation in the category of your numerous office-bearers without function" and wanted to be sure that he would be "exempt as a contributory in case of a winding-up." The concluding words threw the mask of gentlemanliness to the floor: "Your Fellowship has not yet become a title of respect, and your Honorary Membership has been rejected with contumely."[40]

In this same letter Clarke made the much-repeated accusation that was to inflame the rest of the controversy: "It will be with the public to give the verdict—whether the charlatanism, puffery and jobbery of the Anthropological Society shall be rebuked or approved." On the one hand, "puffery" and "jobbery" were two words right in their place in a newspaper that had made a crusade of the fight against puffery in literature. The theme dated perhaps as far back as Francis Bacon's *Great Instauration* and its famous line about time, which Bacon described as "a river which has brought down to us things light and puffed up while those which are weighty and solid have sunk."[41] Yet as readers of the *Athenaeum* could note, the words proffered by Hyde

Clarke also came straight from the stock exchange. As one Anthropological fellow declared, such were "terms familiar to those who may have occasion to use them in their experience of Joint-Stock Companies."[42]

Connections between what was happening in anthropology and in the stock exchange were again brought to the fore in subsequent meetings devoted to trying to expel Hyde Clarke from the Anthropological Society.[43] While the findings of a committee mandated to report on the society's accounts were being reviewed on October 28, A. Bendir, a member of the investigation team, embarked on a diatribe against Hyde Clarke's character. This followed a dramatic scene in which Bendir asked Hyde Clarke whether, given the new evidence, he would consider apologizing. He then paused for effect and, when Hyde Clarke remained silent, continued by stating that "well, [he] hardly expected Mr. Clarke would apologize."[44] Bendir then alluded to a recent mention of Hyde Clarke "in the daily press of September 30th 1868" concerning a meeting of the shareholders of one "Railway or Financial Company" and to an unnamed informant who had described Hyde Clarke's character as "always in hot water." According to the article, Hyde Clarke had participated in the shareholders' meeting, and the shareholders listened to him "with much impatience."[45] The search engine of the British Newspaper Archive shows that the article almost certainly came from the *London Standard*, which on October 1 (rather than September 30th, 1868), ran an article about a meeting of shareholders of the Smyrna-Aidin Imperial Ottoman Railway. The article stated that during this meeting Hyde Clarke had been listened to "with great impatience."[46] And thus, once again, following the tracks of anthropology takes us back to a railway line.

## RAIDERS AND MANAGERS

As we saw Hyde Clarke had been a salaried senior executive—secretary—of the Smyrna-Aidin line, monitoring the (in)activity of the company in Smyrna and occasionally negotiating the renewal of the concession with Ottoman authorities. He had been the man on the spot and as a result became knowledgeable in the matters of the companies.[47] As was often the case in similar schemes, Hyde Clarke had received part of his payment, perhaps all of it, in the form of the stocks or debentures of the company. Of course, insiders such as Hyde Clarke knew how the game was played, and it is highly unlikely that he received them at face value. The important element, however, is that this made of Hyde Clarke an informed stockholder who could either associate with the directors or employ his knowledge to mingle with activist shareholders.

The 1867 audit of the Smyrna-Aidin railway commissioned by the Board of Trade had not settled tensions between shareholders of the Smyrna-Aidin railway and their directors, much to the contrary. The publicity it gave to abuses provided opportunity for Hyde Clarke and a group of wealthy associates to reinvent themselves as defenders of public morality. As we learn from the press, on June 2, 1868, a group of shareholders led by the financier Philipp Ellissen organized a rebellion.[48] Along with Ellissen were six or seven individuals that included Hyde Clarke. The group also included one William Percival Pickering and one Mariano Vives, who would bring lawsuits against the directors of the Smyrna Aidin Imperial Ottoman Railway.[49] These "investors," later described by their adversaries as "self-appointed" sought to interfere with the board of directors. From mid-August onward, they published articles in the press and wrote letters to Foreign Affairs Secretary Lord Stanley containing, according to the court report of the case that opposed the activist shareholders to the directors, "imputations on the council of administration, which [we]re couched in strong and offensive language."[50]

While later censored by the court, the accusations had the immediate effect of cooling off any political support the Foreign Office might have provided to the directors had there been any such intention. In early September 1868, when Hyde Clarke was battling his exclusion from the Anthropological Society, the shareholders' rebellion approached Daoud Pacha, the Ottoman minister of public works. A former governor of Lebanon, Daoud Pacha was depicted by the Western press as a modernizer sojourning in London in order to address the prevarication that was said to characterize Turkish railway lines. During their interview with him, the shareholders made adverse reports on the behavior of the company directors. They argued that the directors "had misapplied the [guarantee, which the Ottoman Government had pledged to provide to the Company] and may do so again." They urged Daoud Pacha not to release the financial guarantee that would end up siphoned by the management. For their part, the directors claimed that the deputation had been led by Hyde Clarke and accused him of being the leader of the rebellion. Hyde Clarke-the-anthropologist insisted that he was by no means an agitator and had acted merely to introduce the deputation and "interpret in case of need" because he spoke Turkish.[51]

The intervention of the activist-shareholders was reported to have injured the company's shares, and this was the background for the electric discussion in the subsequent half-yearly meeting of shareholders on September 30, 1868, a meeting in which Philipp Ellissen, Hyde Clarke, and their allies intended to carry out a motion against the directors of the Smyrna-Aidin company.[52] Against accusations that the activists had caused the recent price

decline observed on the London Stock Exchange and thus injured investors
in the Smyrna-Aidin company, Hyde Clarke insisted that their intervention
with the Foreign Office and Daoud Pacha had no effect on the price of securities because the facts they were complaining about were already known to
everybody.[53] Such was the motive for the "impatience" with Hyde Clarke
mentioned in the *London Standard* and repeated by Bendir.

This context, evidently known to Bendir, explains why he saw a parallel between the impatience that prevailed among the shareholders of the
Smyrna-Aidin railway and the impatience that prevailed among the Cannibals. As he put it: "the fact is, [Clarke] was only doing to this Society what he
had been and was now doing in others." In drawing this comparison, Bendir
was not at all emphasizing the inappropriateness of a behavior financial in a
society learned. It is not that Bendir believed that anything was intrinsically
wrong in dealing with a learned society as if it were a corporation. Much to
the contrary, Bendir outlined the *continuity* of Hyde Clarke's character—
"always in hot water"—in the corporate and the scientific scenes. Far from
being distinct scenes ruled by different laws of motion, they were part of but
one drama. The Smyrna-Aidin shareholders' meeting *informed* what was
happening in the columns of the *Athenaeum* and at the Anthropological
Society. Cannibals felt they were the victims of a bear raid, pure and simple.
Indeed, this theme of Hyde Clarke as the specialist in hostile takeovers does
emerge repeatedly in the controversy. It is alluded to when Hunt mentions
that Hyde Clarke had to leave an amalgamation meeting in June "after half
an hour" because he was reorganizing the Statistical Society. It came again
on October 29, when Bendir stated that "the Ethnological, Horticultural,
and Statistical Societies were also benefited by Mr. Clarke's activity and energy." In fact, Hyde Clarke himself did not dispute his involvement with societies, and in his responses to Bendir merely challenged the negative connotation, asking rhetorically what could be the character of a committee that
said he (Clarke) "had been for many years concerned with disturbing scientific societies and financial bodies" when his real intentions were pure and
well known. [54]

His own take on the virtues of the sort of reorganizations he favored appears in the transcription of the meetings that accompanied Hyde Clarke's
attempted launch of a Shareholders Protection Association in December
1868 discussed in chapter 1. This was an enterprise, as we noted, very similar to Gerstenberg's Foreign Bondholders' Corporation. Its ostensible purpose was to discipline directors, but the true purpose was to extract a surplus
from shipwrecking ("reorganizing") societies. The power that such a body
would have by merely setting itself against such and such a company would

have been enormous. To allay fears, Hyde Clarke emphasized that share-holders in sound companies need not worry, for the goal of the Shareholders Protection Association was not at all the pulling down of companies but, "on the contrary, it would be the means of establishing many companies which had elements of good in them, upon a firm footing."[55] We start to understand better what Hyde Clarke was up to with his *Athenaeum* letter.

It is of course interesting, even fascinating, that Hyde Clarke tried to launch this scheme to unite all shareholders against abusive directors at the same time the Anthropological controversy was taking place. Indeed, in the line of defense he gave for the social utility of the Shareholders Protection Association, we recognize the arguments Hyde Clarke would use in the fall of 1868 against Hunt, Bendir and the other members of the Anthropo-logical Council during the acrimonious debates at St. Martin's Place. As he claimed, his revelations had been *useful* to the Anthropological Society because they had led to the commissioning of the report, and the report clarified the operation of the society for the public, and that was good for the society at large. Destruction was good citizenship. The parallel that comes to mind is the managerial prescriptions of modern accounting firms or investment banks in the context of the reorganization of firms and in-deed, as we have argued earlier, Hyde Clarke's background had been forged in the railway mania in which modern accountancy firms were born.[56]

With so much overlap between things scientific and things financial, we may not even be surprised by the discovery that among Hyde Clarke's associates in his projected Shareholders Protection Association was one "H. Brookes," most probably Henry Brookes, the one fellow of the Anthropo-logical Society who had supported Hyde Clarke against the Cannibals dur-ing the General Assembly meetings in September and October of 1868. This was the same Brookes who also joined in the *Athenaeum* controversy by publishing a letter in which he asserted his opinion that the Anthropologi-cal Society's fellows should feel obliged to Hyde Clarke and his efforts be-cause it was "impossible for any one, not being of Dr. Hunt's clique, either to admire or approve of [their] mode of *managing* the affairs of a scientific society" (my italics).[57] At that moment, however brief, the history of an-thropology and finance are made of but one single thread.

## HYDE CLARKE'S CORNER

In summary, the beating of the Cannibals had the quality of a multipronged public relations program. First, Hyde Clarke had reviewed the weaknesses of the society. It was a new society that had been aggressively growing through canvassing, resorting to the standard corporate trick of portraying things in a happy state and borrowing generously until that state would materialize. It is not for us to judge whether the enthusiasm was warranted. Suffice to say that, if the Ethnological Society could live with 250 members, couldn't the Anthropological Society survive, even with the 400 alleged by Clarke, and pay off its moderate debts? The pro-Cannibal *Pall Mall Gazette* commented on the "difficulties which beset the starting of any new organ of public opinion—the jealousies that are aroused, the susceptibilities that have to be studied, or the financial side of the question, which requires to be considered from more than one point of view. Even the Ethnological Society, at one point of its existence . . . was, so to speak, in a comatose condition."[58] But this ambivalence provided the opportunity to organize a different and equally persuasive narrative. To be sure, Hyde Clarke deftly timed his attack at a point where the growth margins of the society were shrinking. The "academic" year 1867–68 (at the Anthropological Society, a year started in November and ended in June, with a long summer break) had seen a dramatic decline in new admissions.

The second prong of the public relations plan was the execution. The decision to work with the *Athenaeum*, as already suggested, was part of the plot. The same article in the *Pall Mall Gazette* insinuated that the *Athenaeum* was simply a supporter of the Ethnological group.[59] Moreover, Hyde Clarke had been a regular contributor to the journal, and there were individuals prepared to vouch for him and open for him the pages of the newspaper. His foe Bendir indeed suggested taking notice of "Mr. Clarke's frequent contribution to the weekly paper which had opened its pages to him."[60] Highly regarded, largely read by the relevant public, and having a tradition for its papers to be reprinted subsequently in other journals, the *Athenaeum* commanded respect left and right. Pim's coauthor Seemann had been glad to see his work displayed there.[61]

Third, Hyde Clarke's bear raid revealed an attention to the detail, the hallmark of the true professional in public relations. Indeed, the letter published on August 15 by the *Athenaeum*, written a few days earlier, followed closely on a circular sent out on July 30 by the treasurer of the Anthropological Society, Dunbar I. Heath, in which he threatened defaulters who would not have acquitted their dues by September 1 with displaying their names in a separate shame list.[62] The measure was in itself conducive of resignation, something which Hyde Clarke, an insider, was aware of. Another aspect of Clarke's attention to detail can be gleaned from the fact that he more than likely studied the constitution of the Anthropological Society. Hyde Clarke was the sort of man who would study a directors' list before dealing with them and their railways. He probably predicted that the society would respond by trying to exclude him, and knew that the credibility of his accusations would hinge on his ability to resist this exclusion, as it would show that he had supporters inside the Anthropological Society. As with many other societies, the rules were in effect biased against exclusion. Clarke thus calculated that he would need a minority group of supporters, and he conspicuously prepared for it, explaining the staging of his associate Brookes, and certainly some others, who helped him secure the needed votes to block exclusion. A last aspect of Clarke's virtuosity with detail was the timing of the attack, planned for August 15 to coincide with the opening of the annual meetings of the British Association for the Advancement of Science in Norwich only two days later, thus preventing the Cannibals from sending a response in the same periodical in due time. The Parliament of Science was transformed into a soundboard for the *Athenaeum* letter, with gossip spreading the derogatory words and growing a full-fledged rumor.

The effect on membership was soon felt, certainly not the hundreds of resignations Hyde Clarke claimed insincerely were already happening, but rather resignations arriving in little clusters, after the eruption of the dispute. At the end of 1868, the Council of the Anthropological Society would report fifty-nine of these. Attempting to conceal the fact, the next report boasted that the sixty-three new members they had signed in "outnumbered departures," though by taking fourteen deaths and resignations into account, the overall numbers for the society did show a decline. The phenomenon was modest but new, and it outlined the damage done by the *Athenaeum* controversy. In fact, the ability of the society to grow was hampered. Of course, rather than acknowledging that this was the effect of Hyde Clarke, the report for 1868 ascribed the evolution to the Council's stricter enforcement of fee payment.[63] The trend continued the following year. At the end of 1869, reported resignations stood at 58 against 40 newly signed fellows, a net deficit of 18 members—or 23 members if we further subtract the 5 deaths reported.[64] Another way to look at the matter is to consider the decline in revenues from annual fees. They went from £976 to £731 between 1868 and 1869, a 26-percent reduction in annual membership revenues. With annual membership at £2.1, this meant a *net* loss of about 115 members or an equivalent rise in late payments.[65]

In the end, the *Athenaeum* controversy is what reversed the fate of the Anthropological Society and engineered its decline, not the end of the Civil War or, even less so, the "Negro and Jamaica" conference and pamphlet. But if this manufactured controversy was a turning point, then Hyde Clarke is a grander character. His historical significance as an anthropologist is not

merely his scholarly contributions nor the political prowess that helped
this outsider secure access to the Ethnological Council and from there rise
to the vice presidency of the Anthropological Institute. It is true that he
was called to contribute the section devoted to economic anthropology in
the *Notes and Queries on Anthropology, for the Use of Travellers and Residents in Uncivilized Lands*, the ethnographical handbook put out in 1874
by the Anthropological Institute under the auspices of the British Association and which modernized the earlier manual published under the auspices
of the British Navy.[66] But his real legacy was not so much the making of
anthropology but rather the unmaking of the Anthropological Society and,
as a result, the descent of the Anthropological Institute. Should we repeat
that this unmaking was achieved using the language and techniques of the
stock exchange? For those who would still doubt the significance of this, a
last episode must be narrated.

## HYDE CLARKE VS. PIM: REMATCH

In 1872, shortly after the Anthropological Institute was created by the merger
of anthropology and ethnology, the same players were pitted against one
another again, not in the premises of the Anthropological Society of London
but at London Tavern in a bondholders meeting. At the time this new episode occurred, Hyde Clarke was emerging as the powerful secretary of the
Council of Foreign Bondholders, which stock exchange broker and colonization companies promoter Gerstenberg had launched in 1868 a few weeks
before Gladstone's triumphant election.[67] Ever since this had happened, Gerstenberg and Hyde Clarke had been making noise, emphasizing through the
media their desire to cleanse the foreign debt air. They were looking for bubbles to burst in order to further establish the corporation's credentials as cus-
todian of investments for widows and little children. This was how Bedford
Pim again came within Hyde Clarke's reach

We left Pim absorbed in Miskitology and promoting mining companies
in Central America in partnership with Berthold Seemann and C. Carter
Blake. The acceleration of foreign debt promotion in the late 1860s provided
this great connoisseur of Central America with further opportunities. By
1872, he had used his Central American Association and the land grants
the association controlled to reinvent himself as Special Commissioner of
the Honduras government for an enormous Inter-Oceanic Railway loan that
Honduras issued for £15 million (£1.2 billion in today's pounds, according to
the retail price index). As we know, crossing through Honduras was Squier's
route, not Pim's, but Pim's company for the crossing through Nicaragua had

failed to find subscribers. The Nicaragua Railway Company had been wound up, and its New Jersey reincarnation was dormant. But before that, the Honduras Inter-Oceanic Railway Company supported by the Brown consortium had also reached a dead end and had been wound up in 1862 as London directors lost heart and the concession expired.[68]

In 1867, however, a coalition of Honduras politicians that included Honduras ministers Don Victor Herran in Paris and Don Carlos Gutierrez in London reinvented the project with a £1 million sterling loan contracted by Bischoffsheim-Goldschmidt, an international banking house with activity in Brussels, Antwerp, and Paris that was to become one of the most notorious villains of the select committee's inquiry. To tempt investors, the loan promised a return of 12.5 percent, a return made more attractive by guarantees from the Honduras government's customs duties and a pledge on the revenue of the mahogany forests owned by the state of Honduras. The issuers reported to stock exchange authorities that the loan had been successful, although it seems likely from a number of sources that it did not sell out properly despite the contractors' apparent efforts to use the little they had received in subscription to play the loan up on the forward market in the hope that, with prices rising, interest in the loan would eventually catch on. This did not discourage Herran and Gutierrez from trying to make capital from the railway. A second loan was attempted in Paris in 1869 for about £2.5 million in partnership with the bank Dreyfus, Scheyer and Co. This did not succeed either, except that Dreyfus, Scheyer had pledged to take up one third of the loan, although it is not clear that they paid for it.

A last-ditch effort was made in 1870 with a third loan for £2.5 million, and by that time the confusion was substantial. Partly issued bonds were trading in unknown amounts in various markets. Contractors had reportedly built unverifiable portions of the railway line and were asking for payment and threatening lawsuits. Coupons were to be paid on the subscribed portions although there were no permanent provisions for that. In a gamble for survival, Hondurans and their imaginative advisors came up in 1871 with an even more ambitious scheme, the so-called ship-railway, a multi-track monster that would carry not the cargo but the ships themselves between the two oceans. This resulted in the fourth Honduran loan, for £15 million, announced to the public on May 22, 1872.[69] And it was at this point that Pim was appointed Special Commissioner for Honduras in charge of promoting the loan. This was taking place in a context of souring relations between the two plenipotentiaries Herran and Gutierrez, whose cooperation collapsed in adversity. As Pim called at Paris for a road show as a Gutierrez envoy, Herran managed to persuade French authorities to throw Pim into jail because,

Pim claimed when interviewed by the select committee, he had refused to provide Herran with a bribe.

The circumstances whereby Pim managed to get involved in this last Honduras promotion again brings to the fore patterns by now familiar. According to Honduran officials, Pim's involvement had been for "greater precaution" and because the gentleman was "perfectly acquainted with the Republic of Honduras and the affairs of the railway" as well as "greatly esteemed in London for his activity, understanding and previous conduct."[70] Beyond this, we saw that Pim's Central American Association, initially focused on Nicaragua, had also gotten involved in Poyais land grants in Honduras. To the extent that Gutierrez wanted to issue a loan in London, the support of the Central American Association machinery could be valuable if for no other reason than to remove the obstacle of possible harmful publicity. This is how, just because he could block it, Bedford Pim found himself in charge of promoting the reinvention of a defunct project initially imagined by his former rival in anthropology and finance Ephraim George Squier. Illustrating the recognizable connection between reputation, foreign debt, and land grabbing, another agent used by Gutierrez was Edward Haslewood, who needs no introduction and whose name comes up in Foreign Office correspondence of the early 1860s under "Poyais Land Grants."[71]

Publicity was prepared and the new bond was duly puffed up, providing context for Hyde Clarke's intervention. On May 23, the day after subscription to the ship-railway opened, he sent a public letter to Don Carlos Gutierrez, minister plenipotentiary for the Honduras legation for London, making it known that the Council of Foreign Bondholders was not satisfied with Honduras. Hyde Clarke's letter raised all kinds of issues in a long list of queries, complaints, and requests for clarifications where accusations of abuse were implied rather than frankly stated. The method was reminiscent of that used in the *Athenaeum*. The letter drew an immediate answer from Gutierrez, who dealt with every issue and concluded that he "would be happy to afford any further information." But while agents for Honduras were puffing the loan in over-the-counter London trading and increases on the price of issue or "premiums" were being reported in the *Times* and the *Daily News*, opponents of the loan pushed opposite stories and took issue with the computations made in the ship-railway prospectus. The *Economist*, for instance, argued that the annual tonnage for shipment around Cape Horn was not sixteen million tons as the prospectus claimed but more like two million. Given that Walter Bagehot, the editor of the *Economist*, was close to Lubbock, Darwin, and their political and intellectual allies, it is hard to resist the impression that this exchange of letters and the gossip that went along with it represented some deeper confrontation between rival groups through their respective public opinion manufacturers, Hyde Clarke and Bedford Pim.

Be that as it may, the credit of the new Honduras issue was wounded, and, faced with limited subscriptions for the loan, Gutierrez canceled the issue a few days later. But Hyde Clarke had smelled blood. Evidently, part of the plan with the ship-railway loan had been to secure the means to pay the interest on earlier debts, and the failure of the loan suggested that Hondurans were at a disadvantage. So Hyde Clarke wrote another letter to Gutierrez, the content of which was probably divulged through mouth to ear, because there were traces of the letter in the contemporary press but initially not of its content. In the letter, Hyde Clarke questioned Honduras's capacity to keep servicing its debts. He questioned the credit of Honduras, whose securities had been seriously depreciated. As had been the case in the *Athenaeum* controversy where the loss of temper had followed Hyde Clarke's second letter, the new message triggered a change of tone. Gutierrez's answer, sent on July 9, 1872, complained that the aggression on the credit of Honduras by Hyde Clarke and the Council of Foreign Bondholders was what caused the collapse: "The premeditated attack made on my Government by unworthy speculators, for their own gain, has been the chief source of the needless panic (for my Government has not yet failed to fulfill its obligations) in Central and South America securities, and has produced this great depreciation of their value of which you speak."[72]

On July 22, in order to respond to public attacks, Pim called for a meeting of Honduras bondholders to take place a few days later, on July 26 at London Tavern. The idea of Pim and his associates was that since a bear market inflicted losses on subscribers of Honduran loans, those subscribers could be corralled into countering Hyde Clarke's campaign and persuaded to keep their bonds in portfolio, perhaps incidentally permitting the leaders to walk out free. On July 23, the day after the announcement was put in the press, Pim sent the bill for his services as special commissioner to Guttierez. One cannot be too prudent.[73] The July 26 meeting was evidently packed with supporters. The ability of adversaries to do the same was checked by controlling access. The *London Standard* announced that entry rights had been issued at the Honduras government financial agency, where they were to be retrieved and "no person [could] be admitted without one of the admission cards."[74]

The stage was set for the remake of the *Athenaeum* controversy with the Honduras railway in the role of the Anthropological Society, Cannibal Pim in the role of the late Cannibal Hunt, the London Tavern in the role of the Anthropological Society's premises on St. Martin's Place, and Hyde Clarke in the role of Hyde Clarke. Pim occupied the chair as special commissioner for Honduras and was surrounded by a group of associates (a "committee") including Lewis Haslewood and William Digby Seymour, an Irish judge and Liberal M.P. with experience in dabbling in foreign mining companies such as the Waller Gold-Mining Company and a contemporary Liberian government debt security.[75]

The minutes of the Honduras meeting at London Tavern were provided by the *London Standard*. The dynamic they reveal was not unlike the perspective the *Journal of the Anthropological Society* had given of the general meeting when Hyde Clarke had challenged the Cannibals, in that the *London Standard* was partial to Pim just like the minutes in the *Journal* were partial to Hunt.[76] The reporter for the *London Standard* wrote about "the large hall, including the gallery, . . . crowded to overflowing."[77] To address the accusations Clarke had made against Honduras, Pim had resorted to the same argument Bendir had used to defend the Anthropological Society—that the company under attack was dealing with the opportunistic behavior of a rogue, in effect a bear speculator prepared to pull down a project for personal gain. Echoing Bendir's declarations about Hyde Clarke in 1868, the *London Standard* quoted Pim as having declared that "it [was] dangerous, in the interest of commerce, that a body of irresponsible individuals should assume a power which would enable them at pleasure to either wreck or elevate an undertaking, the promoters of which had or had not succeeded in persuading Mr. Gerstenberg and Mr. Hyde Clarke of the *bonâ fide* of their project." Just like the Cannibals had been exhorted to stay together and sustain the brunt of the aggression, Pim concluded by urging upon Honduras bondholders "the absolute necessity of sticking together shoulder to shoulder, and supporting their own property."[78] Just as had happened at the Anthropological Society's meeting, Hyde Clarke's attempts at arguing his case were repressed, and the *Morning Post* of July 27 subsequently ran a summary mentioning Hyde Clarke's "ineffectual attempt" to speak up, an attempt made ineffectual by "the impatience of the audience."

While this was not yet the end of the Honduras railway nor of Pim's promotion of the Honduras railway, it was the beginning of the end, just as had been the case for the *Athenaeum* article, which had not ruined the Anthropological Society overnight but had planted a deadly arrow with a slow poison and eventually jeopardized the Anthropological Society's membership list. Hyde Clarke's attack on the Honduras Ship Railway sent Honduras securities vacillating, and in the next weeks, they plummeted. According to some contemporary reports, including the select committee's own analysis

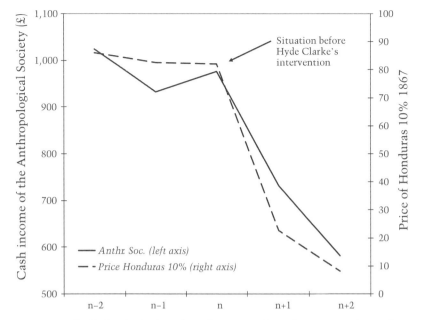

FIGURE 11. Hyde Clarke, the Anthropological Society, and Honduras. *Note*: This graph compares the evolution of the Anthropological Society's cash-raising capacity and the price of the Honduras loan before and after the intervention of Hyde Clarke. The data is organized on an annual basis, taking as benchmark (n) the moment immediately preceding Hyde Clarke's intervention (this coincides with mid-August 1868 for the Anthropological Society and late May 1872 for the Honduras railway). Since there is no Anthropological Society income data for sub-periods, the fall of 1868 was proxied by annual income for the year 1868.

and Jenks's early statement, the Honduras affair was a focal point in the reversal of the market for foreign debt, a statement that is at least consistent with the enormous attention it received in the select committee hearings. Indeed, as visible in figure 11, the general price decline for the speculative state securities examined by the committee had coincided precisely with the attacks against Honduras made by Hyde Clarke and the Council of Foreign Bondholders.

In trying to reconstruct the identity of the man who brought down the Cannibals, this chapter has uncovered a continuity between railway and anthropological promotion. The stellar rise in the number of Anthropological fellows that Hunt had reported in the *Journal* conjured the image of an Anthropological Society bubble and invited accusations of puffing and jobbing. Not only was there an intersection between financial machination and anthropology, but anthropology and railway promotion were taking place in the same linguistic space as well. To a public versed in the delicate art of company abuse ever since the railway mania burst, Hyde Clarke explained that the Anthropological Society was yet another under-capitalized, overblown scheme resting on unsustainable arithmetic. Hunt the "turkey-buzzard" was accused of inflating the fellowship list, just like a London Stock Exchange jobber making mock transactions to puff up a company or foreign government debt. In other words, the Anthropological Society episode can only be fully understood if it is located on a chronology that started with Hyde Clarke's statistical and journalistic involvement in the railway mania, the role of the mania in inspiring developments in accounting, and Hyde Clarke's later involvement as secretary of the Council of Foreign Bondholders. It is at this juncture that the vicissitudes of anthropology take their full significance.

This conclusion enables us to make sense of the superficially incongruous mingling of anthropologists in the adventures of finance and of financiers in the adventures of anthropology. Otherwise, how can we explain that, of all the machinations examined by the select committee, the one subjected to the most intense scrutiny involved so many anthropologists? How to explain that this Honduras railway route had been originated by an anthropologist, Squier, who had drawn a beautiful chart showing how the railway line wound around Indian tribes, losing the Miskito Indians as it twisted and turned through jungles? How to explain that this line was, in 1872, securitized and distributed with the help of another anthropologist who found the Miskito again? How to explain that the Honduras route was finally shorted by yet another anthropologist who was running the day-to-day operations of the most formidable foreign bondholder protection machinery of the time, and one that would be in the forefront of the subjection of Egypt? How to explain that all three anthropologists at one point had been members of the Anthropological Society of London and that, literally a few weeks before the dispute over the Honduras Inter-Oceanic Railway occurred, the two British citizens had been elected in the Council of the new Anthropological Institute so that technically they were colleagues there? And what should be said of the fact that the packing of the Honduras bondholders meeting by Pim in 1872 was happening at a time when the meetings of the Anthropological Institute were themselves packed and counter-packed by the rival factions of anthropology and ethnology of which Bedford Pim and Hyde Clark, respectively, were prominent leaders?[79]

Another important takeaway from this chapter is the discovery of a social function for the multi-plated man. Blanchard Jerrold hinted at this role when he described the manufacturers of public opinion as "experts in the highest skill, who [could] manipulate the various agencies or engines by which public opinion is made." He did not get into the more graphic details, perhaps because he was himself one of those experts, but the similarity of the skills deployed by Hyde Clarke in the Smyrna-Aidin railway case, in the Anthropological Society case, and in the Honduras ship-railway case underscores that reference to the diversity of material, intellectual, and social interests during the Victorian era is deeply misleading.[80] Hyde Clarke may have been multi-plated, but he was single-tasking. He was a fixer, and it is his performance in this job (the job of exposure?) that ensured his success and social promotion.

Fixers work for a fee. In the case of Hyde Clarke slaying the ship-railway loan, he was acting as secretary of the Council of Foreign Bondholders and thus did a paid job, because the Council provided compensation each time he was involved in some government bond negotiation. He might also have benefited on the side by trading on the insider knowledge that his position provided him with. What about his fee for slaying the Cannibals? In the middle of the *Athenaeum* controversy, H. Brookes, Hyde Clarke's planted supporter, had asked rhetorically, "What other motive could he (Mr. Hyde Clarke)—a life member—have [than the welfare of the society]?" implying by that that a life member must have been loyal and disinterested. Anthropological Fellow Charles Harding, a Cannibal, replied sarcastically, "place!" In other words, he predicted that Hyde Clarke would be rewarded for his services.

From the way Hyde Clarke was later elevated to powerful positions inside the Anthropological Institute, of which he became a vice president, it is tempting to conclude that Harding was right. It is likewise tempting to read the fact that he was asked to contribute articles on economic anthropology to the prestigious *Notes and Queries* of anthropology published in 1874 by the institute as yet another form of reward. This was a striking success for a newcomer in a clique of ethnologists who accepted only the purest blood. Hyde Clarke became a regular member of all the important social-scientific events of the time. On the occasion of Darwin's death in 1882, he was part of the deputation sent by the Anthropological Institute. The deputation included people such as Lubbock and Galton and was described in detail by the *Times*, for the ceremony of science feeds also on the dead.[81]

Hyde Clarke's subsequent role in British imperial policy followed on these events. The *Athenaeum* and Honduras controversies had only been preludes and the big affair of the early period of the Corporation of Foreign Bondholders would be the debt crisis of Egypt, which led to the addition of the "gem of the Ottoman empire" to the British Empire.[82] The Egyptian bonds that were defaulted upon in 1876 had been issued in London during the 1860s by the house of Frühling and Goschen, headed by the father of the prominent Liberal politician George Joachim Goschen, and the effects of the default might have been enormous in terms of loss of face. But with Hyde Clarke managing the public relations of the Corporation of Foreign Bondholders, nothing was ever said about the *bona fides* of Goschen, whose help to defend the bondholders was asked and who responded in the most gentlemanly way.

Owing to military intervention and subsequent coercion of the Egyptians, the debts were indeed eventually paid back twenty shillings on the pound, and, given the high rates at which they had been issued, the investment turned out to have been for British capitalists one of the best deals of the time. So why complain? Hyde Clarke, who had experience with the Ottoman Empire (of which Egypt was a dominion until the British takeover) and spoke Turkish (the language of the Egyptian ruling elite), was on the front line throughout. This may explain why, as we saw, he defined himself to the clerk from the British Census who visited him at the time of the British invasion as "Secretary of a Corporation." Clarke would occupy this function until 1884, when he became consulting secretary of the corporation, and a member of its Council until his death.

It occurred on March 1, 1895. On the occasion, Hyde Clarke was up to his own standard as a man of many coincidences. His funeral was announced by the *Morning Post* simultaneously with the death of Ismaïl Pasha, former khedive of Egypt. The article in the *Post* mentioned some of Hyde Clarke's many doorplates: his "active part in Oriental and Colonial Politics" and his prominence as a member of the "Masonic Body." A reference was also made to his role in the Newspaper Press Fund, an organization supporting journalists and their widows and children, of which Hyde Clarke had later become treasurer, enabling him to mingle with the highest strata of the business of charity. It is no coincidence that this powerful lobbyist and manager of reputation was in charge of what was essentially a social insurance system for media people. The *Post,* however, did not mention the Smyrna-Aidin Ottoman Railway, or the Anthropological Society of London, or the Honduras loan.[83]

After that, Hyde Clarke all but disappeared from the radar of historians. He did not even get an entry in the *Dictionary of National Biography*, ironic given that the first edition relied on readers of the *Athenaeum* to provide the basic elements of the social rating system, which the *Dictionary* reflected.[84] Evidently, Hyde Clarke had spent too much time raking the gutters of Victorian gentlemanliness for his memory not to be a threat. The modern *Oxford Dictionary of National Biography* keeps up with the tradition of the *Dictionary* by continuing to ignore him stubbornly, for the pen must be invisible to the reader. This is, perhaps, the subconscious reason for his survival as a pioneer of the study of cycles.

CHAPTER ELEVEN

# Subject Races

It is now time to return to the themes and questions discussed at the outset of this book and provide an interpretation of the role of anthropology in the process of colonial subjection as it shaped fin-de-siècle British imperialism. In so doing, I elaborate on the pun in the title of the book edited by Peter Pels and Oscar Salemink, *Colonial Subjects*. My take is the following: I want to stress the relation between knowledge and property. Knowledge, I argue, is a form of property as reflected by the ambiguous meaning of the word *subject*. Imperial ownership is about the technologies that permit knowing the empire you own. As a result, forms of imperialism and forms of knowledge correspond tightly with each other. Accordingly, previous scholars such as Michel Foucault and Bernard Cohn have emphasized the link between knowledge and power.[1] What I suggest adding is a more detailed attention to the ways knowledge is owned beyond the knowledge bureaucracy. This perspective differs from the conventional focus on the ways knowledge is being produced, and I argue that ownership determines production. In order to underscore the importance of establishing property rights in knowledge as a process that influences imperialism, this final chapter disentangles the forces at work in the respective amalgamations of the Anthropological Institute and of the Corporation of Foreign Bondholders. The result is a narrative of the process of subjection at play in the late 1860s and early 1870s and of the interplay between these two congruent sites: finance and anthropology.

## BEYOND BIPARTISANSHIP

Ideological aspects of the rise of an anthropological monopoly in the 1870s have been previously discussed from the vantage point of the new bipartisan approach to the science of man that developed precisely at that time. An insightful essay by David K. Van Keuren approached this evolution through the personality of Augustus Lane Fox (Pitt Rivers) and his promotion of museology. As van Keuren emphasized, Pitt Rivers ("one of the most interesting figures of British anthropology") is known for his active conservatism and, at a later stage of his life, his no less active advocacy of scientific bipartisanship expressed in characteristic Victorian gibberish:[2]

> The student of science can be neither exclusively Liberal not exclusively
> Conservative. He is bound to apply both principles in their proper places.
> He must be Conservative in the sense of recognizing that it is only by
> preserving the root, the stem and the branches of our social system as it
> now exists that civilization can be expected to bring forth its periodical
> shoots of progress. He must be Liberal in desiring that those periodical
> shoots may be left to flourish and adjust themselves in the free air and
> light of Nature, unfettered by oppressive laws over legislation.[3]

Augustus Lane Fox is indeed a relevant starting point because his career and sociological trajectory can be taken as a metaphor of the evolution of British anthropology. As explained earlier, the politics of the anthropological disputes in the 1860s can be mapped into a Liberal-Conservative debate over empire that developed at this historical juncture and which Disraeli exploited. Lane Fox himself personified the intersection between the two groups. A Conservative born to a lower branch of the Pitt family and certainly not a rich man to begin with, he undertook the proper career in the army and was thus located on the upper reaches of the group where the Anthropological Society canvassed its members. Because of his birth, he likewise came from a milieu where the Ethnological Society recruited its own members and as a result was one of the few joint members of both societies.

Lane Fox went on half-pay retirement in the 1860s and began to devote himself to anthropology and ethnology, joining the Council of the Anthropological Society in January 1867, the same year as Bedford Pim, likewise a military officer on half-pay retirement, though of course one who had been compulsorily retired. Not unlike British anthropology, which would be rebranded and gentrified by amalgamation, Augustus Lane Fox was gentrified in 1880 by inheriting the estates of his great-uncle George Pitt, second Baron Rivers, in 1880, and became Augustus Pitt Rivers, founder of the prestigious Museum of Anthropology in Oxford.[4]

There are various ways to think of Lane Fox's advocacy of bipartisanship in science in general and in anthropology in particular, and they are not

mutually exclusive. Van Keuren's own interpretation emphasized the na-
scent ideology of science as above the parties because of course (as indicated
by Pitt Rivers himself in the same speech as above), "scientific opinion is
based upon such broad induction that it is not amenable to party tactics."
Another interpretation emphasizes that bipartisanship was the natural consequence of the merger that occurred in the early 1870s. To the extent that
the science of man was unified as it was from elements scattered across the
political spectrum, then bipartisanship was of the essence. This interpretation is also a logical consequence of both the Cannibals' and the Ethnologists' emphasis on the importance of guiding policy makers. If scientists as
a whole wanted to be influential, then they had to settle their scores. However, given that this more "civilized" approach was only adopted after a period of war, it is important to spend some time unpacking the logic of the
processes of amalgamation that took place in the early 1870s. This will underscore the tight link between science, finance, and a crisis in imperial
management that occurred at this precise time. It also explains why both
finance and anthropology were restructured and placed under the yoke of
the same group of men. Reader, bear with me—you are getting to the point
where the curious case of the joint destinies of foreign finance and science
is about to be solved.

## INFORMAL EMPIRE AND ITS COMPLICATION

In previous chapters, I have referred to the crisis that developed in the 1860s
within the policy formula on which British imperial management rested—
the formula known to modern scholars as informal empire. As I now explain, the encounter between finance and anthropology and their mingled
destinies can be better understood by taking a closer look at the logic and
problems in the management of informal empire. The starting point is the
1860s, when a crisis that had been simmering for ages became evident as
a result of the organization of the scientific interest, which triggered a scientific uprising that shifted important boundaries in the ownership of foreign policy.

We have already seen that the history of anthropology in the 1860s was
that of a rebellion. Agents of empire organized, gained strength, and—instead
of being enzymes of British dominion spread over the world to nimble new
territories—turned against a senior policy maker in London and used their
learned societies and the highbrow media as mouthpieces against him. One
tangible effect of the rise of the learned society was, I have argued, the new
ability imperial agents had to oppose frontal resistance to the orders from

London. They no longer remained in the fold of the Foreign Office. They had other options, other backers too, including the financial entrepreneurs in the London Stock Exchange. Reflecting this crisis in the relative power of the agents and offices in London, we can point to the declining efficacy of the remedies adopted against rogues. Cases surveyed in this book suggest that this was either because the sanction lacked the expected deterrent effect (as in the case of Consul Cameron and Captain Pim, who knew better and were prepared to face the sanction) or because applying the sanction triggered a riot from both supporters and opponents (as was the case for Governor Eyre, who had a scientific journal explain why repression was an "act of mercy"). Illustrating this rebellion is Pim's endorsement of Governor Eyre in *Negro and Jamaica*, the work of an instinctive recognition by one anthropologist of his common cause with another. We have seen the somewhat convoluted language the Admiralty used with Pim through the pen of Alexander Milne, who stated that he could not "of course" prevent officers on duty from buying land, but that on the other hand the precise land that Pim had bought and the precise project he had sought to promote interfered with British diplomacy, whose interests Pim had been sent precisely to uphold.

Informal empire, therefore, is not a turn of phrase to designate British empiricism and lack of a system. It precisely refers to a refined, sophisticated, and admirable system that evolved over time and enjoyed its apogee in the early phase of the Victorian era under Liberal (or eventually known as Liberal) foreign affairs ministers such as Palmerston, Clarendon, and Russell. For years, the vagueness of the road map and the existence of a private motive by the agent had been exploited successfully. The ambiguity and imprecision of agents' instructions was a virtuous arrangement in the hands of successive foreign secretaries. The agent was given a general framework and template, and within these broad guidelines, he had a large discretion. One benefit from this arrangement was limited cost, an issue that became an obsession of Liberal policy makers—it was empire on a penny.

Among the known benefits from this time-honored vagueness of the imperial agent informal road map was that, as already explained, the bureaus in London could pick what they liked most in what the agent brought back. In a sense, the British imperial system had made its own the age-sanctioned arrangement between merchant bankers and their agents abroad—sharing benefits and shunning failure. For their part, the agents, because they wanted to retain their valuable affiliation to the consular service, were encouraged to use their autonomy with adequate wisdom. The dispatches and archives were there to track their previous moves, construct their character, and, if

needed, apply sanction. The higher echelons, those that carried more substantial revenues, could only be reached by a long career of proper behavior
(or political connection, but that was another matter).

It was this very technology that was being shattered by the scientific
phenomenon, which revealed that this might have been too cheap an empire. Significantly, the policy debate on the cost of empire came up at this
time in relation to the discussion of the indemnification of British consuls.
Subsequent committees of Parliament undertook this debate with inten-
sity and passion between 1857 and 1872. Unsurprisingly, Richard F. Burton,
Anthropological Society of London founder and a British consul, was a star
witness of some of these committees and attracted much interest when he
declared that his predecessor in Santos (Brazil) "lived over a spirit shop and
washed his own stockings." It was really a question, Burton said, of how one
chose to live, but one had to give up any hope to "live on par with foreign
officials."[5]

In return for these unattractive terms, consuls enjoyed (and expected to
enjoy) substantial freedom. The more enterprising among them could use
the allowance as a subsidy and develop their own trade—scientific, commercial, financial, or otherwise. As a result, the consular service was a support
for social ascent, which explains the continued traction it could have for
men such as Burton. It is no surprise that as the number of imperial agents
expanded, they naturally looked to institutions that could represent them
best in their fights with the bureaus. One example of this is the way underpaid consular agents turned to groups such as the Anthropological Society.
We see this in Burton, a man who could by now avoid the more risky assignments. For British consuls, societies such as the Anthropological fulfilled the
functions of a trade union of sorts. Other possible formulas, again unsurprisingly promoted by actors in this tale, reflect this broader trend. Among his
other schemes, the indefatigable Pim was involved in the United Kingdom
Pilots' Association and in the London Seaman's Mutual Protection Society,
which he used in the 1870s, along with his newly acquired law degree, to harass the lords commissioners of the Admiralty who had retired him. He was
also the editor of a journal, the *Navy*, which regularly complained about the
poor shape of British ships and the technological shortcomings of the British
fleet.[6]

The dreamlike efficiency of economical informal empire could easily
morph into a full-scale nightmare if local agents of British imperialism, either consuls or connected merchants, nonetheless ran loose and turned
against British interests at large. The anxiety was that the broker would go

native, raising the specter of the agency operating in reverse. It is easy to understand in this context why Burton made people uneasy. The man knew the locals too well, learning not only their language but their sexual habits. Illustrations of this rising untrustworthiness of the agent are found in the travels of Bedford Pim, who offered his Nicaragua scheme to the French emperor, or when Cameron entered into pledges with Tewodros though he had no authority to make them. Private interests sought to appropriate British policies for personal ends. This *per se* was not the problem. In fact, the whole construct of empire rested on the assumption that agents would always try to do this but that in the end, it would cause a big share of the bargain to fall into John Bull's money box. The problem was that an inadequate response in London could end up steering British policy into dangerous waters. And it seemed that increasingly over time, the agents were able to provoke the wrong answer.

In the early to mid-1850s, the problem was recognized as a full-scale case of moral hazard that might commit British policies to unwanted actions. A typical circumstance, which we observed for both Cameron and Pim, was one where Britain would find itself committed to protect some population. In 1857 the *Examiner* already complained that "we are always getting into quarrels by protecting somebody; We were nearly involved in war with the US by protecting those dear Mosquitoes; We are now at logger-heads with Persia to protect Herat and the precious Afghans."[7] In summary, if informal empire is understood not as synonymous with an amorphous, serendipity-driven, opportunistic British foreign policy but as a specific technology conscious of the need to prevent political moral hazard and the capture of British foreign policy by sectional interests, then we can see what was highly challenging in the organization and rebellion of the agents of British imperialism.

Following the Abyssinia campaign, the danger was recognized in Liberal circles and the Liberal press, beginning with the *Economist*, whose editor Walter Bagehot had joined the Ethnological Society of London in 1865. Bagehot began, between two dinners with Darwin, to articulate the template of a modernized Liberal imperial administration. An illustration was the reaction of his newspaper to the publication of official documents on British policy in Abyssinia under Consul Walter Chichele Plowden. They were released because his successor Charles Duncan Cameron had stated that in his business with Tewodros, he was merely following his predecessor's steps and that they both had received official sanction. But a more rigorous interpretation of the evidence, the newspaper found, was that the vague instructions under which Plowden had operated had enabled him to pledge the British government to "interfere in regions of which the Foreign Office had never heard, in which there are no British subjects, and with which we neither have, nor by possibility can have, the trade Mr. Plowden was employed to protect." Relying on the endorsement by default that Clarendon had provided to Plowden, the newspaper continued, Cameron had been able to read a policy where there was none. This was what had led him to claim that his predecessor had "assumed protection of the tribe of Bogos and others in the neighbourhood."

The newspaper went on to criticize the whole consular agency institution and the leeway it left to the individual consul. The *Economist* advocated that a precise road map ought to be provided to *formally* designated *envoys*:

If the Foreign Office wishes for any reason to acquire influence in a barbarous State, let it send an envoy, with a limited function, precise instructions, and definite object—not an agent, who is sure to try to make trade, sure also, to connect himself with politics beyond his sphere, and apt to imagine himself a representative in all departments of the whole nation which pays his salary. Such men in such places can do no good to our interests, for they must recognize that there is nothing to do, or, by doing something, pledge the country to obligations it has not desired, a policy it has not considered, or enterprises from which it would instinctively draw back.[8]

The result of the Plowden-Cameron imbroglio had been an unwanted war as the Abyssinia campaign, just decided upon at the time of Bagehot's writing, demonstrated. "We are at this moment pledged in some vague and therefore highly dangerous way to protect tribes whose names no European ever heard, who are as much beyond our reach as if they lived in the Moon, from undescribed dangers, arising from undefined Mahomedan oppressors," the Liberal weekly complained. And the problem that had surfaced with Abyssinia and was about to result in so much wasting of money, Bagehot further reasoned, was no mere accident. The agency as it had been designed under informal empire had in fact the potential to produce such outcomes again:

There is reason to believe that conduct of this kind, conduct which inevitably leads sooner or later to what are called "complications," that is, to occasions of wasting money and strength, is by no mean without

precedent in our remoter Asian and African consulates; and it is certain
that it will not remain without imitation. The Consuls selected for such
posts are usually energetic and adventurous persons, eager to advance
themselves and their country—to which we may remark, they are devotedly faithful—impatient of the wretched disorder about them, and
aware of the weakness of all the authorities immediately around them,
authorities whom as they think, England could upset at a word.[9]

There was sarcasm in interjecting a reference to those energetic and adventurous persons being "devotedly faithful" to their country. The agents
of empire were often these academic rangers who haunted jungles, deserts,
and learned societies, and they had arranged for themselves powerful arenas
where they could perform their acts of gentlemanliness and thus the "devoted faithfulness" to which Bagehot referred. The dry humor revealed a
deep dilemma. On the one hand, given how imperial management was conceived, the article concluded, accidents such as the Abyssinia hostage crisis
were normal and would recur for "is it possible that it should not periodi-
cally produce those results?" In sum, if nothing were done about it, there
would be other opportunities for Conservatives to exploit. On the other
hand, the growing power of learned societies was not something that Liberals, who used them a lot, were willing to challenge. Unable and unwilling
to break rogue agents individually, for it would inevitably create epic strug-
gles pitting powerful offices and popular scientists who were sitting at the
top of the academic food chain and controlled a number of mediums, Liberal reformers of the Bagehot breed were led to see the problem where it
could be addressed—not with the inevitable indiscretion of agents, but in
doing away with the lack of precision in the instructions agents received
and the lack of control of the political, ideological, and symbolic resources
that were left at the agent's disposal. One could not prevent Pim from trying to promote the science of Miskito, but one could control anthropology.
    This approach effectively called for a radical transformation. Gone were
the days of the convenient outsourcing of many functions of empire to private interests. The chain of command between the offices in London and
the agents on the ground had to be tightened. Imperial management was to
be integrated inside reinforced bureaucratic-political machinery that would
no longer be exposed to the activism of learned societies and rogue explorers. But if precision was in the essence of government, then informal empire
was no more. In summary, the Abyssinia hostage crisis stood at the apex of
an evolution and became a case in point, sealing the fate of informal empire.
Faced with the challenge that Disraeli had created by deftly throwing in the

highly divisive matter of imperial management, advanced Liberals realized
that safety would only be reached by abandoning Russell and the rear guard
of traditionalists and beginning to think of a transformation in the structures
of empire that would make it more concentrated, more responsive to the
impulses at the center; in short, more formalized empire and less scientific
laissez-faire. Seen from this vantage point, the eventual conversion, in the
1870s, of Liberals such as Gladstone from the older rite of Clarendon and
Russell, who treated empire as something generally dangerous and always
costly, to the new rite of increasingly outspoken eulogy looks like an inevitable evolution rather than a case of political dialectic.

## TAMING THE VULTURES

As this was happening, a parallel evolution took place in the realm of bondholding where, just as had happened in relation with consuls, the dysfunc-
tion of informal imperial management became evident. The business of foreign government loans had been previously ruled by a typical formula of
informal empire summarized in the notorious Palmerston Circular. The *Circular Addressed by Viscount Palmerston to Her Majesty's Representatives
in Foreign States, Respecting the Debts Due by Foreign States to British Subjects* stated that while defaulting foreign governments should expect nothing but stern condemnation by Her Majesty's government, that government
would nonetheless never go to war on behalf of the holders of defaulting
securities. This was informal imperial management at its clearest. Investors
were encouraged to lend to those countries with which Britain traded, and
they could expect the government to be on their side in some loosely defined way. They could expect British policy makers to look favorably upon
foreign lending and negatively upon breaches of commercial faith. They
could expect to have local consuls passing on to local governments Her
Gracious Majesty's legitimate concern with protection of the wealth of her
citizens.

Yet at the same time, boundaries were set to how far British authorities would go. A clear hierarchy was set between higher policy ends and the
defense of foreign investments. The effect was to ring-fence government policy from the encroachment of private investors. If intervention was politically valuable, then it could be considered—but defense of the investors
would never be a motive for automatic intervention. There again, empire
rested on a fundamental ambiguity, fundamental because it was the tool for
the administration of empire. In line with Prime Minister Canning's early

nineteenth-century statements about the money lost in foreign investment being a British investor's own funeral, the circular stated that Her Majesty's government did not bear any liability for residual uncertainty that existed in foreign investment, while nonetheless introducing an element of discretion and opportunism left to circumstances and the wise judgment of policy makers. In sum, the circular and the policy regime it defined were about preventing a form of moral hazard that would have pledged the government to unwanted conflicts, something that reminds us immediately of what we have seen for the consular service, and the protection of foreign populations.

As should have been the case, the subsequent period was marked by opportunistic disputes in the interpretation of the circular. On the one hand, bondholder activists used the words the circular had devoted to the need to enforce commercial faith to secure interviews with secretaries of state for foreign affairs and splashed in the press the words of comfort that the minister had proffered (or was said to have proffered) as if this meant that some terrible sanction was about to fall upon the delinquent foreign government. On the other hand, borrowers understood the bargain behind the violence of language and that there were limits beyond which the punishment would never go. A few years after the circular was made public, one Latin American politician caused quite a stir in the London Stock Exchange when he claimed in his own parliament that default on foreign obligations was perfectly feasible for it would never trigger military intervention by Britain, as explicitly stated in the Palmerston Circular.

However, just as had been the case for anthropologists who had been in effect encouraged to take advantage of the decentralization of informal empire and transformed themselves into a powerful academic lobby and public media, the group of extremely wealthy vulture investors who specialized in bankrupt states became in the 1850s and in 1860s—in tandem with the expansion of foreign lending—an even more powerful clique than before. Around the middle of the decade, the decline of John Diston Powles provided scope for ever more enterprising individuals to create even more powerful vulture coalitions, leading Isidor Gerstenberg in July 1866 to float his suggestion for creating a Council of Foreign Bondholders, as it was initially designated. This formal organization was intended to take care of monitoring foreign governments' debts on behalf of the widow and orphan and also "assist" bondholders in negotiations with defaulters. The scheme was eventually launched in November 1868, causing substantial commentary. With the London foreign debt scavengers henceforth a consistent group, having in its fold scientific, financial, and some political interests, the possibility of a disproportionately powerful bondholders organization capturing now Parliament, now the Foreign Office, now the Navy, had become material. To the extent that such a powerful monopoly would be created, not only would the situation result in more financial problems (since such an institution would inevitably end up encouraging more debt issues to generate more plunder, rather than purifying the financial air), but it would drag policy makers into unwanted conflicts. Walter Bagehot's *Economist* raised the concern and objection in the next Saturday issue in an article called "Mr. Goschen and the Foreign Bondholders":

> We need not say that the various objects pointed out by Mr. Gerstenberg in his speech of Wednesday . . . are wise and legitimate objects. But there is one very great danger attaching to the scheme; and without guarding sedulously and habitually against this danger, it may well be thought by many that the plan of associating the bondholders in a powerful body organized, may threaten more to the peace and commerce of the world than it will do for the special interests intended to be thus protected. This danger is that the association, once formed, and driven on, as it is sure to be, by keen apprehensions of loss from the bad faith of borrowing Governments, may prove too strong and too ungovernable to restrain its operations within the bounds of wise statesmanship, and may be induced to press the claims of the suffering bondholders on weak Cabinets till it induces them to intervene with demands of a nature certain to end in war.[10]

As is evident, the newspaper's policy recommendation was a carbon copy of the position it had taken in the dispute over consuls. Indeed, as the article concluded, "If in any selfish panic [the organized bondholders] are led away to embroil the Government of any great country in the task of enforcing their claims, it would have been a thousand times better for the commerce of the world that they had never formed such an Association at all." Just as it had recognized scathingly that the consuls were devotedly faithful, the newspaper described Gerstenberg's proposal as "wise and legitimate." But in both cases, the red flag was being waved. Blind endorsement of these rogues would inevitably end up in distortions of foreign policy, perhaps even an undesirable war.

Indeed, just as the threat of learned societies was tangible, bondholders had already begun to embroil the policy-making machinery, since Gerstenberg had managed to secure the presence of George Joachim Goschen, as chairman of this first meeting, in order to give some officialdom to his council. This young, ascending politician (he was thirty-seven) had entered Parliament in 1863 as Liberal M.P. for the City of London, a distinct group that included persons such as Lionel de Rothschild. Goschen had spent a short season as vice president of the Board of Trade, interrupted by the fall of Lord Russell in 1866 and, following the electoral triumph of Liberals in 1868, was to enter Gladstone's new government as president of the Poor Law Board. He would later continue a stellar ascent that would have him closely involved in the British takeover of Egypt, and he would eventually come to occupy the position of chancellor of the Exchequer. Gerstenberg's ability to secure a serious figure from the upcoming administration demonstrates the power of the bondholders' clique.

One explanation for Goschen's presence at the meeting is that in previous years, the House of Frühling and Goschen, a prominent London merchant bank of German origin that was run by Goschen's father, had originated a succession of loans to Egypt, thus squarely participating in the foreign debt mania that is background to this tale. The parallel between the issue of the loans and Goschen's progression in the ranks of the Liberal Party is evident. A first loan had been made in 1862 and Goschen had entered politics in 1863, sponsored by Kirkman Daniel Hodgson, a Liberal M.P., a "splendid specimen of the British Merchant," and a representative of that group in Parliament. An interesting aside is that this East India merchant had joined the Ethnological Society in 1863. He would be a prominent member of the Select Committee on Loans to Foreign States.[11] Another loan was made in 1864, and Goschen became paymaster general in 1865. He entered the Cabinet in January 1866 as vice president of the Board of Trade in the month that followed Russell's move from secretary of state of foreign affairs to prime minister. Considering Goschen's stellar elevation, the ways of Victorian Britain, and the manner in which Liberals had been thus far running empire, the connection appears evident. As had been the case when Clarendon had used the House of Brown to secure for Britain a majority interest in Squier's Honduras railway, the House of Goschen had probably been used by Palmerston and Russell to back their commercial conquest of Egypt, which was now within reach thanks to a new *entente cordiale* with the Ottoman Empire following the Crimean War. Egypt was the promise of security on the road to India and an enormous prize in a time of cotton scarcity.

One suspects that Gerstenberg knew as much and took advantage of the context to secure Goschen's participation, possibly threatening him with some kind of embarrassment. If evidence of Goschen's delicate position is needed, we have his own father's letters, written in a state of panic when in 1866 a third loan was undertaken by Frühling and Goschen in the now-difficult context of the Overend Gurney Panic. Goschen's father reported nightmares and wrote of a deep concern over the risk of "failure" of Egyptian loans, which would cause havoc at the precise time George entered the Cabinet. But even if all this were mere literary speculation—for of course we shall never find a trace of Gerstenberg's offer to Goschen, which the latter could not refuse—it remains that organizing a crusade against speculative loans was an evident embarrassment for this City M.P., cabinet member, and former top official of the austere Board of Trade when Frühling and Goschen's Egyptian loans had been made at 8.5 percent, 7.5 percent, and 8 percent in round numbers, well above the return on British debt securities.

What was embarrassment for Goschen must have been trouble for the Liberal Party writ large. Were these financial cannibals about to jeopardize the Gladstone administration in the same way anthropologists and their allies had spoiled Russell's? In order to limit the extent to which the Liberal Party would be embroiled by bondholders, Goschen insisted upon setting tight boundaries to the competence of the intended body, reiterating the essence of the Palmerston Circular and stating in front of the assembled capitalists that "the Englishman lends his money to a foreign Government and gets high interest, because he incurs a risk."[12]

And then the *Economist* was commissioned to dwell on the small print. In its already mentioned issue, it devoted a large space to praising Goschen's "sagacity as a statesman." Goschen was a statesman for having recalled, in his speech before the bondholders' first meeting, that the British Navy was not the property of the bondholders.[13] And he was sagacious for having deftly used another Gerstenberg's stage to remind the market of a few wisdoms. Just as the tipped participant in a society meeting, Goschen made his honest and disinterested remarks. Just as the journal of a learned society, the *Economist* played its own part to perfection and informed the public of how the event should be read, and why Goschen was admirable.

It is not of small consequence to the interpretation of the event that the whole performance was happening exactly two weeks before the landslide vote that restored the Liberal Party to power under Gladstone. Undoubtedly, with the looming election, the bondholders and some city interests wanted some clarification as to how to vote, and for this, they needed to know the take, which the Liberals would adopt toward their efforts at ganging up together. While the Liberal policy makers' answer was characteristically noncommittal, it still showed open-mindedness. After all, Gerstenberg's suggestion for a bondholder council, if properly managed, was an opportunity. Over the short run, it could help Goschen and the Liberal Party to demonstrate goodwill toward the bondholders. Over the longer run, such an institution would help to govern foreign debt markets while at the same time keeping the vultures at arm's length. And indeed, in 1866, when discussing the nightmares he had regarding Egyptian loans and his son's career, Goschen's father stated that he desperately hoped for what he called a "safety valve, some way to escape, someone, or rather some *association* to fall back upon. [He would] not have a moment's rest till [he'd] be assured upon this point" (my italics).[14] If there were to be a bondholder tiger, at least it would dance to a tune set by Her Majesty's government.

## HEAVEN AND HELL AMALGAMATED

In summary, the existence of interests that escaped the reach of Liberals and their connection with the stock exchange did not leave many options to the incoming Gladstone administration, and it explains the traction for the bipartisan solution. Reliance on friendly financiers to address a situation, as had been the case when Clarendon had sought to freeze the operations of Squier in Honduras with the help of the House of Brown, was an ad hoc solution but hardly a policy. For every Rothschild in the Liberal Party, there would be a Baring in the Conservative. The politicization of empire created a potential for disruption. Conservative foreign finance would end up fighting Liberal capital export just as we saw Conservative science fighting Liberal science in the midst of the Eyre controversy.

The limits of a sectional approach to imperialism had been amply played out in the previous years. It was just not possible that each party should try to develop its own machinery for foreign financial policy in order to outmaneuver the other. There could not be a Liberal, allegedly humanitarian, *penny ante* imperialism, and an aggressive, vulture-like Conservative variant that would avowedly bankrupt foreign countries. What the 1860s had demonstrated to policy makers was that the worst-case scenario was precisely the intermixture of and competition between the two brands of imperialism, since this inexorably undermined their controlling efforts.

As Lord Cromer, former administrator of Egypt and a Baring, would summarize it in his infamous 1908 treatise *On the Government of Subject Races*, the "art" of imperial management marched on two legs by which the previous dichotomy may be read. There was what he called respectively the "commercial school" and the "school of philanthropy." The first represented this mercantile and commercial spirit that had been both British greatness and British policy "since Pitt." The other emphasized the "grave national responsibilities which devolve on England, and . . . the lofty aspirations which attach themselves to her civilizing and moralising mission." Both had been important elements for British policy makers because, in Cromer's mind but without Cromer's soapy prose, the first had permitted British commerce to share the benefits from plunder and exploitation while the second, for instance through the concern over slavery, had created an atmosphere that legitimized the Navy's sneaking into other people's territories.[15]

But if policy makers were to rely alternatively on mercantile or philanthropic motives, then they could not ignore anthropological science, which provided pronouncements on the character of peoples and in doing so made statements about British foreign policy. These pronouncements had implications for the degree to which a given people deserved attention or care or protection and also influenced judgments on foreign governments' willingness to repay their debts. Thus anthropological science was relevant for both tools, the mercantile and the philanthropic. And the danger, which the 1860s had abundantly illustrated, was that scientific syllogisms, transformed into policy instruments, would be used *against* the government; Miskitos were Indians, Britain protected Indians, *ergo*, it was essential to keep a close eye on proceedings in Nicaragua—and perhaps construct a railway too.

Just like there could not be two standards of financial faith, there could not be different scales for measuring humanity. The budget, the government, the resources had to be put in front of the skull, the skin, and the virtues. But then, upon reflection, the ability of anthropological science to affect commerce and values was a blessing, if properly managed. It could also serve as counterpoise for the tendencies of the mercantile and philanthropic instincts to run loose. To quote again Lord Cromer, both could also become "dangerous" if unrestrained. The philanthropic school had displayed on occasion "violence" and want of "mental equilibrium." The commercial school could also be nefarious "when allowed to run riot," and there had been "occasions . . . not infrequent when the interests of commerce apparently clash[ed] with those of good government."[16] Beyond not leaving science to an inchoate battlefield, there was value in organizing it. Evidently, it was not so much that the public—neither the well-intended public nor the investing or trading public—would have been unable to cope with competing standards; rather, what was at stake was a problem of interference between scientific considerations and imperial governance. Ultimately, there could be only one finance and one science, and neither could stand in the way of policy making. The practical question must have simply been one of finding a way to make the amalgamation of interests happen. The tight interrelation between the chronologies of finance and science enables us to hypothesize what might have taken place. As already emphasized, the Conservative campaign had come first and can be read in the activism of anthropologists, itself related to the frantic activity that took place in the stock exchange. In the middle of the year 1866, two weeks after the formation of a Conservative cabinet under the Earl of Derby, with Disraeli at the Exchequer and Lord Stanley as foreign secretary, Isidor Gerstenberg, the Ecuador Land Company promoter and all-time vulture, launched his foreign bondholding initiative. He sought to secure Barings' endorsement, which made sense given the firm's specialization in Latin America and its political connections with the incoming administration. But Barings had reservations about some elements in the bondholding clique, whom the firm found to be dangerous if left unchecked. Barings provided background help, but officially they refused to endorse Gerstenberg's proposal. The alternative scheme that emerged was Philip Rose's Foreign and Colonial Government Trust in the spring of 1868, which followed on the back of Disraeli's arrival in Downing Street in February of that year.

The summer of 1868 saw the counteroffensive of Liberals, led by Gladstone, who would triumph in the general election at the end of that year. This was, not incidentally, when Clarke was sent out to break the reputation of the Anthropological Society, a predominantly Conservative group with connections with the foreign debt market at the London Stock Exchange. Hyde Clarke's deft attack on the public image of the Anthropological Society was not coincidentally timed at a significant political juncture. Indeed, the campaign against the Anthropological Society began literally a few months before the election that enabled Liberals, under Gladstone, to defeat Disraeli and reclaim power at the House of Commons. The launching of the Corporation of Foreign Bondholders in November 1868 under the auspices of Gerstenberg and Goschen occurred, too, a few days before Gladstone's election, reflecting Liberal courting of the stock exchange vote. The reason why Hyde Clarke was so ubiquitous in the media at the time of the *Athenaeum* controversy is that this was a turning point in British politics and required mobilizing the agencies of public opinion; behind Hyde Clarke's raid, we can thus discern the hand of the Liberal Party-dominated nexus of humanitarian organizations, science, and banking, whose corners

were the Aborigines' Protection Society, the Athenaeum Club, and seg-
ments of the City.

Put in this light, reasons for the bear raid on the Anthropological Society become straightforward, and those reasons again had much to do with
the initial successes of the Cannibals. Their achievements threatened to
jeopardize a patiently crafted machinery that provided connections between
morality, money, and knowledge. The Cannibals in fact claimed as much
regarding where the attack was coming from, repeating during the *Athenaeum* controversy that Hyde Clarke supposedly verbally boasted of his
role in the Royal Statistical Society in 1868. For those who do not speak
Victorian, it might be worth adding that the Statistical Society was then
presided over by Gladstone, the leader of the Liberal opposition. After the
merger of 1868 had failed, and at a time when Disraeli's ambitions were
becoming all too obvious, it was useful for the Liberal interest to break the
back of this specific, dangerous public relations vehicle. The Anthropological Society's loss of credibility that Hyde Clarke sought to produce resulted
in a loss of membership, which enabled the smaller Ethnological Society to
achieve a merger on equal terms and neutralize it through bipartisanship.[17]

Another aspect of the Liberal victory in 1868 was the deal that their
leaders had cut with Gerstenberg and a number of members of the vulture industry a few days before the general election took place. As we saw,
during the foundation meeting of the Council of Foreign Bondholders in
November 1868, vulture investors had been told by City M.P. George Joachim Goschen that, as far as the Liberals were concerned, they could proceed with their scheme of a foreign bondholders organization but with a number of provisions. In particular, Liberal policy makers would not let British
foreign policy be held hostage by the vultures' market operations. This move
was significant because it signaled that Liberals, if elected, would organize
the activities of foreign country wreckers, rather than put an end to them,
thus ensuring the cooperation of insiders. Of significance as well is the fact
that at the end of that year, Hyde Clarke was formally involved with the
newly created council, of which he would become secretary. Whether Gerstenberg appointed him by his own free will or was led to appoint him, we
can only speculate.

The completion of the Lubbock-led takeover of the Anthropological Society was accomplished by stages between 1870 and 1872. The amenable
counterparts from the other side were Lane Fox, Beddoe, and Robert des
Ruffières. They were co-opted. Others, like Bedford Pim, were gradually evic-
ted. Likewise, 1872 saw a number of efforts at bipartisan supervision of
the foreign debt market. Following on the tracks of Philip Rose's Foreign

and Colonial Government Trust, a second investment trust was created in January 1872. The prospectus of the Government Stock Investment Company splashed the names of Anthony John Mundella, a Liberal M.P. who had entered Parliament with the party's triumph of 1868, and Robert N. Fowler, a banker and moderate Conservative who also entered Parliament in 1868. Fowler was an early "life member" of the Ethnological Society and was heavily involved in humanitarian politics, where he often found himself allied with Liberals.[18] He was heavily invested in the subject of slavery and spoke about the mistreatment of indigenous people, though he argued in favor of Jamaica's Governor Edward Eyre in the aftermath of the Morant Bay rebellion.[19] Fowler was the treasurer of the Aborigines' Protection Society, to which belonged the sons of famous abolitionist T. F. Buxton, especially Charles Buxton, the Liberal M.P. who had resigned with vehemence his permanent membership of the Anthropological Society following the "Negro and Jamaica" lecture: after a Disraelian foreign investment trust, a humanitarian one?

The heart of the emerging bipartisan coalition for the management of foreign debt was a redefinition of what constituted the contours of good and bad in both finance and anthropology. As they lost ground in the Anthropological Institute, the more extremist Cannibals such as Pim retreated in the bond market, where they were hunted down by Hyde Clarke and the Council of Foreign Bondholders. In the summer of 1872, a decision was made to involve the corporation in poking the foreign debt bubble. It is unclear who made it, whether Gerstenberg and Hyde Clarke together or whether they were themselves the instruments of high politics, but this led to the blowing up of the Honduras bubble through a campaign led by Hyde Clarke against Pim's Inter-Oceanic Railway-Ship. The event heralded the reversal of the market. Contractors and underwriters ran for cover, especially when an international monetary crisis in 1873 had the Bank of England steeply raising interest rates. Given the role that some important individuals and institutional actors had played in the puffing of foreign loans (think of Church and the Geographical Society, or of Goschen and the House of Frühling and Goschen), the end game was bound to be messy.

Such was the context in which the select committee's investigation on loans to foreign states took place: illustrating the dangers at hand, in the first days of February 1875, a lawsuit about the misappropriation of £10,000 in the Honduras loan of 1872 was initiated. It pitted against one another George Briscoe Kerferd (a former British consul in Central America who had acted as trustee for the Inter-Oceanic Railway loan), Edward Haslewood, Bedford Pim, the government of Honduras, and the financiers who had dealt in Honduras securities.[20] This promised many exciting revelations, and it called for the staging, in haste, of the select committee, which by contrast enabled the management of consequences and a more controllable release of information on wrongdoings.

From the minutes of the public debates in Parliament that occurred when the inquiry was launched upon Liberal M.P. Henry James's motion in mid-February 1875, it is possible to feel the extent of the underlying tension. There was probably ardent political bargaining over the countries to include in the list. When Henry James spoke before the House of Commons in favor of the setting up of the select committee, he indicated that the inquiry should focus on "Honduras, Costa Rica, San Domingo, and Paraguay," not because they formed "the whole of the States that have made default," but because James preferred to place before the House examples rather than "including the entire case."[21]

The response of Chancellor of the Exchequer Sir Stafford Northcote contained interesting remarks about the need to avoid interfering either with the relationships to foreign states (thus the focus on smaller countries?) or with the course of justice and lawsuits already started (an attempt at excluding Honduras from the list?). The exclusion of Bolivia perplexed observers. A letter to the editor of the *London Daily News* by "another bondholder" seconded the motion by a previous reader to have the circumstances of the Bolivian loan "investigated by the present Commission."[22] What was the reason for a significant carve-out that put George Earl Church's deal out of the purview of the committee's investigations? As Charles E. Lewis, the Conservative M.P. who sat on the board of the Foreign and Colonial Government Trust would state flatly before the select committee, this loan "had not been before the Committee" although there had "already been a default in payment of the interest."[23]

The panic triggered by the release of information damaging to the character of so many gentlemen was to cause a curious episode that saw the return in British politics of the so-called "privilege" of Parliament. The privilege of Parliament was a provision that enabled lawmakers to prevent the circulation of information released during hearings. It still existed nominally, although it was a dead letter. In April 1875, as the hearings of the select committee were attracting considerable attention, the Conservative majority voted in bloc to call for a "breach of privilege" and sanction two newspapers, the *Times* and the *London Daily News*. The occasion was the examination of the circumstances that had led Bedford Pim, now a Tory M.P., to be sent to a French prison upon denunciation to the French authorities by Victor Herran, minister of Honduras in Paris. This had occurred as Pim had been in France promoting the Honduras loan. The object of the breach of privilege was a letter by Herran that had been "in no respect, verified or identified as having been written by M. Herran; and it was full of libels of the most serious character against a Gentleman who had been examined on oath before the Committee, and who has the distinction of enjoying a seat in this House."[24] The Conservative M.P. Charles E. Lewis, member of the Foreign and Colonial Government Trust, had motioned the breach of privilege. Banker and Ethnological Society Fellow Kirkman Daniel Hodgson translated the Herran letter from French to English.

The Liberal press thundered against the measure. The satirical paper *Fun*, which had run the ogre of foreign loans cartoon, listed among the "things [they] don't like": "Conservative Reaction," "attempts to burke the Foreign Loans inquiry," and "'gentlemen and scholars,' of the gutter-raking variety." They had evidently figured out Pim.[25] Leaders of the Liberal Party went into roars clearly intended to scare the ogre of foreign loans. This committee, the *Economist* pounded, "is extremely inconvenient to many persons, and a great effort is being made to stop or hamper it." The paper argued that breach of privilege had existed initially to protect freedom of speech inside Parliament against the king and also to create some liability for misrepresentation of matters discussed either in Parliament or in parliamentary committees, but it was unheard of in recent and less recent history that such a rule should be used to prevent the circulation of politically relevant information toward the outside. This long-forgotten rule had rightly fallen into oblivion and should be left to sleep.[26] It was now an established tradition that material revealed in select committees would be divulged immediately. Liberals were not duped by the subterfuge of Lewis's motion. They suspected that behind the block support that Conservatives had given to Lewis' motion was the hand of Prime Minister Disraeli. The resulting outcry was so loud that Disraeli, faced with a parade of Liberal Party grandees, was eventually forced to back away and sacrifice Pim. As a result of the maneuver, accusations made against Pim received much publicity and enormous political coverage too, enabling writers like R. L. Stevenson to joke about "salt-water finance" without running the risk of being sued for libel, since in Britain, one spoke lightly of politics but carefully of business.[27]

## THE BIPARTISAN MOMENT

What the breach of privilege episode established once again (for it had been readily established by the Eyre controversy) was the deleterious effect of partisan dispute in matters of foreign and colonial interest. To the contemporary modernist minds who had British grandeur and trade at heart and to those practical intelligences who had a stake in the management of empire, the only solution was to draw the contours of a new bipartisan consensus along lines moderate Conservatives such as the would-be Pitt Rivers could recognize as their own. The dispute, as the quotes from Lane Fox at the beginning of this chapter showed, would not be between left and right but between good and evil, between good anthropology and bad Cannibals, between the bona fide export of capital and the vultures. This sheds light on the arrival of John Lubbock at the top of two structures that had evolved by the early 1870s, the Anthropological Institute and the Corporation of Foreign Bondholders. Lubbock was a member of the Liberal Party and a scholar, but he was also the archenemy of the Anthropological clique, the man who in the late 1860s and early 1870s frequently dined with Bagehot, Lowe, and Gladstone, the powerful member of the British Association for the Advancement of Science, and who sat on the councils of so many learned societies. In Parliament, explains his biographer Mark Patton, Lubbock represented both the science and banking lobbies. These two lobbies were discovering the need for bipartisanship, explaining Lubbock's proximity to a number of Conservative elements and why, his biographer explains, he never rose to a ministerial position in the Liberal Party. In effect, he would eventually distance himself from partisan politics.

Just beneath Lubbock, the Council of the Anthropological Institute drew in equal proportion from the two tribes of scientists when it was founded in 1871 and included many of the key protagonists of this tale: Balfour, Lane Fox, and Clarke. There was Pim as well because he was such an insider of the Anthropological clique, but with the help of Clarke's second attack in 1872 and the subsequent bear market in Pim's bona fides, he would be driven out of the institute's executive. Markham also was to be involved in the second council after prominent Cannibals were expelled so that he could play the role of Murchison, working on joint lobbying of the Geographical Society and Anthropological Institute.[28]

The art of co-opting is also observed in the way the Council of Foreign Bondholders was manned after its incorporation by the Board of Trade in 1873. Again, the first council featured more than one of the characters we have encountered in this book. In particular, just as had happened for anthropology, the council combined Liberal- and Conservative-leaning men (although the Liberal element was predominant) such as Balfour and Philip Rose. As we have seen, the presence of Fowler, a "humanitarian-Conservative," reflects well the conspicuous search for consensus characteristic of this bipartisan moment.

But the higher ranks of the corporation are perhaps the most fascinating to describe because they summarize to perfection the tangled history of capitalism and anthropology. The corporation was initially chaired by Isidor Gerstenberg, who owned the Ecuador Land Company with the Anthropological Society's Council member and Jivaro specialist William Bollaert. As Walter Bagehot had recommended in 1868 in his *Economist* article "Mr. Goschen and the Foreign Bondholders," Isidor Gerstenberg had to be put under close political watch. As a result, when the corporation was created, Gerstenberg's freedom was severely restricted and he was flanked by John Lubbock as vice-chairman. Lubbock would rise to higher responsibilities within the corporation. The other vice-chairman was Thomas Matthias Weguelin. We have already met with Weguelin in chapter 7. The son of a London merchant banker of Russian origin, he was a former governor of the Bank of England, and a director in a number of foreign companies such as the Bahia and San Francisco Railway Co. and the Trust and Loan Co. of Canada. He had started a colony in Gran Chaco in 1870, in an Indian territory Argentina had stripped from Paraguay following the end of the War of the Triple Alliance. Most importantly as we indicated, Weguelin had been a member of the political consortium led by William Brown of Liverpool, which had bought from Squier in 1857 his concession and goodwill for a road through Honduras.[29]

Finally, in line with his competence as a man with many doorplates, Hyde Clarke was appointed as secretary to run the Corporation of Foreign Bondholders and its relations with the media. This man who spoke Turkish, the language of the ruling Ottoman elites in Egypt, would come to play a central role in what was to be the most important job of the corporation in the late 1870s—the assistance it lent to British governments in adding Egypt to the British Empire. And thus it is that the four men who ruled the Corporation of Foreign Bondholders had played a hugely important role in the recent history of British anthropology.

## REIMAGINING LUBBOCK

Bipartisanship is a sweet-sounding word, but it is just the other name of cartelization. And the simultaneous cartelization of science and finance narrated in this chapter had significance for the way they were to be governed. Because Lubbock's chairing of the Anthropological Institute has left far fewer traces, the financial counterpart is an interesting parallel to explore. In other words, we may gain an insight into Lubbock's governance of anthropology by looking at a chapter of his life financial, that of the accusations to which his rule at the Corporation of Foreign Bondholders was eventually subjected. If the entanglements of science and finance are now taken for granted, then the lessons from finance should give ideas about science.

Over time, as Lubbock came to play an increasingly important role in the management of the Corporation of Foreign Bondholders, he became the *bête noire* of one W. H. Bishop, an ebullient investor in Austrian bonds whom Hyde Clarke and Gerstenberg had co-opted in the Council of Foreign Bondholders in the late 1860s and was thus a pillar of that institution. Through his dedication, Bishop had become a cornerstone of the corporation and eventually joined its executive body, the Council. There, he developed a concern with the way the corporation was managed, and this concern focused on the operation of the market for the corporation's voting rights. A few months before it came into existence, the Corporation of Foreign Bondholders had issued debentures known as "certificates of permanent membership." These certificates were nominative, and each one conferred a single voting right. This setup was intended to prevent one single individual or group of individuals from cornering the market by purchasing several rights, although it was still possible to use dummies and influence decisions that way. From his position as a member of the council, Bishop had secured from clerks, he said, worrisome information. He began having misgivings with the way the certificates were being traded and transferred by Lubbock. He argued that when one such debenture was coming on the market because someone sold it or died, they were set aside and sold to Lubbock's relatives. And since Lubbock consistently frustrated Bishop's attempts at investigating the matter, Bishop eventually made public accusations of nepotism and gerrymandering, stating that there had been "personal enrichment at the corporation expenses" and even that Lubbock had falsified the minutes.

Lubbock's biographer adopts an agnostic view on an episode which he calls spiritedly "the sins of Saint-Lubbock." Yet Bishop was not any man, which explains why the controversy could not be ignored. He was one of the partners of the then-powerful accounting firm Turquand, Youngs, Weiss, Bishop & Clarke, predecessor of the modern firm Ernst and Young.[30] The transfer books available in the archive of the Corporation of Foreign Bondholders, which I have examined, do support Bishop's accusations.[31] I discovered many Lubbocks among the purchasers of the corporation's certificates of permanent membership after 1880: John Birkbeck, Henry James, Montagu, Frederick, Neville, and Norman, to name a few. This passion the Lubbock family displayed toward holding the securities of the Corporation of Foreign Bondholders (to say nothing of Lubbock's friends whom I have not tried to count) supports Bishop's accusations.

The question, of course, is what we should make of it. Could it be, as Bishop insisted, that Lubbock had in mind insider trading or the scavenging of foreign debts? This may be so, and I do not see there is evidence to reject the inference that there was a little bit of this on the margin. Yet if the previous discussion of the descent of Lubbock at the Anthropological Institute and the corporation were admitted, I'd be more inclined to interpret the shady transactions in the corporation's stock as having had a political significance. As argued, Lubbock came at the corporation with a political mandate. That Lubbock was never seriously harmed by Bishop's accusations and campaign and was eventually returned to the corporation suggests that indeed there was an understanding that Lubbock was acting according to some blueprint. In 1900 he was made First Baron Avebury in recognition of his good offices, and an upset Bishop ended up publishing his recriminations in a pamphlet.

The most likely interpretation of the sins of Saint Lubbock is that Lubbock understood his job as one of making sure that things would be run smoothly at the corporation because the corporation had a direct impact on the direction of British foreign finance and policy. A major concern, as said, was that its voting rights would fall in the wrong hands—in the hands of spirited vultures. It was important for finance to be dull. Policy makers knew the threat posed by enterprising financiers, just as they had learned about the dangers posed when anthropological controversy fell in the hands of voracious and independently minded scholars. To make sure that such institutions as the Corporation of Foreign Bondholders or the Anthropological Institute be put to legitimate use, it was important that the right persons be placed in charge and that they defend the institution against outsiders. It was important to ensure that bankrupt countries would not come under the influence of vulture investors in the City. It was important that not anybody could patent some anthropological or geographical truth and start distributing it to the investing public, in deliberate obstruction of British foreign and imperial policy. This was ensured, among other things, by the concentration of votes with Lubbock, who would report to the right person and make the right decision. The bankrupt nation would now fall in the hand of the proper offices, explaining the need for John Birkbeck, Henry James, Montagu, Frederick, Neville, and Norman, who were somehow the dummies for this faceless power. The irony of course is that words that Herbert Spencer had used in the 1850s to describe the tendency of insider "capture" in British railway companies could well be applied to Lubbock's role in science and foreign finance:

> Not only are the characteristic vices of our political state [i.e., restricted democracy] reproduced in each of these mercantile corporations—some even in an intenser degree—but the very form of government, whilst remaining nominally democratic, is substantially so remodeled as to become a miniature of our national constitution. The direction, ceasing to fulfill its theory as a deliberative body whose members possess like powers, falls under the control of some one member of superior cunning, or wealth, to whom the majority become subordinate, that the decision on every question depends on the course he takes.[32]

Such was the context therefore that had led to Lubbock being put in the armchair, and the reason for the eventual triumph of his clique, now a "circle."[33] Lubbock's arrival at the head of the Anthropological Institute and the Corporation of Foreign Bondholders coincided with a successful attempt, Liberal-inspired but which had co-opted the more amicable contesting elements, at drawing away previous owners and promoters of the science of man and the foreign debt vultures industry. The result put an end to the democratic tendencies of the previous period, which Pim's buccaneering approaches to finance and science epitomized. The manhunt against Pim and the traffic in Corporation of Foreign Bondholders' voting certificates were the two sides of the same coin.

This transformation put an end to a certain form of knowledge and to a certain subtype of the stock exchange modality—let's call it the bubble or puffing modality. The age of brokers such as Gerstenberg, who was soon to die in unexplained circumstances, was giving way to the age of bankers and their tendency to produce less information and trade on relationship. However, for the preservation of the Victorian scene and to the confusion of later students who have written about the Corporation of Foreign Bondholders as a market institution, such an evolution could not be advertised. Instead, it was spoken of in terms of credit and good faith and, as Pitt Rivers did, of truth established in a bipartisan way. This was better than to discuss the significance of Lubbock's rule of the corporation, the shady dealings in certificates, or his operations at the Anthropological Institute, whose control the ethnologists took, it will be recalled, by bankrupting it. But why bring the focus on such matters? Weren't these the venial sins of Saint Lubbock?

More important were the results, and in view of those, much could be forgiven the sinner. Now, with science and bankruptcy of the foreign state monopolized, imperial conquest could resume in a more orderly way—or begin, pure and simple. For as previous authors have remarked, a new form of conquest began after the 1870s with different tools and purposes. With the Anthropological Institute and the Corporation of Foreign Bondholders, the destitute territory would be adequately subjected: governed, studied, and vice versa. Races were becoming subjects—topics and subalterns—and dealt with in a dispassionate, scientific way. The risk that the stock exchange would reinvent them in *independent* ways was checked. A Miskito king would be just what he could tolerably be within the bounds set by the Anthropological Institute. Railways across his territory would need to call first at the Corporation of Foreign Bondholders. The institute could explain why he was (or was not) a proper ruler. Should trouble come, those same institutions would be set in motion. The art of bankrupting foreign countries, formerly a business left to vulture investors, had become a policy tool, made legitimate by science, in the hands of the government. Egypt was to be the first prize, because, as Lord Cromer would have it some years later, "civilisation must, unfortunately, have its victims, amongst whom are to some extent inevitably numbered those who do not recognise the paramount necessities of the Budget System." And inevitably among those victims was what Cromer could call, with all the depth of anthropological science, "the Oriental."[34]

CONCLUSION

# Catharsis:
# The Displayed and the Hidden

This book has been about the Anthropological Society, the London Stock Exchange, and empire, and in telling their tangled stories, it has brought to the fore the importance of white-collar criminality, of trust, and of truth. The three, I have argued, are intimately related because truth and trust involve similar technologies while white collar-crime consists of their abuse. In this claim and in the evidence reported to support it is found the motivation for a new perspective on the history of anthropological science, and beyond that, of human knowledge.

The prestige of the financier and the prestige of the scientist have hidden affinities. The parallel is not solely heuristic. I have argued that between knowledge and the stock exchange there exist revolving doors courtesy of the ever-present problem of the valuation of truth. In essence, the reason for the connection between the stock exchange and the production of knowledge is precisely this problem of trust, which conjures up a set of technologies among which prestige reigns supreme. Focusing on the entanglements between white-collar criminality at the stock exchange and scientific disputes makes this point with particular acuity.

This perspective underscores the insufficiency of the conventional story of the Anthropological Society—that it was founded by James Hunt, a thirty-year-old racist doctrinaire, and that it was eventually defeated by ill health, well-deserved lack of luck or popularity, and the benevolent, philanthropic ethnologists. Could anyone have believed even for one second that such a young man could by himself threaten the bases of British intellectual elites and officialdom? In fact, the history of British anthropology reveals much more powerful trends and forces at work behind Hunt. As shown, a more complete perspective should recognize that it was founded by Hunt *and* Richard F. Burton, the latter a pure product of the many institutions that were put to work by the machineries of British expansion. The birth of the Anthropological Society must therefore be understood as a figment of transformations in the production and administration of trust, which I have argued was a crucial element in shaping late-nineteenth-century empire.

To repeat, Burton was a linguist, a buccaneer in science and society, an employee-turned-critic of the East India Company, and a prizewinner of the Royal Geographical Society who had felt betrayed by Roderick Murchison in the controversy over the sources of the Nile. And he was a British consul too, straddling the line between mercurial and political interests. From his experience with Murchison, he learned the value of the institutions of science—their usefulness as instruments of commoditization, their usefulness for owning or disowning a discovery, and how this ownership was part of a process of securitization. And beyond that, Burton had observed how the Royal Geographical Society was embedded in a vast system of production of information that responded to the needs of European capitalism—a machinery whose ultimate source of impulse and motion was the London Stock Exchange.

The partnership between Burton and Hunt produced a collective action technology that glued together the scattered interests of informal empire. The Anthropological Society stood for the rebelling agent. This was something that those natural agents of imperialism—individuals such as "Governor" Eyre, "Consul" or "Captain" Cameron, and of course "Captain" Pim—immediately recognized. This view is unlike George Stocking's *Victorian Anthropology*, which disposed of Burton in two lines about his latent homosexuality, a curious lash that may reveal more about Stocking and his time than about Burton.

And thus another theme that has emerged from my story is the idea that the fight between the Anthropological Society and the Ethnological Society principally underscores the dialectic of politics and science. In the middle of the nineteenth century, policy makers understood that cannibals—taken to mean the voracious and useful enzymes that had been set free throughout the world to enrich themselves and Britain—had become ungovernable. Informal empire (as explained in this book, the name of the process whereby these cannibals were unleashed) became a victim of its own success. British authorities faced an ecological problem at their headquarters and realized that cannibals could not be left roaming the City. Agents were escaping from the fold of the Admiralty, the Foreign Office, and the Navy that had created and nourished them.

Those "cannibals" were taking over the media, they were taking over Parliament, they were, like Pim, getting degrees in law and becoming legal experts. We saw that as early as in the mid-1860s Disraeli's political cunning sought to harness these myriad individual rebellions. He used the cannibals and their appetites in his drive to political and symbolic triumph in the Abyssinia campaign and the Reform Bill. But beyond Disraeli's own attempts, and as a result of them, it also came to be understood at that point by the Liberal establishment that a politically useful science was a *managed* science. This was especially so given the threat to flood the stock exchange with toxic foreign railway debts and the meeting rooms of learned societies with toxic ideas. In an age of commercial and financial expansion, anthropology, a knowledge that shaped the imagination that Britons should have of distant peoples, could not be left to its own devices. And because scientific men were so restless and because science enabled one to do so many things, it had to be elevated above partisan debate so as to be better constrained. Ironically therefore, the reason why anthropological science became bipartisan was profoundly political.

Hunt underestimated this (or perhaps understood it too well) and overestimated the likelihood of his success, and it partly explains his fate and why the Victorian machine got rid of him. Anthropology was too valuable for too many people for Hunt to be left undisturbed. Statements about peoples that Britain owned or would come to own could not be left to the uncontrolled scientist, and especially not to scientists who claimed to be inspired by scholars from Germany, Switzerland, and (even less so) France, a nation whose thriving stock and money markets came to threaten those in London at this one juncture in the nineteenth century.

And thus someone was found to puncture the Great Anthropological Bubble. That person was Hyde Clarke, and neither the importance of this character nor that of his achievements have been properly recognized thus far, as attested by his almost complete absence from the indexes of books on the subject. To repeat the point one last time, the unmaking of the Anthropological Society was the work of a stock exchange specialist using the techniques and language of the stock exchange. The resulting institution was put under the austere guidance of Sir John Lubbock, a political insider, an eminent representative of the banking lobby in Parliament, and himself a specialist of finance. Unlike Hyde Clarke, he appears everywhere in book indexes and commemorations. Lubbock's family having been handsomely indemnified from its ownership of slaves following abolition, he was evidently the right choice to deliver stern lectures in matters of truth and philanthropy.

At the Anthropological Institute that resulted from the Ethno-Anthropological merger of 1871, just as at the Corporation of Foreign Bondholders that resulted from the merger of rival bondholding groups, Lubbock was surrounded by members of the capitalist and scientific upper world. Gone were the years of the roving stock exchange filibusters and buccaneering anthropologists. Gone were the days when the science of man had a variety of politics. And this is how as early as in the beginning of the 1870s Liberals and Tories were already cooperating in the new institutions of empire, and the Anthropological Institute was one of them. The institute was created simultaneously with the Corporation of Foreign Bondholders and for similar purposes—both were bipartisan instruments to manage Britain's creditor position. In this early entanglement of finance and science can be read much of the subsequent history of anthropology, including, I suspect, its subsequent links with *Imperium Britannicum* and *Imperium Americanum*.

The victory of ethnology—in effect, as we have seen, the acquisition of the anthropology brand by the gentlemen from the Athenaeum Club—illustrates the resistance of incumbent forms of knowledge—the political survival of the intellectual Ancien Régime and its ability to reinvent itself with the help of the new agencies. It shows how, in what John Stuart Mill called suggestively the "marketplace of ideas," status remained paramount, aside from truth. In fact, status *was* truth. It is not a matter of small importance that within the British imperial knowledge system traditional approaches to science and business for which social status was key displayed so much resilience. On the back of a worldview that placed tremendous faith on people and things being bona fide, the gentleman scientist managed to retain his position, even when the trade was "professionalized."

An illustration of this is provided by the experience of King Léopold II. In fact, as the reader will have realized, Léopold's well-known later traffics, which rested on exploiting a private colony (the Free State of Congo) that came to look more like a death camp after initial scientific and humanitarian pledges, were nothing but the implementation on a larger and more successful scale of the sort of schemes Pim and his contemporaries had mulled over—as Léopold's interest for George Earl Church's Amazonian enterprises proves. The difference in their relative success boiled down to a matter of social status—Pim, who was chagrined he could never become K.C.B. versus Léopold, the blue blooded explorer-king, Victoria's cousin, and a Saxe Coburg who convened humanitarian and scientific conferences in Brussels on anti-slavery and geographical topics. In terms of *bona fides*, Pim was no match for Léopold.

I suggest this to the modern student of Victorian anthropology—think less of the relevance of Darwinism as a dematerialized system of ideas and more of its institutions, powers, and forms of exploitation. Deconstruct, but then reconstruct and in doing so, keep an eye on the trafficking propensities of British society. Keep in mind that the expansion of the Anthropological Society's membership in the 1860s coincided with the expansion of foreign investment and that there are correlations between the material rise of membership and the immaterial rise of ideas. Remember that this was the age of Trollope as much as that of Darwin and that *The Way We Live Now*, unlike the *Descent of Man*, is openly critical of the results of human evolution. At the end of my foray, I find that conventional discussion of the subject of races and racism in Victorian Britain ought to be recast from a renewed look at the political economy of scientific discussion. We have come across stock exchange supremacists who had ideas about sciences. We have come across scientists eager to make their discipline valuable.

If one lesson has emerged from this investigation of the jungle of white-collar cannibals, it is that an attempt to understand any science, especially the social sciences, that starts with a select group of "genuine" scientists and then somehow inductively defines the contours of the discipline is a non-starter. The prestige of science is precisely what the politicians of science want us to look at. One can see how previous scholars, by keeping in mind such questions as the greater or lesser prestige of such and such anthropologist from the period, constructed a universe where Lubbock was well in sight except when he was trafficking in Corporation of Foreign Bondholders' voting rights. This was also a universe where Pim was nowhere to be seen except as a puppet in a gruesome horror show suddenly displayed for a one-night stand at St. James's Hall and then hidden again forever—the horror. What was missed as a result—totally missed—was the commonality of the technologies in which Pim and Lubbock were conversant.

In bringing to the fore Lubbock's own machinations, I am not trying to debase poor Sir John, who does need me to do so. I understand that even his own biographer has difficulties thinking of him as a likable or admirable person. I am simply arguing, to reiterate the claim in the introduction and with evidence now to back my views, that anthropology was made of both those whom previous authors have preferred, such as Lubbock, and of the cannibals. Excluding any on the basis of status would be bad history. It would be history of the elite, by the elite, for the elite. I have added the gaping interstice, the "subaltern anthropologist," and I cannot see how those cannibals other scholars have discussed in greater detail were any more real than the main characters in my tale. Each time I have dug deep enough, I have come across connections between the two groups. This is no surprise, for all of them dealt with the question of value, directly or indirectly. All were subjected to the stock exchange modality. Taken together, and helped by the various financial, political, and bureaucratic machines that alternately controlled them or were controlled by them, they all contributed to an accumulation of information that came to define the profession and its uses. They did so, of course, through voyage, organization, publication, fact gathering, discussion, and through the lecture in St. James Hall, but they also did it through the forward market, through puts and calls, through the bubbles of finance and bear raids, and through the making and unmaking of reputation. *Ecce homo*: soldier, sailor, diplomat, politician, attacked by fierce Indians charging naked at midnight, contracting with a king of Miskito, conspiring with financiers, employing Hyde Clarke to destroy the credit of Hunt, caught by the vulture fund, recapitalizing the Anthropological Institute. All these were acts of anthropology.

That anthropologists (and ethnologists) had roots in the stock exchange implies that they were shaped by it in a fundamental way. If the key theme that has gradually emerged from the narrative in this book is to be distilled in one expression, perhaps a correct characterization would be to say that anthropology was one of the *techniques of globalization* that developed in an age when the control of the West expanded and operated through the capital market. It is because the West had capital markets and because imperial conquest raised problems of value and valuation that British anthropology took the form it took. Anthropological science followed the capital pull just like the capital followed anthropological knowledge. The result was the embedding of anthropology inside a network of correspondents or members located in Real del Monte or Massowah, itself reflecting preoccupations with bullion or cotton.

By telling the history of anthropology as a technique of globalization deeply interwoven with the stock exchange, my account has brought to the fore the problem of trust, a problem that is common to science and finance. This approach shifts the conventional focus away from Stocking's cultural interpretation of the anthropological dispute of the 1860s, a perspective that had many merits but also the defect of locking the past in a far, remote, and now inaccessible Victorian Britain. It was also a perspective that saw in racism both the nature and the curse of Victorian society and that read extirpation of racism, however hesitant and tentative, as salvation. If only things were so simple!

The main dilemma with which Victorians grappled was elsewhere. Every trust implies the possibility of betrayal, and thus in a sense that was not a dangerous thing, because betrayal had been discounted. And if "every thief had his consul" (and his financier and his anthropologist), this was no accident but deliberate purpose. The thing that became most dangerous for the overall order was the suspicion that the appearance of trust was for the purpose of betrayal. If that were so, then all the organizational capital that had been expended to accumulate information and process it into knowledge would have been debased in one stroke. What if scientists and financiers were *mala fide* and all appearance was treachery? What would remain of values and of the mercantile economy? Such was the suspicion that reference to the bona fide man and invocation of his rituals served to ward off. So, one had no choice but to take all behavior at face value until proven wrong, or the foundations of value would collapse.

To see how frantic the era was about this subject, we need only to remember how several higher judges handled Church's case, stating (against all that was known about him) that they were not impressed by the *mala fide* approach, and that as far as they were concerned the failure of Church's projects was "a common misfortune," a work that "has proved to be beyond the strength of those who undertook to perform it." What was so valuable to these men that they were prepared to compromise common sense? Referring to the well-known Victorian hypocrisy would be a description but not an explanation. The material danger was to the snowballing effects an indictment of Church would have caused. Since truth only existed through its performance, one at least had to assume that the staging was sincere. The defense of the credit and the mercantile order was worth living in absurdity. And if matters ran out of control, one could always send Hyde Clarke.

The result was schizophrenic. One the one hand, there were those mercantile predators who roamed the world in search of plunder. Without them, for better or worse, ethnological or anthropological information would never have accumulated. Their methodology was hyper-empirical and had much to do with that of the hunter, an expert in the game's watering spot, but also a jolly bragger and amateur of displays. This did not diminish the need for the scene—much to the contrary, as it called for a kind of catharsis of the criminal. And so, on the other hand, work was done that would transform the criminal into something useful. Institutions and rituals were conspicuously developed to eliminate the risk that the staging of credit would be perceived as a burlesque play, explaining the use of dust, stilted ways, and ponderous looks. Those who laughed were threats to be eliminated without mercy.

It fell upon Lubbock to do the job, both at the Corporation of Foreign Bondholders and at the Anthropological Institute. It is ironic to discover, at the end of the day and given the accusations of "jobbing" against Hunt, that the ultimate jobber—the ultimate invisible hand of science and finance—was John Lubbock. And it is even more ironic that it fell upon a partner in the forerunner of the modern accounting firm Ernst and Young, to discover Lubbock's misdeeds. But this is far from merely ironic. It is a subject of the greatest importance that the system of international credit, just like the system of anthropological knowledge that evolved at this juncture, while presenting themselves as boring, anonymous, automatic, in one word scientific, had at their very heart, although carefully concealed, something that was the complete opposite of boring, anonymous, automatic, let alone scientific. As this book has shown, a substantial mileage will have been achieved once we are able to go beyond naive beliefs in the alleged dichotomy between trust and corruption.

In the end, such could be the contribution of this book: to encourage a harder look at the way views of "the rest" were shaped in "the West" at a critical time for the formation of colonial systems. Some have argued that this can only be done if we de-provincialize the "rest" and they are certainly right. Conventional cultural approaches, with all their virtues and insights, have proposed critical readings of racism or orientalism as floating about in the Victorian pea soup. They have dissected Victorian literature, art, photography, botany, feminism, morals, homophobia, and anthropology. Alternatively, they have emphasized all that was left out from these projections. They have discovered voices from below, traces of rebellion and indigenous views of the world that the post-colonial writer is to recover.

But my focus on white-collar crime suggests that a decisive element is left out by such useful approaches, one not outside of the colonial project but very much part of it. In fact, it is the colonial project's beating heart. As I have argued repeatedly, and as was clearly seen in the case of the British consular service whose combination of underpay and mercurial culture was a successful invitation to insider trading, the extent to which the British Empire rested on organizing criminality cannot be underestimated. This, I think, should be put against modern studies of the disorder in the post-colony, which have politely suggested reassessing the way colonial and post-colonial discourse produced the delinquent southern state and made use of it. They have shown how northern propaganda criminalized poverty and race and, more generally, any entity that it branded as subaltern. But what if, in addition to recovering the voices from below, we were to listen more carefully to the voices from above—which for Westerners are truly the voices from within—and pore with a more patient eye over evidence found in the margins of the deceptive transparency of the academic paper, accounting book, and stock exchange quote? This book has shifted the spotlight from the indigenous, after all not so delinquent, to the predations and depredations that were taking place inside the imperial machinery. Science was no exception.

The profound corruption of Victorian mercantile ways, so worrisome to contemporary observers, has been identified in trade. If one looks more closely, it can be seen in science. In fact, the extent to which that corruption was concealed from the public scene and replaced by a dispute over the inferiority of foreign races is striking. The obsession with the character of others was a good cover for the fact that the problem of character was *the* problem within this capitalist society. This book has shown that modern ideas about corruption in the global south (from Massowah to Nicaragua) are the mirror image of that blind spot—rampant white-collar criminality in Victorian Britain. I am impressed by the fact that students of the post-colony still find it necessary to restore the honor of the indigenous. Isn't it still a way to take at face value the prejudices of the powerful? But then, modern science is heir to a mode of knowledge accumulation that was started during the years studied in this book, and this explains the difficulty we have in extricating ourselves from the original sin. Can one consider the origins of a social science and avoid the risk of its devaluation?

In this danger we may find a reason why crime has been—to use the fashionable expression—"provincialized" in narratives of science in Victorian Britain. This book will have succeeded if it has opened the possibility of a new, more difficult story of the center, not a story that would find reassurance in the blackening of a remote and by now less harmful past while making poignant but vague apologies about evil inflicted. This would be a story that recognizes our deep link with this past and, as a result, acknowledges the threats this very past still holds in store for us. The silenced voices of the abused indigenous tell us about how the evil was experienced, not how it was produced. It is altogether a more difficult and dangerous thing to investigate the implacable logic that caused the harm to be made. It requires recognizing the profound relationships that have existed, and still exist, between white-collar crime, science, and power, relationships that have been understudied, ignored, underestimated, and in sum just plain misunderstood. This could be a valid agenda in the study of imperialism, old and new.

# MOSQUITO

M any lands have products rare

O f these Mosquito takes full share.

S uch splendid trees are seldom found,

Q uite three hundred inches round.

U nder the soil and on the tree

I nfinite wealth for industry,

T ropical birds and flowers vie

O n every side to charm the eye.

Sophia Soltau-Pim, *Job and Fugitive Pieces* (London: Gee, 1885), 37

## PRINCIPLES OF SOCIAL EDITING: TWO
## PORTRAITS OF BEDFORD PIM

The following two texts—one an article written in 1877 by the novelist R. L. Stevenson and published in a minor magazine called *London*, and the other the entry for Pim in the *Dictionary of National Biography* (1896)— provide strikingly different perspectives on Bedford Pim. Stevenson's is a witty and well-informed account, which opens with an ironic emphasis, in the title, on Pim's plates. As the writing suggests, Stevenson had the risk of a libel suit in mind, and thus all the attacks on Pim's character were made on the basis of material released publicly during the Select Committee hearings as well as on a number of other public occasions.

The contrast with the *Dictionary*'s entry, attributed to C. H. Coote, a contributor to both the *Dictionary* and the *Encyclopaedia Britannica*, is striking. The *Dictionary* emphasizes a totally different set of events and portrays Pim as "a true-hearted sailor of the old school brave, generous, and unselfish." Coote claims to have worked with family papers and printed sources, but the article contains many factual errors. Even the chronology of Pim's career in the Navy is flawed. For instance, it gives September 1851 as date for his promotion to lieutenant, although his Navy files show him being promoted on October 1, 1850.

Yet today, it is the substance of this early "official history" of Pim which defines more or less the content of modern reference dictionaries such as the *Oxford Dictionary of National Biography* (reviewed by Andrew Lambert) and the *Dictionary of Canadian Biography* (reviewed by L. H. Neatby). Lambert's entry followed Coote's and "modernized" it by removing a few "normative" words on the dubious theory, one suspects, that this would make the entry more scientific. The result is of course the exact opposite. L. H. Neatby largely follows the substance and language of Coote's early contribution, again with strange edits, these ones completely random. For instance, the sentences, "He reached England on 6 June 1851. In the following September he was raised to the rank of lieutenant" have been rewritten as: "The *Herald* reached England in June 1851 and was paid off; Pim was promoted lieutenant on 6 September." Neatby also ends the entry with a quotation of the earlier version, provided with quotation marks but without indicating the source, thus lending to the judgment an air of authority and respectability: "He was the best type of Victorian adventurer, 'a true-hearted sailor of the old school, brave, generous and unselfish.'" It is regrettable that modern descendants of the British Empire no longer speak nor understand Victorian.

### "OUR CITY MEN. NO I.—A SALT WATER FINANCIER. CAPTAIN BEDFORD CLAPPERTON TREVELYAN PIM, R.N., F.R.G.S., ASSOC. INST. C. E., M.P., LATE SPECIAL COMMISSIONER FOR HONDURAS"

#### R. L. Stevenson (*London*, 1: 9–10, 1877)

The character of the British seaman, his courage, tenderness, and childlike simplicity, are among those things which make us most proud of our native land. The heart of the playgoer has often swelled within him as he saw Mr. T. P. Cooke point his toe in a hornpipe, or triumph in broad-sword combat over many dark men in petticoats. The novel-reader cherishes the memory of Commodore Trunnion, Midshipman Easy, and Peter Simple; and there is nothing we value higher in all our annals than the exploits of Drake, Raleigh, Benbow, or Nelson. In fact, we are proud of our Navy as it was and our sailors as they were and could have been content to let them remain unchanged unto the end of time. But the Navy and the naval character, like all sublunary things, are subject to the action of Great Natural Laws; and in these later days the action of Great Natural Laws has introduced some curious and unprofitable innovations–such, for instance, as Tea-kettle Ships and Salt-water Financiers. It is the particular distinction of Captain Bedford Clapperton Trevelyan Pim to represent Salt-water Finance. The gallant Captain has seen service in his day. He is an old Arctic hero, and has searched for Sir John Franklin. He wears a Crimean medal. He received three wounds in China. He has written books, pamphlets, and reports. And nowadays he is a member of the Inner Temple, and rolls his quid in Parliament upon the Government benches. So far so good. To complete the portrait it must be borne in mind that he is simply the Ideal Seaman. It might be thought that he was a man of some capacity, not without a dose of what worldly people would call shrewdness; but this is to take up the Captain by the wrong end. A man of warm feelings, if you will, rugged, hasty; using strong language when he meets with unsailorlike conduct from the landsharks among whom his lot is cast; but tender and true at heart, referring with tears in his sympathetic voice to the poor people who "embarked their little all, the savings of years of toil and industry" in the Honduras Loans, and not ashamed to confess his weaknesses, his untutored parts, and his body stiff with old age and scarred with honourable wounds. And he is specially modest as to his intellectual endowments. When he writes a book he must remind his readers that he has spent "fewer years at school than among the icebergs of the Arctic regions or under the scorching rays of tropical suns." And when he is pressed before the Committee as to his relations with Messrs. Bischoffsheim and Lefevre, he owns at once, with a hitch and a frank smile, that he does not think these clever people cared to speak about finance with the likes of him. Altogether, a simple, hearty, hale old salt as you would wish to see.

Only, why did he trail his pig-tail through all the narrows and shoals of Foreign Loan Finance? What an incongruity, what an anachronism was there! And how are we to know our way any longer through the labyrinth of the world, if Paladins, and Patriots, and Pims, and all we think most highly of, are to keep getting into the strangest of false positions, flaunting arm in arm with Sir Giles Overreach, or sitting cheek by jowl with Fraisier in his Cabinet? What on earth, we ask again, what on earth took the gallant and distinguished officer on board the galley where he was captured by Sir Henry James? And we are glad to have a thoroughly satisfactory answer. It appears that he has given a good part of his life to the question of Pacific and Atlantic intercommunication. And so, when ever he heard the Honduras Railway was in a pickle, although he had no interest in the concern but that unselfish interest common to generous and far-seeing spirits, the Captain consented to accept the position of Special Commissioner, and a salary of fifteen hundred a-year. He seems to have made no inquiries before he plunged into the transaction. It was enough for him to learn there was a ship in distress. It is one of the most chivalrous instances we have met with of that spasmodic philanthropy common to all who sail the troubled waters of Finance.

This was how the ancient mariner got in. We do not propose to inquire too closely how he got out—whether in triumph by the captain's barge, or at midnight, perilously clambering over the ship's hinder parts. We have no desire to criticise the Captain's conduct in detail, or offer him our aid in determining the precise amount he received for his services, or the precise nature of the services themselves. We shall simply point out a few details in his behaviour, for the guidance of bondholders and the elucidation of Saltwater Finance.

And first, here is a little circumstance that cannot fail to gratify all who love him, as it shows there is life in the old dog yet, and his heart is green though his body be somewhat declined into the vale of years. About a month after his appointment as Commissioner it fell to him to address and lead a meeting of Honduras Bondholders. Some of us might have liked to satisfy our own minds, before we proceeded to satisfy the minds of others, particularly when these others stood to lose "the savings of years of toil and industry," and we ourselves risked precisely nothing on the event. But that was not the way with Captain Bedford Clapperton Trevelyan Pim. His warm heart beat too fast for such calculating ways. He had promised to reassure the bondholders; and, shiver his timbers! As he almost told Sir Henry James, there was no information grave enough, no disclosure sufficiently scandalous and shabby, to make him go back from his word. He would say what he had to say. He was a man of honour.

But—and this is what specially tickles us—on the day before the meeting, as he sat in the office with Don Carlos Gutierrez, a waggish thought came into the Captain's head. He didn't wish for information—no, not he; or, of course, he might have asked for it; but it struck him it would be funny to play at being a real business man writing real business letters. And so the light-headed sailor scratched off a little note, in which, after some thoughtful oratorical prelusions, he requested a fact or two as to the financial prospects of the loan. And then he handed it over the table to the Don, with his tongue in his cheek and a wink of his merry blue eye. It was so fresh and spontaneous that even the grave half-breed could not find it in his heart to blame his companion's levity; nay, he even entered so far into the joke as to write an answer with the funniest affectation Spanish seriousness, and hand it back again to Pim. We wonder if the bondholders who listened next day to Captain Bedford Clapperton Trevelyan Pim had heard the story of this little playful correspondence, and instead of the Captain's sprightly consolations, had been sent away each one with a copy of Don Carlos's answer in his pocket—we wonder, and we can't help wondering, whether they would have gone home from the London Tavern in equally good spirits.

The second and last episode in the Captain's Special Commissionery on which we shall touch, was referred to by Mr. Digby Seymour, in impas-

sioned nonsense, as "the attempt in the sacred name of law which was suffered in the name of Captain Bedford Pim." In July, 1872—that is to say, at
the date of the meeting at the London Tavern; previous loans to Honduras
had already been expended in paying their own interest, rigging the market,
and building some portion of the first third of the proposed railway—things
were in a sorry plight. But, courage! Long Tom Coffin has still his harpoon
under his arm! Ten times he visited the Continent; and finally a new loan of
two millions sterling.

This statement contained a trifling mis-statement, which the Captain's
knowledge of French did not permit him, it appears, to rectify. What did he
know about their d–d lingo, the old sea-dog? Just at the moment when all
looked so well, there occurred a difference between the Captain and MM.
Herran and Lepelletier, the Honduras representatives in France, about—we
regret to say—about a bribe. Such conduct, the Captain says, we will easily
understand, "was a sort of thing that English sailors are not accustomed
to." There was a quarrel; and Captain Bedford Clapperton Trevelyan Pim, as
near as we understand his innuendo, seems to have taken God's name in
vain: a circumstance which, with the natural piety of sailors, he subsequently
regretted.

And now for the fifth act of the melodrama. The wicked attorneys denounced the honest tar to the French Government: and the tar "was summarily arrested at his hotel, lodged in the cells like a common felon, dragged
before the judge twice with a chain round his wrist, and incarcerated for
forty-six hours under circumstances which he disdains to excite our feeling
by narrating." That is an extract from his own eloquent speech to the second meeting of Honduras Bondholders, over which he presided. The picture
is moving and not uninstructive, perhaps not displeasing. Where there is
such a prodigious lot of naval bounce and bluffness there should be something like circumspection in a man's walk and converse.

"PIM, BEDFORD CLAPPERTON TREVELYAN (1826–1886)"

C. H. Coote (Dictionary of National Biography, edited by
Sidney Lee, Volume 45 [Pareira to Pochrich], 306–7,
[London: Smith, Elder and Co., 1896])

Pim, Bedford Clapperton Trevelyan (1826–1886), admiral, born on 12 June
1826 at Bideford, Devonshire, was son of Lieutenant Edward Bedford Pim,
who died of yellow fever off the coast of Africa in 1830, when he was engaged in the suppression of the slave trade, in command of the *Black Yoke*,

tender to the *Dryad*. His mother was Sophia Soltau, eldest daughter of John Fairweather Harrison. Pim was educated at the Royal Naval School, New Cross, and entered the navy in 1842. He served under Captain Henry Kellett [q.v.] on the *Herald* from 1845 till 1849. In that year he was lent for duty on the brig Plover, and, wintering in Kotzebue Sound, Alaska, made a journey in March and April 1850 to Michaelovski in search of intelligence of Sir John Franklin. He reached England on 6 June 1851. In the following September he was raised to the rank of lieutenant.

At this period Pim proposed an expedition in search of Franklin to the north coast of Asia, and offered to survey the coast. After receiving a grant of 500£ from Lord John Russell, unlimited leave from the admiralty, and recommendations to the authorities in St. Petersburg, he went to Russia in November 1851; but the Russian government refused to sanction his project. On board the *Resolute* he left England on 21 April 1852, and served under Sir Edward Belcher [q.v.] in the western division of his Arctic search expedition. In the following October, when the *Resolute* was in winter quarters off Melville Island, a travelling party discovered in a cairn on the island the information (placed there by McClure the previous April) that McClure's ship, the *Investigator*, was icebound in Mercy Harbour, Banks Land, 160 miles off. It was too late in the season to attempt a communication; but on 10 March 1853 Pim was despatched as a volunteer in charge of a sledge for Banks Land. The journey was accomplished in twenty-eight days ; and on 6 April Pirn safely reached the vessel, only just in time to relieve the sick and enfeebled crew (see McClure, Sir Robert John Le Mesurier).

In January 1854 Pim was appointed to the command of the gunboat *Magpie*, and did good service in the Baltic. He was wounded at the bombardment of Sveaborg on 10 Aug. 1855, for which he received a medal. In April 1857 he was appointed to the command of the *Banterer* in the war with China, being severely wounded at Sai Lau, Canton river, 14 Dec. 1857. He was invalided home in June 1858, and promoted to the rank of commander. In June 1859 he was appointed to the *Gorgon*, for service in Central America. While stationed off Grey Town he originated and surveyed the Nicaraguan route across the Isthmus, through Mosquito and Nicaragua, which now bids fair to supersede the ill-fated Panama route. While on the station he purchased a bay on the Atlantic shore, now known as Gorgon or Pim's Bay. For this he was somewhat harshly censured by the lords of the admiralty in May 1860. Returning to England in June, he retained the command of the Gorgon, and took her to the Cape of Good Hope in January 1861. On his way home he exchanged into the *Fury*. The following June he retired from active service; his name, however, remained on the navy list. He became captain on the retired list in 1868. Pim made three journeys to Nicaragua, in March 1863, October 1863, and November 1864, in reference to his transit scheme. After he had obtained additional concessions, in November 1866 a company, called the Nicaraguan Railway Company, Limited, was registered; but the necessary capital was not forthcoming, and it was dissolved in July 1868.

Pim now turned his attention to the law. On 20 April 1870 he entered as student of the Inner Temple, and on 28 Nov. of Gray's Inn, being called to the bar on 27 Jan. 1873. He was admitted a barrister of Gray's Inn *ad eundem* the following month. His practice was almost exclusively confined to admiralty cases, and he went on the western circuit. At Bristol his name became a household word among seamen, he represented Gravesend in the conservative interest in Parliament from 1874 to 1880, but failed to retain the seat at the following general election. He was elected F.R.G.S. in November 1851, and an associate of the Institute of Civil Engineers on 9 April 1861. He laid before the institute, on 28 Jan. 1862, his mode of fastening armour-plates on vessels by double dovetail rivets. He was on the first council of the Anthropological Institute, 1871–4, and remained a member of the institute up to the time of his death. He was raised to the rank of rear-admiral on the retired list in 1885. He died at Deal on 30 Sept. 1886, in his sixty-first year, and a brass tablet and window were placed in his memory at the west end of the church of the Seamen's Institute, Bristol, by the pilots of the British empire and the United States of America in 1888. He was a true-hearted sailor of the old school brave, generous, and unselfish. Pim married, on 3 Oct. 1861, Susanna, daughter of Henry Locock of Blackheath, Kent, by whom he had two sons.

His published works include: 1. 'An Earnest Appeal . . . on Behalf of the Missing Arctic Expedition,' 1857 ; 5th edit, same year. 2. "Notes on Cherbourg," with map, 1858. 3. "The Gate of the Pacific," 1863. 4. 'The Negro and Jamaica,' 1866 (special No. of 'Popular Magazine of Anthropology'). 5. "Dottings on the Roadside in Panama, Nicaragua," &c., 1869 (in conjunction with Berthold Seemann). 6. 'An Essay on Feudal Tenures,' 1871. 7. 'War Chronicle: with Memoirs of the Emperor Napoleon III and of Emperor-king William I,' 1873. 8. 'The Eastern Question, Past, Present, and Future,' 1877–8. 9. "Gems from Greenwich Hospital" 1881. He also contributed an article on shipbuilding to Bevan's "British Manufacturing Industries" 1876.

.

## PIM'S TRAVELS: FROM THE *GORGON* TO THE *FURY*

I established the following summary of Pim's travels using the ships' logs of the *Gorgon* and the *Fury* as well as information from two books narrating episodes that took place during these voyages (*Dottings along the Roadside* by Pim and Seemann, and *Cruise in the "Gorgon"* by Devereux). When in doubt, the ships' logs command greater authority. My coverage starts with Pim taking over the *Gorgon* in April 1859 and ends with his return to London aboard the *Fury* in March 1861, at which point he was "compulsorily retired."

### 1859

*From Britain to Bermuda*

April 27: Pim takes over the *Gorgon*, a new ship log is started.

April to August 1859: Coastal navigation in Britain.

August 30: Off Plymouth.

September 5: Steaming past Cork.

September 6: Alongside coal wharf at Haulbowline.

September 7: Off Haulbowline steering for Madeira archipelago.

September 15: Reaches Funchal (Madeira).

September 15: Leaves steering for Halifax.

October 10: Entering Halifax harbor.

October 20: Leaving Halifax.

October 25: At anchor along Bermuda dockyards.

*Caribbean and Mosquito Coast*

November 5: "Made fast to buoy off Jamaica Dockyard."

November 6: Coal wharf (episode narrated in *Dottings,* chapter 13, p. 215).

November 10: Port Royal harbor.

November 18: Steaming out of Port Royal harbor.

November 22 1: Arrives at Greytown, meets with *HMS Racer* (see also *Dottings,* p. 229).

November 23 to December 1: At single anchor, outer roads, Greytown; Dec. 1 "Ship rolling heavily."

December 2: Leaving Greytown to Corn Islands.

December 8: Leaving Great Corn Island for Pearl Cay Point.

December 10: Pearl Cay Point to Bluefields Bluff.

December 11: "The King Musquito [*sic*] came on board [& Valet Wm Nute (?)]."

December 12: Bluefields Bluff to Monkey Point.

December 15: Monkey Point to Greytown "Left the ship for shore. King Musquito [???]."

December 20: [at Greytown] "King of Musquito [on (?)] visit board."

December 21: Greytown to Bluefields. "The King of Musquito left for shore. Saluted with 21 fires."

December 22: Bluefields to Monkey Point. Christmas at Monkey Point, logging.

December 30: Leaves Monkey Point.

December 31: Arrives at Greytown; The *Gorgon* spends January in Greytown.

## 1860

January 16: "Commander Pim left the ship to proceed up the River San Juan de Nicaragua for the purpose of communicating with Mr. Wyke consul."

January 29: "Ret'd on board Com'd B. Pim."

February 2: Greytown to Bluefields Bluff.

February 3: Bluefields to Great Corn Island.

February 5: Leaves Great Corn Island.

February 6: Arrives at King Cay.

February 10: King Cay to Great Corn Island.

February 12: Great Corn Island to Monkey Point.

February 12: Monkey Point to Greytown.

February 15: Leaves Greytown, southbound.
February 23: Anchor at Limon Bay (Atlantic mouth of today's Panama Canal).
February 26: Leaves Limon Bay.
February 28: Arrives at Greytown.
March 21: Leaves Greytown for Monkey Point; Last farewell to new property.
March 21: Leaves Monkey Point for Great Corn Island.
March 22: Anchor at Great Corn Island.
April 4: Leaves Great Corn Island.

*Jamaica and Bermuda*
April 11: Anchor at Pedro Banks (sand and coral bank, 80 km south of Jamaica).
April 13: Reaches Port Royal, Jamaica.
May 9: Proceeding out of Port Royal.
May 12: Anchor at Cape Dame Marie, Haiti.
May 13 and 14: Anchor at Matthew Town, Bahamas.
May 16: At anchor at Cay Verde (120 km off north coast of Cuba).
May 27: Arrives in sight of Bermuda, at anchor.
May 28: Enters dock yard Camber, Bermuda.
May 29: Moored at dock yard Camber, Bermuda.
June 11: Leaves Bermuda.
June 26 and 27: At anchor in Horta Bay (Azores).

*Britain: Refitting and Troubles*
July 14: Arrival at Spithead, off Portsmouth.
July 1: Enters Portsmouth.
July 20: In Portsmouth refitting in Steam Basin begins (an area where timbers were softened with steam so that they could be bent and shaped).
August 8 (until August 16): Steam basin in Portsmouth.
August 9 (until September 4): *Gorgon* in dock, Portsmouth.
September 5 (until September 30): Steam basin in Portsmouth, new boilers installed.
September 30 (until October 10): Several days for which Pim's handwriting predominates. Various assignments (October 1: "Gig in blue Victory"; participation in court martials and Navy examinations).
October 11: *Gorgon*'s refitting continues.
November 7: *Gorgon* proceeds out of Portsmouth.
November 8 (to 14): Plymouth.

November 15: Plymouth to Spithead (off Portsmouth).

November 16 (to 25): Enters Portsmouth harbor.

November 26: Anchor at Spithead.

November 27 (to 29): in the Hamoaze (Estuary of Tamar in Plymouth Sound).

*Heading to Cape Town via Brazil*

November 30: Off Plymouth Sound, heading to Porto Santo (Archipelago of Madeira); The Gorgon has *HMS Swift* (a packet brig launched in 1835, which was destined to be used as a mooring vessel) in tow and will carry it till it reaches the Cape, its final destination.

December 13: Rounds the north end of Madeira Island to Funchal.

December 22: Proceeds out of Funchal Roads, heading to Gran Canaria.

December 26: Areynaga, Gran Canaria, in sight.

## 1861

January 1: At anchor, Porto Grande, Sao Vincente (Saint Vincent), Cabo Verde.

January 6: Tarrafal Bay, Santo Antonio, Cabo Verde; Leaving on the 7, heading to Fernando de Noronha (off Brazil's coast), then Cape Frio.

January 31: At anchor in Rio de Janeiro harbor; Captain Pim goes off for a trip on the Don Pedro II Railway, in construction: "The officials, although very courteous at first, appear to disrelish the idea of our worthy captain's curious determination to examine their establishment; but at last consent and kindly pass him onwards, allowing him to travel on the engine—a peculiar vanity of his. Now Captain P—, clad in Arab coat, fez, and umbrella, waves farewell; the engine gives a snort and a puff, and she is off." (W. Cope Devereux, *A Cruise*, 28–29).

February 6: Steaming out of Rio de Janeiro harbor, heading to Tristan da Cunha (volcanic islet in the South Atlantic, midway between Africa and Brazil).

February 23: Tristan da Cunha in sight; On the 24, *Gorgon* departs for Cape of Good Hope.

March 7: Anchor at moorings in South Simons Bay (near Cape Town); "7:30, cast off the 'Swift'"; Stays there until the change of captain.

## THE *FURY*

*Pim Removed*

April 1st: Pim changes ship; Commander J. C. Wilson takes over *Gorgon*
    and does "read his commission for the ship"; "[illegible] Pim to the
    Fury"; (Devereux describes a transfer taking place the day before,
    with Pim accompanied by his officers. He thus concludes the de-
    scription of the scene: "Captain P—takes with him our sincere love
    and respect, and leaves an impression which time cannot efface,"
    entry "March 31st, W. C. Devereux, *A Cruise*, p. 48. The entry for
    April 1st reads: "New Commander reads his commission and we
    have a fine opportunity of finding fault with him; All I intend to say
    of him is, he is considered a smart officer, loves his profession, and
    knows more of a ship than most men."

April 3: Table Bay, Cape Town.

April 18: Saint Helena.

April 27: Clarence Bay, Ascension.

May 14: Tarrafal Bay, Saint Vincent.

May 24: Santa Cruz, Madeira, May 25, heads to Cape Finisterre.

June 19: Portsmouth; page signed "Bedford Pim"; Pim's Navy file men-
    tions that service ends with that day and ship.

FIGURE 12. *H.M.S. Gorgon*, by Sir Oswald Walters Brierly. Source: Wiki Commons.

# THE DEMOGRAPHICS OF CANNIBALS

The identification of members of the Anthropological and Ethnological societies and the production of membership statistics is not straight-forward. Early researchers have grappled with the issue, as illustrated by Chapman's discussion of Pitt Rivers's association with the Ethnological Society when he reported being unsure about the date when Pitt Rivers joined the Ethnological Society, and after strenuous deductions concluded (rightly) that "it is likely that Fox was one of 21 new members elected in 1860–61" (Chapman, *Ethnology in the Museum*, chapter 4, footnote 1).

This supplement reviews the sources for measuring the number of eth-nologists and anthropologists and discusses previous attempts to quantify them. It emphasizes the importance of membership registries, on which my research drew. Using the sources described in this supplement, Royal Anthropological Institute archivist Sarah Walpole is putting together an online database of fellows that will prove tremendously useful for future researchers.

## LEONE LEVI

The first attempt ever to count the students of man was made by the nineteenth-century jurist, economist, and statistician Leone Levi. In 1868 (coinciding therefore with the Clarke scandal), he contributed a paper to the British Association for the Advancement of Science meetings titled, "On the Progress of Learned Societies, Illustrative of the Advancement of Sci-ence in the United Kingdom During the Last Thirty Years" (*Report of the Thirty-Eighth Meeting of the British Association for the Advancement of*

*Science*, pp. 169–73, 1868). More than a decade later, he contributed an update, "The Scientific Societies in Relation to the Advancement of Science in the United Kingdom" (*Report of the Forty-Ninth Meeting of the British Association for the Advancement of Science*, pp. 458–68, 1879).

Counting members was far from straightforward. As indicated in the text, there were various groups of members: the ordinary fellows, the permanent fellows, the honorary fellows, the foreign correspondents and the local secretaries. In his articles, Levi provided no explanation of how he went about gathering numbers for their membership. Nor was he apparently interested in methods or definitions of membership. One suspects he merely asked the secretaries of the various societies and assumed he would get *bona fide* answers. For 1867, Levi reported 219 members for the Ethnological Society and 1,031 for the Anthropological. In the later update, the figure for the Anthropological Society of London in 1867 was reported again and chained to the number given for the successor Anthropological Institute of England and Wales, as if the latter followed the former. This was as if the Ethnological Society had never existed and the Institute was the descendant solely of the Anthropological Society. A figure of 462 was reported for Anthropological Institute membership in 1878.

It is unclear how the figure of 1,031 for 1867 was established. It does not square with available evidence or commentary. In his rejoinder to Hyde Clarke's first *Athenaeum* letter, Hunt stated in the summer of 1868 that, while the list of fellows on the books of the Anthropological Society had amounted to about 400 in January 1865, "they now number more than seven hundred" (recall that it was in the spring of 1865 that the 500th fellow celebration had taken place). I suspect Levi may have been communicated this figure by a publicity-conscious member of the Anthropological Society, and that such a number was obtained by counting *all* fellows signed in since the beginning of the society (about 840 according to my own reckoning of the number of fellows since inception), plus honorary fellows, foreign correspondents and local secretaries, which stood at roughly 200 at this date. Some local secretaries were fellows and had been counted twice. The figure of 1,031 was clearly inflated, since it did not account for departures.

## David K. Van Keuren

The other notable statistical foray is found in the work by D. K. Van Keuren, "Human Science in Victorian Britain: Anthropology in Institutional and Disciplinary Formation, 1863–1908," an unpublished 1982 Ph.D. dissertation from the University of Pennsylvania with an excellent chapter providing statistical material on the sociology of the two rival societies. In particular, Van Keuren provides a chart showing the rise and fall of the Anthropological Society bubble (p. 51). The chart underscores that the collapse of the society had little to do with the end of the American Civil War (and thus makes one wonder about the reasons for the persistence of the myth). As Van Keuren indicates, "after Hunt's death, a disastrous decline of membership was initiated which finally forced amalgamation with the Ethnological" (p. 53). I find myself in broad agreement with Van Keuren, but since he does not explain the details of his computations either, it may be worth reviewing the subject further.

## Sources

There are six main sources for counting membership: The annual reports, the minutes, the membership lists (a rich collection of these being held in the Anthropological Institute archive), the accounts, material in the Royal Anthropological Institute's archive, and the accounts with the publisher Trübner. As explained below, these various sources provide different perspectives on membership.

(1) Annual reports were printed in the respective societies' flagship publications at the date of the anniversary meeting (in January for the Anthropological Society and in May for the Ethnological Society). They often contained useful indications concerning outstanding membership, new fellows, and departures through death or resignation. This might appear as a useful source, but these were never rigorous, nor systematic. During its first years of existence, the Anthropological Society's annual reports gave elements that permit tracking the dynamics of regular fellows through elections, totals, and departures, but after 1865, the numbers reported became confusing and incomplete. Similar information published by the Ethnological Society is absent in the early 1860s, and scant afterwards. Because the Ethnological Society was the smaller of the two societies, the absolute size of the errors is smaller, but not their relative size.

(2) At the beginning of each meeting, new members were voted in and their names were listed in the meeting summary published in the *Journal of the Anthropological Society of London*. A list of new fellows, local secretaries, honorary members, and foreign correspondents can be constructed from such material. However, this list is silent on resignations.

(3) Membership lists were printed and circulated (see RAI Archive, especially A2, A8, A31). Because the publishing of names could trigger complaints (as in the case of Buxton, a permanent member who insisted,

successfully, that his name be withdrawn), they represent a reasonable indication of membership. The lists also recorded the various categories of
membership (fellows, honorary members, corresponding members, etc.). In
this book, we have exploited three lists for the Anthropological Society
which are found in the back matter (with separate, Arabic numbering) of the
volumes of *Memoirs Read before the Anthropological Society* (1865, 1866,
and 1870). The last volume provides the list as it stood in August 1869, a
few days before the death of James Hunt. These lists make it possible to
establish membership on a given date.

The lists circulated by the Ethnological Society of London are found in
the *Transactions of the Ethnological Society of London*. Volumes 2, 5, 6, and
7 give memberships for 1863, 1866, 1867, and 1869 respectively. They seem
to correspond to the existing situation of membership on the date of the
annual meeting, in May. Additional interesting information in these lists
includes a mention of members who had contributed a paper; the member's
address or club; occasionally the member's profession; and, in the 1869 list,
the date when the member had joined.

(4) It is possible to use the balance sheets to determine the number of
ordinary fellows in any given year, because we know the cost of an individual subscription (an annual fee of £2.2) and the total revenue from annual
subscriptions reported in books (adjusting for permanent membership). In
some years, the information is detailed enough to determine the number
of fellows subject to the fee as well as the late payers. For instance, in the
last issue of the *Journal of the Anthropological Society of London*, vol. 8,
p. lxxiv, a line records "subscriptions received 1869." In other years, however, the material from the published accounts is more difficult to exploit.

(5) The Royal Anthropological Institute Archive (A6 Membership Subscription Ledgers, 1863–1872) provides an interesting document regarding
members and payment of their dues, and it could be exploited to further the
evidence reported in this book. However, the sheer number of members and
complications resulting from the many layers displayed in this document
have prevented full exploitation at this stage.

(6) A last potentially interesting source that was explored consists in the
accounts of publisher Trübner (Publication Books and Publication Account
Books). However, their investigation yielded discouraging results. The surviving material concerns the circulation of the *Anthropological Review*
beyond the copies sent to fellows (this represented a circulation of about
200 copies per issue). It says nothing on those copies of the *Anthropological
Review* that were sent to fellows, and which must have been recorded in a
separate account with the Anthropological Society of London. Likewise, the

accounts are silent regarding the circulation of the *Journal of the Anthropological Society of London* to fellows.

## Some Estimates

Using these sources, fairly reliable numbers for various groups of members can be established at the dates when selected membership lists were circulated. Information from other sources was also used to complete this information and the basic output at benchmark years is shown in table 1. "Paying members" is to be understood here not as members who actually paid their dues but as members who were *expected* to do so. This eliminates honorary fellows, corresponding members, and, for the Anthropological Society, those local secretaries who were not members.

TABLE 1. Society Membership

| | Ethnological Society | | | | Anthropological Society | | | | |
|---|---|---|---|---|---|---|---|---|---|
| | Paying members | Nonpaying members | | Total | Paying members | | | Nonpaying members | Total |
| | Permanent and annual fellows | Honorary fellows | Corresp. members | | Permanent and annual fellows | Local secretaries | Honorary fellows | Corresp. members | |
| 1862 | 155 | n.a. | n.a. | n.a. | 0 | 0 | 0 | 0 | 0 |
| 1863 | 180 | 44 | 28 | 252 | 198 | 28 | 25 | 17 | 268 |
| 1865 | n.a. | n.a. | n.a. | n.a. | 522 | 60 | 25 | 40 | 647 |
| 1866 | 219 | 47 | 31 | 297 | 701 | 68 | 29 | 42 | 840 |
| 1867 | 218 | 47 | 31 | 296 | n.a. | n.a. | n.a. | n.a. | n.a. |
| 1869 | 230 | 39 | 32 | 301 | 636 | 125 | 32 | 66 | 859 |

*Sources:* Ethnological Society, *Transactions of the Ethnological Society of London*, vols. 2, 5, 6, 7; *Journal of the Anthropological Society of London* 1 [1863], xxiv ff; *Memoirs of the Anthropological Society*, vols. 1–3.

*Note:* For the Ethnological Society, the author has estimated the number of paying members for 1862 by subtracting the number of members signed in 1862 from the total membership in 1863 according to volume 6 of *Transactions of the Ethnological Society of London*, which gives year of membership. (This calculation assumes no resignations or deaths that year and no resignations or deaths before 1867 among those who joined in 1862. The two errors, in two different directions, should be small and hopefully cancel one another.) The remaining Ethnological Society totals correspond to data from May of the specified year. For the Anthropological Society, the author has estimated membership for 1863 from data in the society journal's first issue and the totals correspond to the situation in late December of that year. The remaining totals correspond to data from April 4, 1865, August 20, 1866, and August 1, 1869, respectively.

# HOW TO PRICK AN ANTHROPOLOGICAL BUBBLE

This supplement explores technical aspects of Hyde Clarke's "bear raid" on the Anthropological Society. Understanding the logic of the raid requires an understanding of membership recording and financial accounting. As explained in text, Hyde Clarke told the story of an unsustainable debt, of a society on the verge of bankruptcy, and voiced his concern as a member about personal liability for the society's debts. The argument was that, since learned societies did not have the limited liability form, their members were individually exposed to collective losses.

## CASH AND CANNIBALS

Because membership was the main source of income, the membership increase reflected on the subscription list was a statement about the society's credit. In principle, each fellow had to pay his dues annually, and at £2.1 per annual subscription, or £2 for simplicity a *bona fide* membership of say 500 members meant an annual income of £1,000. This equivalence between membership and cash is evident in the Hyde Clarke letter and the rejoinders, which take figures on membership and figures on revenues as synonymous. As a result, the question of *bona fide* membership and the question of *bona fide* accounts were two sides of the same coin. Deciding on whether Hunt was puffing and jobbing, therefore, boils down to a problem of membership reporting. This might appear quite simple, but as it turns out, it was far from straightforward.

Indeed, the simple initial mathematical link was blurred by several factors. First, some members compounded their subscriptions by paying a larger amount of cash, about £20, though the exact sum appears to have been adjusted for the interest rate and the time of year the subscription was paid. Those interested in actuarial computations will see that compounding was given a big incentive since a life subscription of £2 per year was perhaps worth something around £40 rather than £20 (since interest rates were below 5 percent). This subsidy had to be understood as an offset of the cost borne out by the compounder, should the society go out of business. Life members brought a substantial contribution when they were signed up. Yet it seems that the windfall was typically treated as current revenue and spent immediately. This profligacy became a burden later. For instance, a 500-member society with 50 life members only earned £900 annually (500 − 50 = 450 × £2). Yet the society was still committed to deliver its publications to all members, including life members.

Another complication that disconnected revenues from membership was the question of whether members were paying or nonpaying. As noted, learned societies had members who were not charged annual fees although they received the publications and had full membership rights. In the Anthropological Society, this was the case with honorary fellows, foreign correspondents, and local secretaries, although the last group was encouraged to join as fellows as well.

Of course, such members brought advantages, so there was a rationale for not charging them fees. Honorary fellows benefited the society with prestige, foreign correspondents with reports, and local secretaries with photographs, artifacts, and information. It may have been difficult to charge these individuals who had been solicited or were contributing goodwill. It remained that such members did cost the society the price of the publications to which they were entitled. The 500-member society in the example above with, say, 50 honorary fellows, 50 corresponding members, and 50 local secretaries could rightly claim to have 650 members (and expenditures proportionate to this number), yet its revenue was only £900, proportionate to its 450 *paying* members. In his first letter to the Athenaeum, Clarke argued that the "four hundred paying Fellows" were subsidizing the "four hundred non-paying fellows" and there was decidedly an element of truth to the remark (Clarke, "Anthropological Society of London," *Athenaeum*, August 15, 1868, 210). But what he did not emphasize was that the very same logic applied to the Ethnological Society.

## BAD PAYERS

Yet even taking all of this into account, one cannot reconcile the amounts received in subscriptions and the number of members that ought to have contributed. On average, the actual amount in the books stood consistently about 20 percent below the amount a society might have hoped to receive from subscriptions. Before jumping to the conclusion that the membership list was doctored, we must consider one straightforward reason for this— late payments. There were inherent delays, such as caused by the distance at which many fellows lived—it may not have been easy to send a reminder to Fernando Po or Shanghai! Some fellows simply forfeited their dues. Efforts at boosting membership resulted in fellows being put on the membership list who did not have a real interest and failed to actually pay their dues when the time came. It might be suspected that some fellows of higher rank or those with political or social connections looked at their participation as a benefit they brought to the society (a bit like honorary fellows) and resisted payment.

In the end, these various phenomena conspired to open a chasm between the membership of fellows who actually paid and the broader membership of all fellows, paying or not, punctual or not. Note that the difficulty of reconciling the number of fellows with the amount of cash that actually rolled in was not specific to the Anthropological Society. It can be observed at the Ethnological Society too. In 1867, for instance, I find 175 paying members, and at £2 per member, revenues from paying fellows should have stood at £350, but they were shown at £285, a difference proportional to that observed for the Anthropological Society.

## DEBTS AND DEFICITS

One thorny issue with the accounts of the Anthropological Society was the manner in which the society recorded the money owed by late payers. In the first two years, unpaid dues were booked as credits, which the society held on late payers. To take our earlier example once again, if 100 members were late in their dues, then the actual revenue of a society with 450 paying members was £700 (£900 less the £200 owed by the non-payers). If the society had spent £900 that year, then it had a deficit of £200, against which, in principle at least, it could claim to hold a £200 credit against fellows. This might have been recorded into a capital account that would keep track of the pending claims and debts, but nothing of the kind was provided with annual returns.

Moreover, if those late payers forfeited their dues, then the society's claim became an irrecoverable asset, and the society would be in debt for £200. In 1863, for example, the society recorded £384 in membership fees with £63 of that total not yet paid. In 1864 the receipts were put down at £709 with £239 of that total still owed, or a total of £302 (63 + 239). Looking at subsequent payments on the dues for these two years, all we can find is about £150, less than half of the £302 due in late payments. The credit on members had become an outright loss.

From 1865 onward, the reporting of membership in arrears on the revenue side stopped, although how they proceeded exactly is unclear. There is the possibility that such questionable practices continued entirely concealed from sight by the absence of proper reporting of membership revenues. Indeed, until 1867, the account books merely indicated "subscriptions," and it is possible that the society used this as a way to borrow. Following Hyde Clarke's accusations, the accounts for 1868 clarified the matter with the use of the expression "subscriptions received" to record membership fees.

## THE TWO BODIES OF THE ANTHROPOLOGICAL SOCIETY

The lack of clarity that characterized the management of Anthropological Society accounts and the efforts that had been made to conceal the existence of a debt were not unique by British corporate standards, but they provided an opportunity for raising questions about the society's *bona fides*. As indicated in the text, the actual amount of the debt may not have been unsustainable. Creating a new outlet was costly, and this cost was reflected in the outstanding debt. The "bipartisan" audit of 1868 set the debt of the Anthropological Society at £670 11s 4d and indicated that it was entirely owed to the publisher ("Report of the Committee of Investigation into the General and Financial Condition of the Anthropological Society of London" *Journal of the Anthropological Society of London*, vol. 7, pp. i–vi, iv). Against this, the society held offsetting assets valued at £1,177, although these assets were not liquid and therefore hard to price. They comprised in roughly equal proportions that fraction of the subscriptions in arrears estimated as good, the stock of books for sale held by the publisher, and physical assets such as the museum, the library, and furniture. The Committee of Investigation deplored the confusing use of subscriptions in arrears and the lack of reporting regarding outstanding debts and liabilities. It recommended that, in the future, a complete balance sheet for the assets and liabilities ought to be published annually. In the Report for 1869 (*Journal of the Anthropological Society of London*, vol. 8, p. lxxvii), the debt was assessed at £834 8s 11d. The report for 1870 was not so precise, stating instead that the Council was "sorry to say that the debt of the Society has not been reduced during the past year." But they were still confident, they added, since the debt "was no more than one year's income," and again offset by the society's asset (see roman numeral "Appendix" to the *Journal of the Anthropological Institute*, vol. 1, 1872, p. xxiii).

The logic of Hyde Clarke's attack becomes clear. What the Hyde Clarke raid had damaged was the earning capacity of the Anthropological Society of London, transforming a previously deferrable debt problem into a pressing issue: a liquidity problem into a solvency problem. Rather than recognizing the need for a debt outlay that was useful to provide for the building of the anthropological infrastructure, he pointed to the spread between the Anthropological Society's membership numbers and the picture painted by the accounts. This was an astute move, because the more the membership had been puffed up, the less defensible financial returns became.

This underscores that there were really two ways to think of the Anthropological Society's bottom line. The first, favored by the Cannibal Club, looked at the society as a kind of fellowship of jointly responsible scholars who had pledged their names and credit and could thus raise money to carry on their agenda. The second, promoted by Clarke, looked at the Anthropological Society as a simple matter of debts and assets and discounted the fellows' goodwill. Clarke's involvement as early "accountant" in the disputes surrounding the railway mania of 1845–47 takes here its full significance. Clarke portrayed the Cannibals as insiders and looters in the same way directors of a railway line had been described during the railway mania. Against the abuses (alleged or real) of a clique of insiders, the dry science of the accountant provided a remedy (Odlyzko, "Collapse of the Railway Mania," 321–22).

This gives context and perspective to the accusations in the *Athenaeum* that members of the Anthropological Society were "subjected to such losses and liabilities as no other society has suffered" and were "oppressed" by a "large debt" (*Athenaeum*, p. 210, August 15, 1868). In sum, by emphasizing his own concern with being associated with the Anthropological Society, Clarke was really calling for a subscribers' rebellion, encouraging them individually to leave the society. It was this, the falling membership, and not the debts that was the essential cog of the bear raid. Connoisseurs would have recognized the essence of any successful short on a leveraged firm, where the leverage itself is used by the bears to bring the company to the tipping point. This provides background to the exaggerated claims Clarke made of a run by fellows resigning "by the hundreds"—this was precisely what he intended to provoke, and he largely succeeded.

# NOTES

## PREFACE

1. George W. Stocking, "What's in a Name? The Origins of the Royal Anthropological Institute (1837–71)," *Man*, n.s., 6 (1971): 369–90, 378.

2. The character is Clements R. Markham, and the remark appears to have been a compliment; see Elizabeth Baigent, "Markham, Sir Clements Robert (1830–1916)," in *Oxford Dictionary of National Biography* (Oxford: Oxford University Press, 2006).

3. Translated as Pierre Gripari, *Tales of the Rue Broca* (Indianapolis, IN: Bobbs-Merrill, 1969).

4. Eugène Ionesco, *Théâtre I* (Paris: Gallimard, 1951). Since 1957 this play has been in permanent production at Paris's Théâtre de la Huchette, not so far from the Rue Broca.

## INTRODUCTION

1. For many references on the subject, see John D. Kelly et al., eds., *Anthropology and Global Counterinsurgency* (Chicago: University of Chicago Press, 2010); or the more recent book by David H. Price, *Weaponizing Anthropology: Social Science in Service of the Militarized State* (Petrolia, CA: Counterpunch, 2011), which discusses the rising use of anthropology by the U.S. military, particularly in counterinsurgency operations.

2. J. F. W. Herschel, *Manual of Scientific Inquiry Prepared for the Use of Her Majesty's Navy and Adapted for Travellers in General* (London: John Murray, 1849). The first edition mentioned John Murray as publisher to the Admiralty. The third edition (1859) stated, "published by authority of the Lords Commissioners of the Admiralty." The Ethnological section had been first drawn up by J. C. Prichard and was later revised by T. Wright.

3. B. Malinowski, "Practical Anthropology," *Africa: Journal of the International African Institute* 2 (1929): 22–38; L. Tournès, "La fondation Rockefeller et la naissance de l'universalisme philanthropique américain," *Critique Internationale* 35 (2007): 173–97; D. H. Stapleton, "Joseph Willits and the Rockefeller's European Programme in the Social Sciences," *Minerva* 41 (2003): 101–14; Volker R. Berghahn, "Philanthropy and Diplomacy in the 'American Century,'" *Diplomatic History* 23 (1999): 393–419.

4. See, for instance, Jack Stauder, "The 'Relevance' of Anthropology to Imperialism," in *The "Racial" Economy of Science: Toward a Democratic Future*, ed. S. Harding (Bloomington: Indiana University Press, 1993), 408-33, 418. A. Kuper, *Anthropology and Anthropologists: The Modern British School*, 3rd ed. (London: Routledge, 1996), 56, quoting Evans-Pritchard, and 94, quoting from Johan Galtung, "Scientific Colonialism," *Transition* 30 (1967): 11-15. M. Fortes and E. E. Evans-Pritchard, *African Political Systems* (London: Oxford University Press, 1940).

5. T. Asad, introduction to *Anthropology and the Colonial Encounter*, ed. T. Asad (London: Ithaca, 1973), 17-18. This theme of the small effect of anthropology on colonialism is also emphasized by Asad in another article, where he writes about how study of the history of anthropology can illuminate aspects of the expansion of Europe's power, a transformation "of which this discipline was a small part"; T. Asad, "From the History of Colonial Anthropology to the Anthropology of Western Hegemony," in George W. Stocking, ed., *Colonial Situations: Essays on the Contextualization of Ethnographic Knowledge*, in *History of Anthropology*, vol. 7 (Madison: University of Wisconsin Press, 1993), 314-24; 314.

6. As illustration, see Benoît de L'Estoile, Federico Neiburg, and Lygia Sigaud, *Empires, Nations, and Natives: Anthropology and State-Making* (Durham, NC: Duke University Press, 2005), where the outcome of the struggle between anthropologists and interested government agencies is presented as open-ended. Likewise, A. L. Conklin's *In the Museum of Man: Race, Anthropology, and Empire in France, 1850–1950* (Ithaca, NY: Cornell University Press, 2013) argues that students of anthropologist Marcel Mauss began to challenge both colonialism and the scientific racism that was used to justify it, enrolling in the fight for the humanist values they had learned from their teachers and in the field.

7. The reader may rightly perceive here a distant echo of Immanuel Wallerstein's *Unthinking Social Science: The Limits of Nineteenth-Century Paradigm*s, 2nd ed. (Philadelphia: Temple University Press, 2001).

8. Londa Schiebinger, "Forum Introduction: The European Colonial Science Complex," *Isis* 96, Special Issue (2005): 52–55. For the eighteenth century, see for instance Richard Drayton, "Maritime Networks and the Making of Knowledge," in *Empire, The Sea and Global History: Britain's Maritime World, c. 1760–c. 1840*, ed. D. Cannadine (2007), 231–52, or his chapter entitled "Knowledge and Empire," in *The Oxford History of the British Empire*, vol. 2, *The Eighteenth Century*, ed. P. J. Marshall (Oxford: Oxford University Press, 1998), 72–82. For the late nineteenth and twentieth centuries, see Helen Tilley, *Africa as a Living Laboratory: Empire, Development and the Problem of Scientific Knowledge* (Chicago: University of Chicago Press, 2011).

9. James Urry, *Before Social Anthropology: Essays on the History of British Anthropology*. New York: Routledge, 1993), esp. chap. 5: "Imperial Anthropology and Institutional Development in British Anthropology."

10. See Tilley, *Africa as a Living Laboratory*. A similar metaphor serves as subtitle for the collection of sociological essays edited by George Steinmetz, *Sociology and Empire: The Imperial Entanglements of a Discipline* (Durham, NC: Duke University Press, 2013). See also George Steinmetz, *The Devil's Handwriting: Pre-coloniality and the German Colonial State in Qingdao, Samoa, and Southwest Africa* (Chicago: University of Chicago Press, 2007).

11. Edward Beasley, *Empire as the Triumph of Theory: Imperialism, Information, and the Colonial Society of 1868* (London: Routledge, 2005); Edward Beasley, *Mid-Victorian Imperialists: British Gentlemen and the Empire of the Mind* (London: Routledge, 2005); Edward Beasley, *The Victorian Reinvention of Race: New Racisms and the Problem of Grouping in the Human Sciences* (London: Routledge, 2010).

12. P. J. Cain and A. G. Hopkins, *British Imperialism: 1688–2000*, 2nd ed. (Edinburgh, UK: Pearson, 2002 [1st ed., 1993]); J. Gallagher and R. Robinson, "The Imperialism of Free Trade," *Economic History Review* 6 (1953): 1–15; D. C. M. Platt, *Finance, Trade, and Politics in British Foreign Policy, 1815–1914* (Oxford: Clarendon Press, 1968). On the history of the role of the city: David Kynaston, *The City of London: A World of Its Own, 1815–1890* (London: Pimlico, 1995); R. C. Michie, *The London Stock Exchange: A History* (Oxford: Oxford University Press, 1999).

13. See B. S. Cohn, *Colonialism and Its Forms of Knowledge: The British in India* (Princeton, NJ: Princeton University Press, 1996), 61. See also C. A. Bayly, *Empire and Information: Intelligence Gathering and Social Communication in India, 1780–1870* (Cambridge, UK: Cambridge University Press, 1996).

14. Pierre-Joseph Proudhon, *Manuel du Spéculateur à la Bourse*, 5th ed. (Paris: Garnier, 1857), 170.

15. One may add to the list a "missionary modality" more recently emphasized by Patrick Harries (not his expression). P. Harries, "From the Alps to Africa: Swiss Missionaries and Anthropology," in *Ordering Africa: Anthropology, European Imperialism, and the Politics of Knowledge*, ed. Helen Tilley and Robert J. Gordon (Manchester, UK: Manchester University Press, 2007), 201–24. See also P. Harries, *Butterflies and Barbarians: Swiss Missionaries and Systems of Knowledge in South-East Africa* (Athens: Ohio University Press, 2007).

16. As one reviewer remarked, there is a connection between this agenda and the one pursued by other scholars who have favored a materialist paradigm for the study of anthropology's history. An illustration is provided by T. C. Patterson, *A Social History of Anthropology in the United States* (Oxford: Berg, 2001). One significant difference of my work is that, rather than looking at macro-conditions of production, I am focusing on the articulation between specific financial and knowledge technologies.

17. In 1827, in the immediate aftermath of a previous foreign debt bubble, William Maginn (himself a professional puffer) invented a pseudo-scholarly Latin terminology for the art of puff: *de arte puffandi*; W. Maginn, *Whitehall, or the Days of George IV* (London: Marsh, 1827), 139.

18. For instance, George Stocking, "What's in a Name," or G. Stocking, *Victorian Anthropology* (New York: Free Press, 1987). For a recent work echoing similar themes, see E. Sera-Shriar, *The Making of British Anthropology, 1813–1871* (London: Pickering & Chattoo, 2013).

19. For a discussion of Stocking's historical methodology, see Kevin A. Yelvington, "A Historian among the Anthropologists," *American Anthropologist* 105 (2003): 367–71.

20. Simon Schaffer, Lissa Roberts, Kapil Raj, and James Delbourgo, eds., *The Brokered World: Go-Betweens and Global Intelligence, 1770–1820* (Sagamore Beach, MA: Science History Publications, 2009) explores the role of "brokers and spies, messengers and translators, missionaries and entrepreneurs, in linking different parts of . . . densely entangled

systems." The more recent anthropology of the stock exchange pioneered by authors such as Karen Ho (discussed later in this chapter) also falls under this heading. For a discussion, among many others, of the role of anthropologists as social or cultural brokers, see Jean-Pierre Olivier de Sardan, *Anthropology and Development: Understanding Contemporary Social Change* (New York: Zed Books, 2005). These works, however, are mostly devoted to a descriptive or normative perspective. In none of these works can one find a theory of cultural brokerage as embedded in a careful study of capital market relations.

21. See William Pietz, "The Problem of the Fetish, I," *Res* 9 (1985): 5–17, and especially "The Problem of the Fetish, II: The Origin of the Fetish," *Res* 13 (1987): 23–45, and "The Problem of the Fetish, III: Bosman's Guinea and the Enlightenment Theory of Fetishism," *Res* 16 (1988): 105–23. On Marx, see Lawrence Krader, "Karl Marx as Ethnologist," *Transactions of the New York Academy of Sciences*, series II, 35 (1973): 304–13.

22. Cohn, *Colonialism and Its Forms*, 16–17 ("Language of Command").

23. Colebrooke's father was a chairman of the East India Company who went bankrupt following the crisis of 1773. His subsequent life and the lives of his heirs were geared toward repaying his creditors and explains why, using his old contacts, he was able to send his sons to India at the peril of their lives but opening as well the possibility of redemption. See Rosane Rocher and Ludo Rocher, *The Making of Western Indology: Henry Thomas Colebrooke and the East India Company* (London: Routledge, 2012).

24. On the historical evolution of ethics in anthropology, see Peter Pels, "Professions of Duplexity: A Prehistory of Ethical Codes in Anthropology," *Current Anthropology* 40 (1999): 101–36. A sociological history of the later professionalization of anthropology is provided by Henrika Kuklick, *The Savage Within: The Social History of Anthropology, 1885–1945* (Cambridge, UK: Cambridge University Press, 1991), esp. chap. 2: "Scholars and Practical Men."

25. Examples include Dr. Tett's keynote lecture in the Anthropology in the World Conference of 2012, sponsored by the Royal Anthropological Institute (www.youtube.com /watch?v=QJKLqWiIh8k), or her summer 2013 radio show, "The Anthropology of Finance."

26. Karen Ho, *Liquidated: An Ethnography of Wall Street* (Durham, NC: Duke University Press, 2009).

CHAPTER ONE

1. This was to study, as the prospectus stated (for instance on the back cover of the first volume of the *Anthropological Review*, in May 1863): "Man in all his leading aspects, physical, mental and historical: to investigate the laws of his origin and progress; to ascertain his place in nature and his relations to the inferior forms of life; and to attain these objects by patient investigation, careful induction and the encouragement of all researches tending to establish a de facto science of man."

2. *Journal of the Anthropological Society of London* 3 (1865): lxxix; Sarah Walpole, "Nuts and Bolts: A Survey of the Pre-1871 Anthropological Institutes," manuscript presented at the Workshop on the History of the Royal Anthropological Institute, London, December 9–10, 2014.

3. Leslie Howsam, *Kegan Paul: A Victorian Imprint; Publishers, Books and Cultural History* (London: Kegan Paul, 1998). See sources at the end of this book for information on

the anthropological periodicals carried by Trübner. The role of Trübner is discussed in greater detail later in the book.

4. *Anthropological Review* 1 (1863): cover page.

5. See Stocking, *Victorian Anthropology*, 252.

6. Fawn M. Brodie, *The Devil Drives: A Life of Sir Richard Burton* (New York: Norton, 1967), 241. This foreword was written by Isabel without Richard's knowledge, since he was at the time in South America surveying the atrocities of a war waged by a Brazilian-Argentine-Uruguayan coalition against Paraguay and had entrusted Isabel with the editing and publication of his book.

7. Hyde Clarke, "Anthropological Society of London," *Athenaeum* 2129 (1868): 210; Stocking, "What's in a Name?" 369–90 has an oft-cited discussion of the controversy. A more recent discussion is Efram Sera-Shriar, "Observing Human Differences: James Hunt Thomas Huxley and Competing Disciplinary Strategies in the 1860s," *Annals of Science* 70 (2012): 1–31, 5. On the *Athenaeum*, see E. E. Kellett, "The Press," in *Early Victorian England*, vol. 2, ed. G. M. Young (Oxford: Oxford University Press, 1934); and especially Leslie A. Marchand, *The Athenaeum: A Mirror of Victorian Culture* (Chapel Hill: University of North Carolina Press, 1941).

8. M. Winkler-Dworak, "The Low Mortality of a Learned Society," working paper, Österreichische Akademie, 2006.

9. The Club used, in algebraic form X, the date of the next dining event: X = 3 would mean that the Club would meet on the 3rd of the month; see Mark Patton, *Science, Politics and Business in the Work of Sir John Lubbock: A Man of Universal Mind* (London: Ashgate, 2007), 68.

10. See Mark Patton's excellent *Work of Sir John Lubbock*; P. E. Smart offers a concise business biography in "John Lubbock," ed. D. J. Jeremy, *Dictionary of Business Biography*, vol. 3, (London: Butterworths, 2004), 873–76.

11. Patton, *Work of Sir John Lubbock*, 109. This was in 1873.

12. *Sheffield Daily Telegraph*, May 11, 1875.

13. Anthony Trollope, *Letters of Anthony Trollope*, vol. 2, *1871–1882* (Stanford, CA: Stanford University Press, 1983).

14. See for instance the article published in 1876 by the *Westminster Review*, "Foreign loans and national debts. A review of the report of the select committee on foreign loans," reprinted as a pamphlet by Trübner.

15. 31/1884/7528; Other names were Gerstenberg, Francis Lycett, Thomas Matthias Weguelin, John Henry Daniell, Samuel Montagu, Roger Eykyn, and Philip Rose.

16. Patton, *Work of Sir John Lubbock*, 106–7; W. R. Chapman, "Ethnology in the Museum: AHLF Pitt-Rivers (1827–1900) and the Institutional Foundation of British Anthropology" (diss., Oxford University, 1981).

17. A. Desmond and J. Moore, *Darwin's Sacred Cause: Race, Slavery and the Quest for Human Origins* (London: Allen Lane, 2009), 330.

18. Stocking, "What's in a Name?," 382.

19. Susanna B. Hecht, *The Scramble for the Amazon and the "Lost Paradise" of Euclides da Cunha* (Chicago: University of Chicago Press, 2013), 159.

20. Wm. Bollaert, along with Gerstenberg, featured among both the promoters and the shareholders of the Ecuador Land Company. If the point were to take issue with the

conventional approach to disqualify members of the Anthropological Society as "unscientific," Bollaert and Seemann (the latter a protégé of Sir William Jackson Hooker) would provide examples that are difficult to discard. Among Bollaert's significant works, see his "Contribution to an Introduction to the Anthropology of the New World," in *Memoirs Read before the Anthropological Society of London* 2 (1866) 92-152. The paper was discussed during a meeting of the Anthropological Society of London, in 1866 (see *Journal of the Anthropological Society of London* 4 [1866], clxxi-clxxv). For details on Bollaert's works, see the four-page bibliography at the British Library: W. Bollaert, *W. M. Bollaert's Researches from 1823 to 1865: Principally on South American Subjects* (London, 1865). An obituary is found in the *Journal of the Anthropological Institute* 6 (1877) 510ff); Samples of Bollaert's large academic production include his *Antiquarian, Ethnological and Other Researches in New Granada, Equador, Peru and Chile* (London: Trübner, 1860); "The Jivaro Indians in Ecuador," *Transactions of the Ethnological Society of London* 2 (1863): 112-18; "On the Ancient Indian Tombs of Chiriqui in Veraguas (South-West of Panama), on the Isthmus of Darien," *Transactions of the Ethnological Society of London* 2 (1863): 147-66; "Maya Hieroglyphic Alphabet of Yucatan," *Memoirs Read before the Anthropological Society* 2 (1865/66); "On the Alleged Introduction of Syphilis from the New World. Also Some Notes from the Local and Imported Diseases into America," *Journal of the Anthropological Society of London* 2 (1864): cclvi-cclxx. On the Ecuador Land Company, see National Archive, Kew, for details, FO 25/134, Ecuador Land Company Colonization project. For lists of shareholders, see Ecuador Land Company, BT 41/215/1225. For evidence on the Ecuador Land Company's prospecting of foreign territory with colonization objectives see BJ 7/919; see also Ulrike Kirchberger, *Aspekte Deutsch-Britischer Expansion: Die Überseeinteressen der Deutschen Migranten in Großbritannien in der Mitte des 19. Jahrhunderts* (Stuttgart: Steiner Verlag, 1999) as well as the discussion of the links between the Ecuador Land Company and the Royal Geographical Society in chapter 5.

21. *Economist*, "Special Dangers of High Commercial Developments," quoted in D. Kynaston, *The City of London: A World of Its Own, 1815-1890* (London: Pimlico, 1995), 265.

22. Charles Stansifer, "E. George Squier and the Honduras Interoceanic Railroad Project," *Hispanic American Historical Review* 46 (1966): 1-27. This paper extends beyond Stansifer's unpublished dissertation, "The Central American Career of E. George Squier" (Ph.D. diss., Tulane University, 1959). Ephraim George Squier was named an Anthropological Society of London honorary member on May 2 1865, and joined as a fellow on November 14 of same year; *Journal of the Anthropological Society of London* 3 (1865). The *Journal of the Anthropological Institute of New York* existed in 1871 and 1872.

23. Stansifer, "E. George Squier," 1. At the time Stansifer wrote, previous writers were diplomatic historians specializing in Central American history such as Mary W. Williams, *Anglo-American Isthmian Diplomacy, 1815-1915* (Washington, DC: American Historical Association, 1916). T. A. Barnhart's recent biography, *Ephraim George Squier and the Development of American Anthropology* (Lincoln: University of Nebraska Press, 2005) discusses Squier's involvement in railroad promotion as a diversion from Squier's career as an anthropologist (214-16). The book claims to build on Stansifer's own critical distinction between Squier's work as a "publicist and scholar." I am not sure that Stansifer makes this

distinction in the way Barnhart understands it. The suggestion is also made that Squier never made money out of the project. This is an unwarranted statement because Stansifer shows that all details cannot be known. The money paid by Brown and company to neutralize the Honduras route cannot have been negligible. Of course, the suggestion that Squier would have lost money on his financial schemes is appealing to the view that conflicts of interest do not pay. Against this, we have the other examples in the book.

24. Private e-mail to author, January 24, 2013.

25. George Bemis, *Report of the Case of John W. Webster, Indicted for the Murder of George Parkman, before the Supreme Judicial Court of Massachusetts* (Boston: Little & Brown, 1850), 16, 62.

26. Hyde Clarke, "On the Epoch of Hittite, Khita, Hamath, Canaanite etc.," *Transactions of the Royal Historical Society* 6 (1877): 1–85.

27. Paul Frank Meszaros, "The Corporation of Foreign Bondholders and British Diplomacy in Egypt 1876 to 1882: The Efforts of an Interest Group in Policy-Making" (Ph.D. diss., Loyola University Chicago, 1973).

28. Search consisted of looking for "Hyde Clarke" in the British Newspaper Archive, and the Times Digital Archive, 1785–2006, for the period May 15 to December 15, 1868.

29. *London Standard*, May 22, 1868, 4.

30. *Morning Post*, June 1, 1868, 7

31. *Morning Post*, June 8, 1868, 3

32. *Leicester Journal*, June 12, 1868, 3; *Birmingham Daily Post*, June 8, 1868, 3. This was distinct from the London and Middlesex Archaeological Society, est. 1855.

33. "Consular Jurisdiction in the Levant," *Times*, August 10, 1868; *Liverpool Daily Post*, August 11, 1868, 4, where Hyde Clarke is quoted saying that consular courts are a "byword to natives, to our own people and to foreign officials especially those who have had a legal education." The letter to the *Times* followed a debate initiated by a discussion in Parliament in July 1868 and to which both the Tory secretary of foreign affairs, Lord Stanley, and the former Liberal under-secretary, Sir Austen Henry Layard, participated (see, e.g., Layard's detailed exposé in "Consular courts in Turkey and Egypt," HC Deb 10 July 1868 vol. 193 cc1024–54). On July 2 and 13, the newspaper had published two leaders on the subject, hostile to capitulations. Hyde Clarke's letter had been preceded by letters sent to the *Times* by the editor of the *Anglo-Egyptian and Levant Herald* (August 6, 1868), "Consular jurisdiction in the Levant" and a responses by A. H. Layard (August 8, 1868). A discussion of the surrounding controversy and issues is found in D. C. M. Platt, *The Cinderella Service: British Consuls since 1825* (London: Longman, 1871), 136–63.

34. *Norfolk Chronicle*, August 26, 1868, 4.

35. *Norfolk News*, August 22, 1868, 23.

36. *London Standard*, August 7, 1868, 2. Hyde Clarke was rejoicing in a public gathering at London Tavern that the "two committees had agreed to send out the same agent."

37. "The Ottoman, Smyrna and Aidin Railway," *London Daily News*, September 4, 1868; *London Standard*, October 1, 1868.

38. *Pall Mall Gazette*, November 4, 1868, 8.

39. *London Standard*, December 18, 1868, 2; *Morning Post*, December 18, 1868, 8. The association comprised men, such as W. P. Pickering, who can be spotted in the Smyrna-Aidin

314

NOTES TO PAGES 31–38

shareholders meetings. Pickering is one of the authors of the Ottoman Railway Company Report, (FO 78/2275, p. 41) mentioned in chapter ten.

CHAPTER TWO

1. George W. Stocking, *Race, Culture, and Evolution: Essays in the History of Anthropology* (Chicago: University of Chicago Press, 1968); Stocking, "What's in a Name?" 369–90; Stocking, *Victorian Anthropology*; Douglas Lorimer, *Colour, Class and the Victorians: English Attitudes to the Negro in the Mid-Nineteenth Century* (Leicester, UK: Leicester University Press, 1978); David Keith Van Keuren, "Human Science in Victorian Britain: Anthropology in Institutional and Disciplinary Formation, 1863–1908" (diss., University of Pennsylvania, 1982). In *The Myth of the Noble Savage* (Berkeley: University of California Press, 2001) Terry Jay Ellingson gives a detailed narrative of the years preceding the foundation of the Anthropological Society. See also Henrika Kuklick, "The British Tradition," in *New History of Anthropology*, ed. H. Kuklick (London: Blackwell, 2008), 52–78; Adrian Desmond and James Moore, *Darwin's Sacred Cause: Race, Slavery and the Quest for Human Origins* (New York: Houghton Mifflin, 2013); Efram Sera-Shriar, *The Making of British Anthropology, 1813–1871* (London: Pickering & Chattoo, 2013).

2. For a contemporary critical statement by an anthropologist, see J. B. Davis, "Anthropology and Ethnology," *Anthropological Review* 6 (1868): 394–99, 395. A recent history of the Aborigines' Protection Society is James Heartfield, *The Aborigines' Protection Society: Humanitarian Imperialism in Australia, New Zealand, Fiji, Canada, South Africa, and the Congo, 1837–1909* (London: Hurst, 2011).

3. The paper was presented before the society on November 17, 1863; see the minutes of discussion of Hunt's paper in *Journal of the Anthropological Society of London* 2 (1864): xv–lvi. It was published in pamphlet form (with a foreword by Burton) as *On the Negro's Place in Nature* (London: Trübner, 1863). In this article, usually regarded as Hunt's manifesto, Hunt inferred from a discussion of bones, types, and so on that "there is greater difference between the Negro and Anglo-Saxon than between the gorilla and chimpanzee," then moved to the conclusion that "the Negro is inferior, intellectually, to the European." A U.S. edition (New York: Van Evrie) was published in 1864.

4. See *Journal of the Anthropological Society of London* 3 (1865): lxxxix, cxiii. The marriage of ethnology and geography in the British Association's Section E went back to 1851.

5. *London Standard*, September 3, 1863; Desmond and Moore, *Sacred Cause*, 338; "Sambo and the *Savans*," *Caledonian Mercury*, September 4, 1863.

6. See *Journal of the Anthropological Society of London* (January 3, 1865): lxxxix, cxiii; Paul Broca, *Sur le volume et la forme du cerveau, suivant les individus et suivant les races* (Paris: Hennuyer, 1861), 15. Broca cautioned his readers against undue inferences. Given that there was no proper measure of male intellectual superiority a "definite ratio" could not be computed between this weight advantage and males' greater acumen.

7. Consider, for instance, the following from Burton: "After being for some years 'paradoxical' in my conviction of the innate and enduring inferiority of a race which has had so many an opportunity of acquiring civilization, but which has ever deliberately rejected improvement, I find that the rising authors are beginning to express opinions far more

decided than mine, and I foresee the futurity of hard compulsory labour which the negro-maniac will have brought upon his African *protégé*." R. Burton, *A Mission to Gelele, King of Dahome, with Notices of the So-Called "Amazons," the Grand Customs, the Yearly Customs, the Human Sacrifices, the Present State of the Slave Trade, and the Negro's Place in Nature*, vol. 2 (London: Tinsley Bros., 1864), 177–211, 181. W. Winwood Reade (Esq., FRGS, FASL), a close friend of Burton, declared in his 1865 paper on the "Efforts of Missionaries among Savages" that all he had "heard and seen" during an "extensive" five months' travel in equatorial Africa led him to conclude that "every Christian negress was a prostitute and every Christian negro was a thief." W. Winwood Reade, "Efforts of Missionaries among Savages," *Journal of the Anthropological Society of London* 3 (1865): clxiii–clxxxiii, clxv. See also his *Savage Africa: Being the Narrative of a Tour in Equatorial, Southwestern and Northwestern Africa, Complete with Notes on the Habits of the Gorilla and the Existence of Unicorns* (New York: Harper Bros., 1864). On the friendship between Burton and Winwood Reade, see Isabel Burton, *The Life of Captain Sir Richard F. Burton*, vol. 1 (London: Chapman & Hall, 1893).

8. Stocking, "What's in a Name?": 381.

9. Stocking, *Victorian Anthropology*, 253. In "What's in a Name?" this is done in one full paragraph, at p. 380.

10. Brodie, *Devil Drives*, 336. Brodie further writes: "His skill in the symbolically rich sport of fencing, and especially his scholarship on the history of the sword, tend to suggest latent rather than active homosexuality, for active homosexuals do not content themselves with the symbol of the phallus, but seek out the real thing." She also indicates: "One should not forget that he was fascinated also with all forms of heterosexuality—which most male homosexuals find utterly repugnant—and that an extraordinary amount of energy went into his 'field research' as well as into translations on the subjects." Archer provides a parallel between Burton's anthropological writing on sexuality and Malinowski's *Sexual Life of Savages in North-Western Melanesia* (London: Routledge & Sons, 1929): W. G. Archer, preface to *The Kama Sutra of Vatsyayana*, trans. Sir Richard Burton and F. F. Arbuthnot (London: Allen & Unwin, 1963).

11. Brodie, *Devil Drives*, 66.

12. Yet Brodie also argues elsewhere in her teeming book that this "rigorous" spouse wrote her own "Guidance as a Wife," where one of the seventeen rules was: "Let him find in the wife what he and many other men fancy is only to be found in a mistress." Brodie, *Devil Drives*, 70, 191.

13. A rare counter-example is Anjali Arondekar, *For the Record: On Sexuality and the Colonial Archive in India* (Durham, NC: Duke University Press, 2009).

14. R. F. Burton, *Scinde, or the Unhappy Valley*, vol. 1 (London: Richard Bentley, 1851), 182.

15. Burton, *Mission to Gelele*, vol. 2, 14. See also P.P. 1863 (3179) LXXIII, *Dispatches from Commodore Wilmot, Respecting His Visit to the King of Dahomey, in December, 1862, and January, 1863. Presented to the House of Commons by Command of Her Majesty, in Pursuance of Their Address.*

16. *Athenaeum* 1926 (1864): 391–93. This is something that Stocking also reproached Burton for ("What's in a Name?" 380).

17. The phrase "How a Hatred of Slavery Shaped Darwin's Views on Human Evolution" comes from the pitch of Desmond and Moore's book. Charles Darwin, *The Descent of Man, and Selection in Relation to Sex* (London: John Murray, 1872); Desmond and Moore, "Cannibals and the Confederacy in London" (chapter 12 of *Darwin's Sacred Cause*); Patton, *Work of Sir John Lubbock*, 62, argues that, because of this infiltration, the conflicts of the American Civil War were "played out on the pages of English scientific journals."

18. The myth of a decline after 1865 was pioneered by W. R. Chapman, "Ethnology in the Museum," 334. He argues that "by 1866, or just three years after the society's foundation, it had already entered decline." One also finds it in Patton, who states that "by the beginning of 1871, with Confederate funds having dried up, and amid allegations of fraud, the Anthropological Society was bankrupt" (Patton, *Work of Sir John Lubbock*, 96).

19. Robert E. Bonner, "Slavery, Confederate Diplomacy and the Racialist Mission of Henry Hotze," 51 *Civil War History*, (2005): 288–316; A. de Gobineau, *The Moral and Intellectual Diversity of Races, with Particular Reference to their Respective Influence in the Civil and Political History of Mankind*, (Philadelphia: J.B. Lippincott, 1856); In the accounts for 1864, Hotze appears among a few other donors for the "Library fund" for £5 5s; one J. W. Conrad Cox gave £10; *Journal of the Anthropological Society of London* 3 (1864): lxxv. Such open gifts would not be the preferred means to provide subsidies, and if they existed at all, they probably would have taken the shape of book purchases of translations of Vogt or Broca. In fact, I do not see the amount sold on that account declining markedly after the Confederate defeat, although the accounts of the Anthropological Society were sufficiently opaque to make it impossible to argue conclusively either way.

20. J. F. Jameson, "The London Expenditures of the Confederate Secret Service," *American Historical Review* 35 (1930): 811–24, 823; For a recent study of Confederate and Union propaganda in London, see Thomas E. Sebrell II, *Persuading John Bull: Union and Confederate Propaganda in Britain, 1860–65*, (Lanham, MD: Lexington Books, 2014).

21. Edward Walford, *Old and New London: A Narrative of Its History, Its People, and Its Places*, vol. 3: *Westminster and the Western Suburbs* (London: Cassell, Petter & Galpin, 1878), 155, devotes a few lines to the admission of the "Duke" to the Royal Society of Literature: "The Society, however, incurred considerable ridicule by having admitted a certain M. Cosprons, a few years since, to read a paper, as a French savant, under the assumed title of 'M. le Duc de Rousillon.' [Cosprons is a village in France's Roussillon region and so Monsieur Cosprons might have come from the area.] The mistake was soon found out, and the 'illustrious' soi-disant duke was never asked to read a second paper." Roussillon appears as vice-president of the Anthropological Society in 1863 and 1864, but not afterward. The minutes of the Anthropological Society meeting (February 28, 1865) mention: "The resignation of the Duke of Roussillon was read, when it was resolved, that the Duke of Roussillon be communicated with to the effect, that his resignation will be accepted by the Council when he shall have paid his subscription for 1865." He was considered resigned in the minutes of December 19, 1865, and the subscription ledgers report that his dues had been paid for all three years of membership (1863 to 1865); Among other spottings of the learned duke, we find a "memoir" read before the Syro-Egyptian Society "upon the discovery of a date of the year 135 of Our Lord, as well as of the pace and sacred cubit of the Hebrews" (*Athenaeum*, 1843, February 21, 1863, 267). Other occasions are frequently related to the duke's expertise "on the Scythians," the title of a paper presented at the Ethnological So-

ciety in 1864 (Council meeting, December 27, 1864) and of a presentation at the annual meetings of the British Association at Bath in 1864. There are occasional mentions of his interventions at the society, such as in the *Reader* (March 26, 1864, 401). In 1867 the duke was still reported by the *London Standard* as participant at the Royal Society's soiree (*London Standard*, April 15, 1867). The same year, he published *Origines, Migrations, Philologie, Monuments Antiques* in two volumes. As Sarah Walpole remarked to me, the election of Cosprons as vice president in 1863 had been met with some resistance. According to the Council minutes (March 24, 1863), the resolution that the duke of Rousillon be elected vice president won by a majority of seven against three (the minority amendment asking that it should be deferred for six months). Could this mean that there was some awareness that he was not quite all he claimed to be? The Duke of Roussillon's episode evokes a parallel with Burton's well-known penchant for disguise and hoaxes and his mingling around the same period with a famous impostor, the "Tichborne Claimant," who alleged to be heir to a baronetcy and substantial estates in England (see Brodie, *Devil Drives*, 244). The Tichborne claimant case was the occasion of a resounding fallout between partners of the law firm Baxter, Rose, Norton and Co., an important character of my tale. I am grateful to Sarah Walpole for having helped me put together this note.

22. Burton, *Mission to Gelele*, vol. 2, 180.

23. Freda Harcourt, "Disraeli's Imperialism, 1866–1868: A Question of Timing," *Historical Journal* 23 (1980): 87–109, 104.

24. T. J. Ellingson, *Myth of the Noble Savage*, 269.

25. Thomas C. Holt, *The Problem of Freedom: Race, Labor, and Politics in Jamaica and Britain, 1832–1938* (Baltimore, MD: Johns Hopkins University Press, 1992); Nicholas Draper, *The Price of Emancipation: Slave-Ownership, Compensation and British Society at the End of Slavery* (Cambridge, UK: Cambridge University Press, 2010).

26. Clare Midgley, *Women Against Slavery: The British Campaigns, 1780–1870* (London: Routledge, 1995).

27. *Anthropological Review* 3 (1865): clxxii.

28. See Bernard Semmel, *The Governor Eyre Controversy* (Jamaica: MacGibbon & Kee, 1962). For a discussion of the Morant Bay affair, see Holt, *Problem of Freedom*; and Gad Heuman, *The "Killing Time": The Morant Bay Rebellion in Jamaica* (Knoxville: University of Tennessee Press, 1994).

29. "Anniversary Address," *Journal of the Anthropological Society of London* 4 (1866): lxxviii.

30. Archive ASL, Office, A5; "Henry Butler, Admiralty, Somerset House to Anthropological Society of London, 31 Jan. 1866—requests ticket for meeting on 1 Feb."

31. Among others, Tim Watson ("Jamaica, Genealogy, George Eliot: Inheriting the Empire after Morant Bay," http://english.chass.ncsu.edu/jouvert/v1i1/WATSON.HTM#21) argues that Pim's lecture would have been published both "in the Anthropological Society of London's *Popular Magazine of Anthropology* and as a separate pamphlet, *The Negro and Jamaica.*" The announcement in *Journal of the Anthropological Society of London* 4 (1866): lxxxix, makes reference to a special issue of the *Popular Magazine* to be put out and containing the text of the lecture. The volume for 1866 of the *Popular Magazine* (Volume I) does not have any such article among the many articles relating to the controversy. Also, the typical length of articles in the *Popular Magazine* is much shorter than the Pim essay.

The most likely interpretation is that *Negro and Jamaica* came out *as* a special issue of the *Popular Magazine* rather than *in* a special issue. This is consistent with my inspection of the records of the publisher (Trübner) and with what was printed on the copies in circulation. For instance the digital copy held by the online library Hathi Trust (www.hathitrust .org), shows a mention of the pamphlet being put out as a "'Special number' of 'The Popular magazine of anthropology.'" In this book, I refer to this version as Bedford Pim, *The Negro and Jamaica* (London: Trübner, 1866).

32. Information communicated by Charles Priestley, correspondence with author, July 10, 2013. From Michael Hammerson, I drew the useful precisions about the Reverend Francis Tremlett of Belsize Park. Pim was one of Tremlett's parishioners, and, Dr. Hammerson suggested to me, played an active role in helping Tremlett in his work. He suspects Pim used his high-level naval contacts to enable Confederate naval personnel, of whom there were many either based in or visiting England, to get access to the latest in ironclad and weapons technology.

33. Pim, *Negro and Jamaica*, 4; the "Negrophilists" were sometimes referred to as "the Negrophilist at Exeter House." Exeter House was a reference to where the public meetings of the Aborigines' Protection Society took place (Exeter Hall), underscoring the identification between Ethnologists and humanitarians in Cannibal language.

34. Pim, *Negro and Jamaica*, 35.

35. Ibid., 72.

36. Markham, *Royal Geographical*, 53.

37. "Occasional Notes," *Pall Mall Gazette*, June 3, 1868, 4.

38. On R. B. N. Walker, see Nora McMillan, "Robert Bruce Napoleon Walker, F.R.G.S., F.A.S., F.G.S., C.M.Z.S. (1832–1901), West African Trader, Explorer and Collector of Zoological Specimens," *Archives of Natural History* 23 (1996): 124–41. See "19th Century Field Collecting," website of the Pitt-Rivers Museum of Anthropology in Oxford, http:// www.prm.ox.ac.uk/Kent/shieweap/shfldcol.html#anchor2738136. See also Edward Walford, *Old and New London*, vol. 3, 155. The same curators prefer to talk of Augustus Pitt-Rivers and his collection of riffles. I am grateful to Sarah Walpole for having discussed the matter of the Anthropological Society's collection with me.

39. Catherine Hall, "The Economy of Intellectual Prestige: Thomas Carlyle, John Stuart Mill, and the Case of Governor Eyre," *Cultural Critique* 12 (1989): 167–96.

40. Augustus Lane Fox (Pitt Rivers), *Primitive Warfare: Illustrated by Specimens from the Museum of the Institution; A Lecture at the Royal United Service Institution* (London: United Service Institution, 1867).

41. Lane Fox (Pitt Rivers), *Primitive Warfare*, 9.

42. Patton, *Work of Sir John Lubbock*, 65.

43. Walpole, "Nuts and Bolts." The two females were Eliza Lynn Linton, a professional journalist, and Adelina Paulina Irby.

44. The definitive treatment of the subject is Evelleen Richards, "Huxley and Woman's Place in Science: The 'Woman Question' and the Control of Victorian Anthropology," in *History, Humanity and Evolution: Essays for John C. Greene*, ed. James R. Moore (Cambridge, UK: Cambridge University Press, 1989). The argument in Richards has been more recently repeated by Sera-Shriar, *British Anthropology*, 116–19, who finds a generational

aspect to the subject of women's admission in Ethnological meetings, with older-generation ethnologists such as Crawfurd and Hodgkin supporting an "inclusive science open to both genders." Once they had passed away, it would have become much "easier for Huxley to reverse the council's original decision to admit women." This may be an optimistic perspective. See Ellingson's portrait of Crawfurd in *Noble Savage*. Also, as emphasized by Richards, women's admission has to be understood as "admission in meetings" not admission as fellows. Women were students at best, not adepts, who might benefit from minimal exposure to anti-creationist ideas but would remain amateurs. In 1869 the Ethnological Council announced that women would be admitted only to extraordinary meetings. On the Anthropological Society accepting women in extraordinary meetings, see *Journal of the Anthropological Society of London* 8 (1870), meeting at St. James's Hall, with Berthold Seemann in the chair (May 19, 1870, clv–clvi): "This is the first time in the annals of this Society that the Chairman is able to welcome at its meeting the fairer portions of mankind; and their presence is a proof that they do not think us quite as black as we have been painted."

45. Apart from a brief episode in 1892 when about twenty women were admitted, the Royal Geographical Society kept excluding women until 1913; opening up of British science to women only began in the early twentieth century. See W. J. Atkinson, "Scientific Societies and the Admission of Women," *Nature* 79 (1909): 488; T. E. Thorpe, "Scientific Societies and the Admission of Women Fellows," *Nature* 80 (1909): 67–68. The reader is also referred to Morag Bell and Cheryl McEwan, "The Admission of Women Fellows to the Royal Geographical Society, 1892–1914: The Controversy and the Outcome," *Geographical Journal* 162 (1996): 295–312; and Joan Mason, "The Admission of the First Women to the Royal Society of London," *Notes and Records of the Royal Society of London* 46 (1992): 279–300. In 2013, London School of Economics Professor Dame Judith Rees (C.B.E., R.G.S.) became the first female president of the Royal Geographical Society. Of course, such attitudes as documented in the paragraph were not limited to individuals who used the same idiom as Carlyle: consider, for instance, the dedication by the Geneva-based German anthropologist Vogt of his *Lectures on Man* (1854) to the (French) anthropologist Paul Broca, lamenting the "admission of ladies" in British Ethnological meetings; Karl Christoph Vogt, *Lectures on Man: His Place in Creation, and in the History of the Earth* (London: Longman, Green, 1854), viii.

46. Edward Caudill, "The Bishop-Eaters: The Publicity Campaign for Darwin and *On the Origin of Species*," *Journal of the History of Ideas* 55 (1994): 441–60. Boyd Hilton, *The Age of Atonement: The Influence of Evangelicalism on Social and Economic Thought, ca. 1795–1865* (Oxford: Oxford University Press, 1988).

47. The presence of the clergy in both the Ethnological and Anthropological societies is evident when one inspects membership lists, although one fails to identify the predominant rural elements that Hyde Clarke imagined in the case of the Anthropological Society; Hyde Clarke, "Anthropological Society of London," *Athenaeum* 2129 (1868): 210; Reade, "Efforts of Missionaries"; and Colenso's response: "Efforts of Missionaries among Savages," *Journal of the Anthropological Society of London* 3 (1865): clxiii, ccxlviii–cclxxxix. There is, however, evidence of a tension on the subject between the leadership of the society and elements of its rank and file. The conflict erupted in the open following the

*Athenaeum* dispute when, one new convert to the Anthropological Society, Dr. Cornelius Donovan, author of a paper on the "Fundamental Principles of Anthropological Science," tried to have a motion adopted by the society to"[view] with regret and disapprobation the fact that a great majority of the Council are professed atheists." Die-hard Cannibals, however, repelled the attempt; *Journal of the Anthropological Society of London* 7 (1869): xvii.

48. J. W. Burrow, "Evolution and Anthropology in the 1860s: The Anthropological Society of London, 1863–71," *Victorian Studies* 7 (1963): 137–54; Thomas Gondermann, "Die Etablierung der Evolutionslehre in der Viktorianischen Anthropologie: Die Wissenschaftspolitik des X-Clubs, 1860–1872," *NTM Zeitschrift für Geschichte der Wissenschaften, Technik und Medizin* 16 (2008): 309–31; Efram Sera-Shriar, "Observing Human Differences: James Hunt Thomas Huxley and Competing Disciplinary Strategies in the 1860s," *Annals of Science* 70 (2012): 1–31.

49. Leone Levi, "Of the Progress of Learned Societies Illustrative of the Advancement of Science in the United Kingdom during the Past Thirty Years," *Report of the Thirty-Eighth Meeting of the British Association for the Advancement of Science, Held at Norwich in August 1868* (London: John Murray, 1868), 169–73.

50. Typically ranging between 1 and 3 guineas per year, they stood initially at about £2 for both the Ethnological and Anthropological societies; the May 1863 issue of the *Anthropological Review* (in publicity section, no page numbering) states the fee to be 2 guineas for a one-year subscription, raised afterwards to £2 2s or £2.1 ("Annual Report" for 1864). This was similar to that for the Ethnological Society (Chapman, "Ethnology in the Museum," 1). Life or compounded subscriptions were £21 and £20 respectively. In 1873 the annual membership of the Corporation of Foreign Bondholders would also cost £2 2s, with compounded life membership at £20.

51. For instance, Charles Harding, "Anthropological Society of London," *Athenaeum* (September 12, 1868): 334.

52. Feb. 1, 1866; Archive ASL, Office (A5);

53. Throughout the process, and especially during the early years, substantial amounts were spent on advertising. While this was principally to advertise the purpose of the society and its meetings, it also served to purchase the goodwill of recipient journals.

54. Dinner announced in *Journal of the Anthropological Society of London*, March 14, 1865, clxxxiii. The description of the dinner and transcription of speeches is in *Anthropological Review* 3 (1865): 167–82, and was also sold or distributed as an independent pamphlet, printed by Trübner.

55. "Fifteenth List of the Foundation Fellows of the Anthropological Society," *Memoirs of the Anthropological Society of London* (1865). Lists circulated by the Anthropological were confusingly called lists of "Foundation Fellows." One could be a foundation fellow and yet not have been around at the time of the foundation. See for instance the back matter of the first volume of the *Memoirs Read before the Anthropological Society*.

56. Between August 1866 and August 1869 membership lists have been taken from subsequent editions of the *Memoirs of the Anthropological Society of London*. In 1866, I find (among the twenty-nine honorary members) nine Britons, eight Frenchmen, four Americans, three Germans, two Swedes, one Russian, one Spaniard, and one Swiss.

57. The present discussion focuses on so-called local secretaries abroad. There were local secretaries in Great Britain too, though reportedly less enterprising.

58. D. K. Van Keuren, "Human Science in Victorian Britain: Anthropology in Institutional and Disciplinary Formation, 1863–1908" (diss., University of Pennsylvania, 1982).

59. Van Keuren, "Human Science in Victorian Britain," 38.

60. Van Keuren considered all officers in the Council including the lower-rank officers (librarian, secretaries, treasurer). My own computation suggests an even larger chasm, with the proportion of Athenaeum Club members within the Ethnological executive closer to 60 percent.

61. For a contemporary history of London clubs, see John Timbs, *Clubs and Club Life in London: With Anecdotes of Its Famous Coffee Houses, Hostelries, and Taverns, from the Seventeenth Century to the Present Time* (London: John Camden Hotten, 1872); Thomas Humphry Ward, *History of the Athenaeum, 1824–1925* (London: Athenaeum Club, 1926); F. R. Cowell, *The Athenaeum Club and Social Life in London* (London: Heinemann, 1975). Benjamin Disraeli had to wait until his political triumph in 1866 in order to make an application, which the club could not refuse. On Disraeli and the Athenaeum Club, see Marvin Spevack, "Benjamin Disraeli and the Athenaeum Club," *Notes and Queries* (March 2005): 68–70. The significance of the Carlton Club as an informal instrument of the Conservative Party under Disraeli and Sir Philip Rose has been emphasized by Robert Stewart, *The Foundation of the Conservative Party, 1830–1867* (London: Longman, 1978).

62. F. G. Waugh, *Members of the Athenaeum Club from Its Foundation, 1824 to 1887* (London: Athenaeum, 1894). Athenaeum Club membership is to be understood as membership at the date when the society's membership list was established, not membership in the life time of the Anthropological and Ethnological societies. Significantly, a proportionately larger number of Anthropological Society members later became members of the Athenaeum. This could be interesting material for a study of social-scientific trajectories. The Anthropological Society list I use in this section was that for August 1866; the Ethnological Society list was that for May 1868.

## CHAPTER THREE

1. On Burton in Fernando Po, see Platt, *Cinderella Service*, 51.

2. Indeed, as some research shows, the sins of the Anthropological Society were quite widespread. Just like the Anthropological, the Ethnological Society was also involved in the art of "puffing"—the name of the game rather than an Anthropological shortcoming: in the May 1867 "Report of the Council of the Ethnological Society of London" (*Transactions of the Ethnological Society* 6, back matter), it is said that "the accession of members has been greater than in any one year for some years past, and the increase of the Society's numbers has been considerably greater than during previous years." However, the report for 1866–67 shows a gross increase of twenty-six Ethnological Fellows, against twenty-three for 1865–66. Likewise, looking at the list of fellows who joined in 1862 we find twenty-five at least (from *Transactions of the Ethnological Society* 6, showing members still there in 1866–67). What was talking of a "considerably greater" number of Ethnological Fellows in 1866–67 than during previous years if not puffing, pure and simple?

322 NOTES TO PAGES 60–64

3. Van Keuren, "Human Science in Victorian Britain," 302. The images have persisted. A recent conference (March 2013) at the Royal Society commemorating the centenary of John Lubbock's death was called "Avebury's Circle: The Science of John Lubbock FRS (1834–1913)." On the Lubbock circle, see Ruth Barton, "Huxley, Lubbock, and Half a Dozen Others: Professionals and Gentlemen in the Formation of the X Club, 1851–1864," *Isis* 89 (1998): 410–44.

4. D. N. Livingstone, "Post-Darwinian British and American Geography," in *Geography and Empire*, ed. Anne Godlewska and Neil Smith (London: Blackwell, 1994), 134–35, for both illustration and quotation.

5. Sir Roderick was the British Association's general secretary 1837–44, president in 1846, and a longtime member of its Executive Council; see Roy McLeod and Peter Collins, *The Parliament of Science: The British Association for the Advancement of Science 1831–1981* (Northwood, UK: Science Reviews, 1981). He presided over the Royal Geographical Society almost continuously between 1843 and 1871 (1843–45, 1851–53, 1856–59 and 1862–71); see Clements R. Markham, *The Fifty Years' Work of the Royal Geographical Society* (London: John Murray, 1881). Anthropologists would later complain about an alliance that was really a takeover and would have deprived the science of man of the modicum of sovereignty it had once enjoyed as a subsection within Section D (zoology and botany), "submerging ethnographical and ethnological papers amongst geographical ones."

6. See "The President's Address," *Journal of the Anthropological Society of London* 3 (1865): lxxxix. For a discussion of the problems of sections within the British Association, see P. Sillitoe, "The Role of Section H at the British Association for the Advancement of Science in the History of Anthropology," *Durham Anthropology Journal* 13 (2005): 1–17.

7. *Athenaeum* 2129 (1868): 210.

8. Markham, *Fifty Years' Work*, 52.

9. Catherine Hall writes about the Anthropological Society as "a new society of which Eyre was a founder member, that combined 'science' and politics." C. Hall, "Imperial Man: Edward Eyre in Australasia and the West Indies, 1833–1866," in *The Expansion of England: Race, Ethnicity and Cultural History*, ed. Bill Schwarz (London: Routledge, 1996), 129–68. The expression *founder member* suggests that he was in the group of initial promoters. Hall must have meant "Foundation Fellow," a misleading term used by the Anthropological Society repeatedly, even after it was created, and splashed in membership lists. It was indeed granted to Eyre in lists published in 1865, for instance. Technically, Eyre's joining as fellow was announced on November 1, 1864: *Journal of the Anthropological Society of London* 3 (1865): ccxlvi. So he joined early, but certainly not as founder.

10. "The rooms of the Anthropological Society, N°4 St Martin's Place now offer a refuge to destitute science." Isabel Burton, *Life of Captain Sir Richard F. Burton*, vol. 1, 402.

11. For an introduction to the Burton-Speke controversy, see Brodie, *Devil Drives*; and Dane Kennedy, *The Highly Civilized Man: Richard Burton and the Victorian World* (Cambridge, MA: Harvard University Press, 2007). A recent study of the East African expedition by Burton and Speke emphasizing indigenous geography is provided by Adrian S. Wisnicki, "Charting the Frontier: Indigenous Geography, Arab-Nyamwezi Caravans, and the East African Expedition of 1856–59," *Victorian Studies* 51 (2009): 103–37.

12. Isabel Burton, *Life of Captain Sir Richard F. Burton*; and Thomas Wright, *The Life*

*of Sir Richard Burton,* 2 vols. (London: Everett, 1906); Brodie, *Devil Drives,* 205 and 233. The chronology of Burton consulships is: in August 1861–November 1863, Burton was in Fernando Po, with a four-month leave from December 1862 to March 1863; in November 1863, Burton left Fernando Po upon appointment by the British government as commissioner and bearer of a message to Gelele, king of Dahomey; in August 1864, Burton obtained another few months' leave in time to participate in the Bath meetings of the British Association for the Advancement of Science (episode of Speke's death in September 1864); in late 1864 or early 1865, Burton received his new assignment in Santos (episode of the 500-fellows banquet in April 1865); he occupied this position, interrupted by several travels in the area, until resigning in April 1868 under a Tory administration. In July 1868, he set off for Montevideo and the Paraná River, on "sick leave" to explore the battlefields of the murderous wars Paraguay was waging against a coalition comprising Brazil, Argentina and Uruguay. He spent the next months, "haggard and thin, and still drinking," wandering in Latin America until news of his new consulship in Damascus reached him in Lima in February 1869 (Brodie, *Devil Drives,* 242–43).

13. See, for instance, the letter Burton sent to the Anthropological Society during the missions controversy after he had left the country for Santos in Brazil. There he lashes out at one J. Reddie, regretting that "this gentleman whose travels extend so little beyond the Admiralty, should have departed from his hitherto admirable habit, by putting his arguments on paper." *Journal of the Anthropological Society of London* (1865): ccxci. Burton's take on the Admiralty contrasts with his views on individual sailors. Supporting statements for the argument in this section are legion.

14. *Journal of the Royal Geographical Society* 29 (1859): xcvii. Quoted in Brodie, *Devil Drives,* 354.

15. Beyond and beneath academic tactics were also Burton's own preferences. His ethnographical explorations vastly surpassed in scope his more purely geographical ones. In his undercover trip to Mecca he had already been faced with such a choice since taking with him his measurement equipment might have given his identity away. He opted instead for carrying a system that enabled him to scribble field notes in the deep darkness of his tent.

16. Blackwood to Delane (editor of the *Times*); David Finkelstein has described in detail the "making of" Speke's *Journal of the Discovery of the Sources of the Nile.* David Finkelstein, *The House of Blackwood: Author–Publisher Relations in the Victorian Era* (University Park: Pennsylvania State University Press, 2012), 67.

17. On the subject of human exhibitions, the reader is referred to Sadiah Qureshi's *Peoples on Parade: Exhibitions, Empire, and Anthropology in Nineteenth-Century Britain* (Chicago: University of Chicago Press, 2011), where this author describes the contribution of such displays to the formation of early anthropological inquiry and the creation of broader public attitudes toward racial difference.

18. T. J. Ellingson, *The Myth of the Noble Savage.* In a similar vein, W. R. Chapman, "Ethnology in the Museum," 334, declared that the Anthropological Society was "in many ways, less a scientific society than a polemical club."

19. Buxton had joined on March 15, 1864, *Journal of the Anthropological Society of London* 2 (1864). See ASL Archive, Office (A5); Letter 167: Charles Buxton, January 29,

1866, resigns from the Anthropological Society; 168: February 5 confirms resignation; "among scientific men it is spoken of with general ridicule" (ibid.); 169: February 6 expresses his "very deep indignation . . . [at] the evident attacks made by Captain Pim"; 170: to Hugh J. C. Beavan, February 26 "after the delivery and publication of Capt. Pim's lecture does not wish to be associated with the ASL." Georges Pouchet, *The Plurality of the Human Race* (London: Longman, Green, Longman & Roberts, 1864).

20. Isabel Burton describes how, when he came back from his trip in Latin America, her husband "went to the Royal Geographical Society meeting, found it slow, and *was not satisfied with his reception.*" Isabel Burton, *Life of Captain Sir Richard Burton*, vol. 1: 458 (italics in original). He also dined with Murchison and met with the former and current Foreign Office secretaries (Stanley and Clarendon).

21. "Annual Report for Year 1867," *Journal of the Anthropological Society of London* (1867).

22. "President's Address," *Journal of the Anthropological Society of London* 6 (1868): xcix ff. Hunt declared: "The history of scientific societies in this country is pretty nearly as follows. During the first three or four years, the novelty of the formation of a new scientific society attracts a large number of persons to join it in the hope of getting 'some sensation.' The novelty, however, soon ceases and the trial begins. The managers, during the first few years of the flourishing existence of such societies, are led to embark in various undertakings on the strength of the large number of Fellows on their list. There is, unfortunately prevalent in our day, a very low state of morality amongst those who, from motives of their own, are the foremost to join a new scientific society; and this class of members, I regret to say, usually decline to pay their obligation for expenses incurred on their behalf. There is also, I understand a prevailing impression abroad that this society is likely to be no exception to what has too frequently been a general rule. Some of our friends I am told, have given this impression too much weight; and some, who have never been friendly to us have made it a ground of attack."

23. "Minutes Special General Meeting 2nd September," *Journal of the Anthropological Society of London* 6 (1868): cxcvi.

24. The membership subscription ledgers in the archive of the Anthropological Institute (A-6), which provide indications of dates of cancellation of membership, contain interesting material on the topic discussed here.

25. "Anthropology at the British Association, 1869," *Anthropological Review* 7 (1869): 414–32, 428.

26. "Death of Dr. James Hunt: Science or Sun-Stroke," *British Medical Journal* 2 (1869): 355.

27. The editors of the new *Journal of Anthropology* were John Beddoe, J. Barnard Davis, Hermann Beigel, the Reverend Dunbar I. Heath, and C. Staniland Wake. Wake's book contained opinions that were as many openings towards ethnologists, as the following tortuous passage, which ends up on a clear anti-polygenist statement, suggests: "It is asserted, however, that, even if the mind of the negro, for example, be essentially superior to that of the anthropoid ape, his intellectual development is so small, that he must be of a species quite distinct from, and vastly inferior to, that of the European. It certainly has not been proved, and it is probably incapable of proof, that this incapacity for civilization, if it exist, would continue under all circumstances, and that it may not be the effect merely

of peculiar conditions of existence, which, if changed, would result in the removal of that incapacity. But if the negro were shown to be utterly incapable of receiving more than the rudiments of civilization, it would be no proof of his absolute and essential inferiority to the European . . . As well might it be said, however, that the mental faculties of the child are *absolutely* inferior to those of the man, because the functions of the brain not being fully developed, the former has not intelligence equal to that of the latter, as that the negro and the European are of different species because the brain of the one is inferior to that of the other. *It may be affirmed without fear of contradiction that all men have similar mental powers, and also the faculty of spiritual perception in common, and, if this be so, they must all have an equal intellectual capacity, the development of which, however, being dependent on physical conditions not equally favourable in each case*"; C. S. Wake, *Chapters on Man: With the Outlines of a Science of Comparative Psychology* (London: Trübner, 1868), 104–5. That Wake would nonetheless find himself subsequently battling the ethnologists suggests that ideas were not at the heart of the fight.

28. Stocking felt that he could contrast the Anthropological leadership, the Cannibal club, with a group of "independents" ("What's in a Name?" 383). My own take is that it is incorrect to read the tensions inside the Anthropological Society as resulting from a conflict of ideas. My narrative suggests reasons for individual dissidence and why a number of previously less visible individuals could emerge as plausible leaders in the context of a merger.

29. Vaux was admitted to the Athenaeum Club in 1870; among other active members of the Cannibal Council or Anthropological authors/participants, Farrar joined in 1873, Duncan in 1877, Bouverie-Pusey in 1885, Beddoe in 1888, Brabrook in 1889. There was a parallel between this and admission in the Royal Society; see *Anthropological Review* (1868): 324.

30. *Journal of the Anthropological Society of London* 9 (1870): iii–vi. The relevance of previous disputes can be read in Wake's insistence on quoting Eyre in "Tribal Affinities among the Aborigines of Australia," *Journal of the Anthropological Society of London* 9 (1870/71), ix ff.

31. Patton, *Work of Sir John Lubbock*, 93, 109. Stocking, *Victorian Anthropology*, argues more cautiously that the [Anthropological] "organization's finances and membership continued to decline" (256) and his narrative next discusses amalgamation, thus implying a chronological rather than strictly causal link. It would seem from the letter Ellen Lubbock wrote to Emma Darwin that a scheme had been suggested whereby Lubbocks, Darwins, Busks and others would raise the money to cover the debts of the Anthropological Institute upon the condition that it would revert to being called "Ethnological"; Ellen Lubbock to Emma Darwin, Darwin Correspondence Project, Letter 8700, "Nov-Dec?" 1873, http://www.darwinproject.ac.uk/.

32. £700 in today's currency amounts to about £50,000 according to purchasing power or £400,000 according to relative incomes. "Report of the Council of the Anthropological Institute for year 1874" in *Journal of the Anthropological Institute* of 1875, 473–75. The list of contributors to the fund shows thirty-nine "pure" ethnologists against seventeen "pure" anthropologists and five former members of both societies: Colonel Lane Fox, J. Milligan, John Beddoe, C. Robert des Ruffières, and Hyde Clarke. One finds seventeen men whose names are neither on the August 1869 list of the Anthropological Society nor on the

Ethnological Society's anniversary meeting list of 1870. Some may have joined during the interval, but mostly these include donors connected to the Ethnological establishment. The largest donor in this group was David Forbes (for £50) a fellow of the Royal Society. In round numbers, the breakdown of contributions is as follows: ethnology, £400; ethnology/anthropology £50; anthropology £150; others £100. Lubbock and Busk contributed £50 each. Darwin's name is not on the list.

33. For details on this, see George W. Stocking, "What's in a Name?" 383–84. See also Stocking, *Victorian Anthropology*, 254–57.

CHAPTER FOUR

1. Thanks to Amalia Ribi for pointing out to me the original rhyme.

2. Motion for a Select Committee, to "inquire into the circumstances attending the making of Contracts for Loans with certain Foreign States, and also the causes which have led to the non-payment of the principal moneys and interest due in respect to such Loans"; HC Deb February 23, 1875, vol. 222 cc772–87. *Fun*, April 10, 1875, 150, 153; The poem "A terrible monster" is dated April 7.

3. J. R. T. Hughes, *Fluctuations in Trade, Industry and Finance: A Study of British Economic Development, 1850–1860* (Oxford: Clarendon, 1960).

4. See David Landes, *Bankers and Pashas: International Finance and Economic Imperialism in Egypt* (Cambridge, MA: Harvard University Press, 1958). For a recent account and case study of the political economy upheavals brought about by railroad construction, see Richard White, *Railroaded: The Transcontinentals and the Making of Modern America* (New York: W. W. Norton, 2011).

5. Ramón de Silva Ferro, *Historical Account of the Mischances in Regard to the Construction of a Railway across the Republic of Honduras; The Failure of the Loans Solicited from the Public to Carry Out That Undertaking, and the Principal Difficulties Which Have Occurred at Each Fresh Combination Attempted for the Completion of the Line* (London: C. F. Hodgson & Son, 1875), 38.

6. *Times*, June 1, 1872.

7. My italics; Earl of Cromer [Evelyn Baring], "On the Government of Subject Races," *Edinburgh Review* (1908), reprinted in *Political and Literary Essays, 1908–1913* (London: Macmillan, 1913), 44.

8. P. J. Cain, "Empire and the Languages of Character and Virtue in Later Victorian and Edwardian Britain," *Modern Intellectual History* 4 (2007): 249–73. At one point in the article, Cain provides the following interpretation of the meaning of character: "The sense of dynamism and progress embodied in these components of 'character talk' owed a great deal to the Scottish Enlightenment view of 'commerce' and civil society as the chief vehicles of moral formation and was clearly reflective of the commercialization and urbanization of Britain over the previous centuries, though it was also strongly influenced by the Evangelicalism of the first half of the nineteenth century that was still a powerful force in 1900" (252). I am struck by the fact that, just like the literature on the Anthropological Society likes to blame the Confederacy, Cain makes of character a "Scottish thing." One author who has come closer to putting the finger on the underlying economics of pres-

tige is Catherine Hall, "Economy of Intellectual Prestige," 193–96. On nineteenth-century affinities between character and reputation, albeit with an American perspective, see Scott Sandage's excellent *Born Losers: A History of Failure in America* (Cambridge, MA: Harvard University Press, 2009).

9. C. M. Reinhart and K. S. Rogoff, "Serial Default and the 'Paradox' of Rich to Poor Capital Flows," *American Economic Review* 94 (2004): 53–58; as well as Reinhart and Rogoff, *This Time Is Different: Eight Centuries of Financial Folly* (Princeton, NJ: Princeton University Press, 2009). Robert Dudley Baxter, *National Debts* (London: Robert John Bush, 1871).

10. Joseph Conrad, *Nostromo, a Tale of the Seaboard* (New York: Harper & Bros., 1904); Edward B. Eastwick, *Venezuela; or, Sketches of Life in a South American Republic, with the History of the Loan of 1864* (London: Chapman & Hall, 1868); Edward Said, *Joseph Conrad and the Fiction of Autobiography* (Cambridge, MA: Harvard University Press, 1966); Edward Said, *Orientalism* (New York: Pantheon Books, 1978). See also Cedric Watts, "Four Rather Obscure Allusions in *Nostromo*," *Conradiana* 28 (1996): 77–80. The Venezuela loan of 1864 had been issued at 60£ and had a 6 percent coupon, thus the 10 percent yield.

11. Eastwick, *Venezuela*, vii. On the role of Philip Rose in the reorganization of the Conservative Party under Disraeli, see Robert Stewart, *Foundation of the Conservative Party*.

12. *Oppenheim v. General Credit and Finance Company*, 1865; National Archive, C 16/291/014.

13. Beyond the sharing of a name, which may be just a coincidence, it appears from the Archives and Manuscripts catalogue of the British Library that Samuel Laing wrote in favor of Laing-Meason's applications for financial support from the Royal Literary Fund. Note this Samuel Laing should not be mixed up with the travel writer who became a member of the Anthropological Society.

14. Harold Perkin, *Origins of Modern British Society, 1780–1880* (London: Routledge, 1969), 442; Martin Joel Wiener, *Reconstructing the Criminal: Culture, Law and Policy in England, 1830–1914* (Cambridge, UK: Cambridge University Press, 1990). From the vantage point of the arguments in this book, a relevant theorizing is provided by Vilhelm Aubert, "White-Collar Crime and Social Structure," *American Journal of Sociology* 58 (1952): 263–71. For a useful recent survey of the matter from a legal history perspective, see Sarah Wilson, "Law, Morality and Regulation: Victorian Experiences of Financial Crime," *British Journal of Criminology* 46 (2006): 1073–90; and Sarah Wilson, *The Origins of Modern Financial Crime: Historical Foundations and Current Problems in Britain* (London: Routledge, 2014).

15. H. Drummond Wolff, *Rambling Recollections*, vol. 2 (London: MacMillan, 1908), 56–66. Drummond Wolff is considered by Jenks to have been an "authority" on the matter, he certainly was a participant; see L. H. Jenks, *Migration of British Capital to 1875* (New York: Knopf, 1927), 274.

16. United Kingdom, House of Commons, *Report from the Select Committee on Loans to Foreign States: Together with the Proceedings of the Committee, Minutes of Evidence, Appendix, and Index*, House of Commons Paper 367, 1875.

17. Karl Marx, *Capital: A Critique of Political Economy*, vol. 3 (Chicago: C. H. Kerr, 1909), chap. 25: "Credit and Fictitious Capital."

18. Walter Bagehot, "Why the Stock Exchange Is Likely to Have More and Greater Frauds in It Than Any Other Market," *Economist*, March 27, 1875, 361.

19. This point of view, which emphasized a combination of middle-class incompetence (on the demand side) and aspirational corruption (on the supply side) may be traced to Spencerian readings of the railway mania of the 1840s; H. Spencer, *Railway Morals and Railway Policy* (London: Longman, Brown, Green & Longmans, 1855).

20. Sir R. Malins V.C., in Rubery v. Grant [1871] R. 139.

21. *Westminster Review* (April 1850); Marmaduke Blake Sampson, *Central America and the Transit between the Oceans* (New York: S. W. Benedict, 1850); Sampson, *Criminal Jurisprudence Considered in Relation to the Physiology of the Brain* (London: Samuel Highley, 1851).

22. "'The Times' behind the Scenes," *Grey River Argus* 16 (1875). The result was that the *Times*, whose City article had indeed, under Sampson, published criticism of Rubery, was condemned for libel. See *Times*, January 19, 1875, "The Alleged Libel against Mr. Rubery" for the text of incriminated articles.

23. D. Kynaston, *The City of London: A World of Its Own, 1815–1890* (London: Pimlico, 1995), 267–69; "Rubery v. Grant," *Economist*, January 23, 1875, 86–87.

24. *Fun*, January 30, 1875, 47.

25. United Kingdom, House of Commons, *Report from the Select Committee*, xlv.

26. Charles Duguid, *The Story of the Stock Exchange: Its History and Position* (London: Grant Richards, 1901), 32, 137.

27. See Robert Blake, *Disraeli* (New York: St. Martin's, 1967) on the puffing pamphlets written at the time. W. F. Monypenny and G. E. Buckle, *The Life of Benjamin Disraeli, Earl of Beaconsfield*, vol. 1 (London: Russell & Russell, 1968). They describe John Diston Powles as the "head of a financial house which had been reaping a rich harvest from the boom, and whose credit was deeply involved in its continuance and justification" (60). At the peak of the bubble, John Murray published several editions of *An Inquiry into the Plans, Progress, and Policy of the American Mining Companies* (London: John Murray, 1825).

28. Thomas Strangeways, *Sketch of the Mosquito Shore, including the Territory of Poyais, etc.* (Edinburgh, UK: Leith, 1822). As to the credibility of the "Strangeways" patronym, Robert A. Naylor finds one Thomas Fox-Strangeways among local elites; Naylor, *Penny Ante Imperialism: The Mosquito Shore and the Bay of Honduras, 1600–1914: A Case Study in British Informal Empire* (London: Fairleigh Dickinson University Press, 1989), 16.

29. Drawing on a history of McGregor by David Sinclair, *The Land That Never Was: Sir Gregor MacGregor and the Most Audacious Fraud in History* (Cambridge, MA: Da Capo, 2004), a Christmas edition of the *Economist* ("Financial Crime: The King of Con-Men," December 22, 2012) has painted Poyais as one "imaginary country" and then turned to "new research" from behavioral economics from Boston University about the traits of the victims ("They tend to be excessively trusting, have a high risk tolerance, and—especially the more educated victims—have a need to feel exclusive, or part of a special group"). For alternative perspectives on the McGregor swindles, see Frank G. Dawson, *The First Latin American Debt Crisis: The City of London and the 1822–25 Loan Bubble* (New Haven, CT: Yale University Press, 1990); and Dawson, "MacGregor, Gregor (1786–1845), Soldier and Adventurer" in *Oxford Dictionary of National Biography* (Oxford: Oxford University Press, 2004). For a promising critical perspective on the subject, see Damian

Clavel, "Gregor MacGregor, the Poyais Bonds and the Latin American Sovereign Debt Market 1822–1823: Financial Fraud or Failure?" (MSc diss., University of Edinburgh, 2013).

30. Naylor, *Penny Ante Imperialism*, 90, 91ff.

31. *Fortune's Epitome* (1833): 121.

32. Bernard Cohen's *Compendium of Finance*, first published in 1822, was dedicated to Alexander Baring, William Haldimand, and N. M. Rothschild, three leading foreign debt underwriters.

33. H. Spencer, "Morals of Trade," *Westminster Review* (April 1859), reprinted in *Essays: Moral, Political and Aesthetic* (New York: Appleton, 1881), 107–48. The production of circulars by key brokers of the market was a traditional feature of the time illustrated by Barings on American states' debts, Jardine's on opium, or Pixley's on gold and silver. Another example was one J. Edouard Horn, who, when he sought to promote Ottoman and Egyptian securities, published and then kept updating (for three years, between 1859 and 1861) a remarkable yearbook for state credit, *Annuaire international du crédit public*, that contained rich statistical information about borrowing governments and was the model for Macmillan's successful *Statesman Yearbook*. J. E. Horn is spotted in Landes, *Bankers and Pashas*, 201.

34. R. L. Nash Jr. was voted a member of the Anthropological Society on May 5, 1868 (*Journal of the Anthropological Society of London*). On Nash Sr. and Jr., see A. Odlyzko, "The Collapse of the Railway Mania, the Development of Capital Markets and the Forgotten Role of Robert Lucas Nash," *Accounting History Review* 21 (2011): 309–45, 340.

35. For details on R. L. Nash and the trade test, see Marc Flandreau and Frédéric Zumer, *The Making of Global Finance* (Paris: Organisation for Economic Co-operation and Development, 2004).

36. This group has received little attention from previous researchers and yet its role is essential, as briefly mentioned in Jenks, *Migration*. The most complete secondhand sources (but devoted exclusively to Mexico) are Michael P. Costeloe, *Bonds and Bondholders: British Investors and Mexico's Foreign Debt, 1824–1888* (Westport, CT: Praeger, 2003); and Richard Salvucci, *Politics, Markets and Mexico's "London Debt," 1823–1887* (Cambridge, UK: Cambridge University Press, 2009). For a study of why bondholders got organized after 1827, see Marc Flandreau, "Sovereign States, Bondholders Committees, and the London Stock Exchange in the Nineteenth Century (1827–68): New Facts and Old Fictions," *Oxford Review of Economic Policy* 29 (2013): 668–96. For a study of interlocking bondholder committees from 1827 to 1866, see Marc Flandreau and Joanna Slawatyniec, "The Dart and Elliott Associates of Victorian Britain: Foreign Debt Vulture Investors in the London Stock Exchange, 1827–1866" (Working Paper, Graduate Institute of International Studies and Development, forthcoming).

37. Giorgio Fodor, "The Boom That Never Was? Latin American Loans in London, 1822–1825" (Discussion Paper no. 5, Universita' degli studi di Trento, 2002).

38. See for instance W. D. Rubinstein, *Men of Property: The Very Wealthy in Britain since the Industrial Revolution* (Rutgers, NJ: Rutgers University Press, 1981).

39. David Morier Evans, *City Men and City Manners* (London: Groombridge, 1852), 139.

40. J. D. Scott, "Hambro and Cavour," *History Today* 19 (1969): 10; Bo Bramsen, Kathleen Wain, and Kathleen Brown, *The Hambros, 1779–1979* (London: Joseph, 1979). On war

and investor risk-taking in the mid-nineteenth century see Marc Flandreau and Juan H. Flores, "The Peaceful Conspiracy: Bond Markets and International Relations during the Pax Britannica," *International Organization* 66 (2012): 211–41.

41. Jenks, *Migration*, 121–22.

42. On Robinson, see N. Draper, *The Price of Emancipation*, app. 4. For members of the bondholders' committee, *London Daily News*, January 5, 1849.

43. For all the quotes in this paragraph, W. M. Mathew, "The First Anglo-Peruvian Debt and Its Settlement, 1822–49," *Journal of Latin American Studies* 2 (1970): 81–98, 96–98.

44. Costeloe, *Bonds and Bondholders*, 230. See the *Times*, February 3, 1869, for a list of the first Council which comprised several of the future members of the Council of the Corporation of Foreign Bondholders. Names with (*) are found in the 1873 Council: Charles Bell, George Worms, Philip Cazenove, Thomas Moxon, Isidor Gerstenberg (*), Philip Rose (*), Mr. H. R. Jameson, Francis Lycett (*), Admiral Sir Provo Wallis (*), Colonel R. B. Beaumont (*), George Cavendish Bentick (*), and H. B. Sheridan; *Journal of the Anthropological Society of London* 2 (1864): clxx.

45. This applies particularly well to Hyde Clarke, who was a senior member of the Imperial Cotton Commission in Smyrna and to Charles Duncan Cameron, the British consul in Massowah. Pim would also imagine cotton development in Central America.

46. Hecht, *Scramble for the Amazon*, 141–63.

47. Edward Haslewood, *New Colonies on the Uplands of the Amazon* (London: King, 1863; published for private circulation).

48. "Commencing at the mouth of the Orinooko, and following the southern bank through Venezuela to the parallel of west longitude 72 degrees in New Granada; thence in a straight line southward to the boundary between New Granada and Equador; thence following that boundary to the parallel 74 degrees; thence southward to Naupta; and thence following the north bank of the Amazon to its mouth in the Atlantic; though very irregular in form, this territory comprises probably 650,000 or even 800,000 square miles—about the size of one sixth of Europe. The whole of this enormous space, though under different governments, is utterly valueless to them—demanding care which it does not receive—and involving responsibilities they dare not assume." Haslewood, *New Colonies*, 4. If we leave aside the confused geographical terminology when it came to "parallels" (the "parallel of west longitude 72 degrees" is a "meridian"), Haslewood was describing an area roughly delineated by the Orinoco on the north and the Amazon on the south, the Atlantic on the east and on the west, finally, the 72th and 74th meridians west cutting across Colombia ("New Granada") and Ecuador. Notwithstanding innumerable indigenous tribes, this area contained territories owned or claimed by Brazil, Venezuela, Ecuador, Colombia, Britain, France, and Holland. As described by Hecht, it would host several of the later settlements by the Brazilian politician and colonial promoter José Maria da Silva Paranhos, Baron of Rio Branco (Settlements 1, 2 and 3 in her "Map of Territorial Settlements made by Rio Branco," Map 1; Hecht, *Scramble for the Amazon*, xvi).

49. Ecuador Land Company Ltd. Prospectus, F. O. 25/34, reproduced in http://www .europa.clio-online.de/site/lang__en/ItemID__324/mid__11373/40208215/default.aspx (Themenportal Europäische Geschichte).

50. Haslewood, *New Colonies*, 5

51. Ibid., 14.

52. Patricia Seed, *American Pentimento: The Invention of Indians and the Pursuit of Riches* (Minneapolis: University of Minnesota Press, 2001). See also Seed, *Ceremonies of Possession in Europe's Conquest of the New World, 1492–1640* (New York: Cambridge University Press, 1995).

53. Joseph Dalton Hooker, "On the Struggle for Existence amongst Plants," *Popular Science Review* 6 (1867): 131–39; Bedford Pim and Berthold Seemann, *Dottings on the Roadside, in Panama, Nicaragua, and Mosquito* (London: Chapman & Hall, 1869), 119.

54. Pim and Seemann, *Dottings*, 57, 59, 83.

55. Ibid., 316.

56. Ibid., 334.

57. *Times*, June 1, 1872.

58. Pim and Seemann, *Dottings*, 59.

59. See Ulrike Kirchberger, *Aspekte Deutsch–Britischer Expansion*; FO 25/34.

60. BT 41/215/1225. Royal Geographical Society membership list in *Journal of the Royal Geographical Society of London* 29 (1859): i–xciv.

61. BJ 7/919. December 19, 1859. Of interest is the fact that the approach, which raised specifically the subject of meteorological information collection, recognized the Board of Trade as a relevant authority. Of course, the Board of Trade was also the body in charge of delivering authorizations to British corporations. The answer by Fitzroy to the secretary of the Ecuador Land Company stated with "thanks" and "consideration" the Board of Trade's concern not to add "to the embarrassments of new and difficult undertakings by asking for meteorological observations at its commencement."

62. Publicity for the Anthropological Society in the reprint of Pim's *Negro and Jamaica*. Bollaert's name is not found in the subsequent list of shareholders (for April 1860). He must have cashed out, once his academic certification had been sold. The degree to which public connections with commercial endeavors could have reputational consequences and the resulting brand management tactics would be worthy of a fuller examination.

CHAPTER FIVE

1. Ranald Michie, "The Social Web of Investment in the 19th Century," *Revue Internationale d'Histoire de la Banque* 18 (1979): 158–75. Max Weber's first discussion of the stock exchange is found in "Die Börse" (1894), translated as "Stock and Commodity Exchanges" in *Theory and Society* 29 (2000): 305–38. Max Weber's "Die Börse" was the first of a series of articles (until 1896) devoted to the subject. On the importance of networks to credit relations see Naomi Lamoreaux, whose seminal *Insider Lending* revisited networks as a positive force in development; Lamoreaux, *Insider Lending: Banks, Personal Connections, and Economic Development in Industrial New England* (Cambridge, UK: Cambridge University Press, 1996). On the relation between bureaucratic control and credit computations in the context of late nineteenth-century foreign debt, see Marc Flandreau, "Audits, Accounts, and Assessments: Coping with Sovereign Risk under the International

Gold Standard, 1871–1913," in *International Financial History in the Twentieth Century: System and Anarchy*, ed. M. Flandreau, C. L. Holtfrerich, and H. James (New York: Cambridge University Press, 2003).

2. W. R. Chapman, "Ethnology in the Museum," chap. 4; Elizabeth Baigent, "Markham."

3. Hecht, *Scramble for the Amazon*, 159. She suggests that Church stands on a par with "his friend the celebrated imperial botanist" Clements Markham and the "widely traveled Sir Richard Burton."

4. Clements R. Markham, "On Crystal Quartz Cutting Instruments of the Ancient Inhabitants of Chanduy (Near Guayaquil in South America)," *Journal of the Anthropological Society of London* 2 (1864): lvii–xli.

5. Neville B. Craig, *Recollections of an Ill-Fated Expedition to the Headwaters of the Madeira River in Brazil* (Philadelphia: J.B. Lippincott, 1907); "Biographical Notice," in *Aborigines of South America*, by G. E. Church and Clements R. Markham (London: Chapman & Hall, 1912), xi–xix. The "Notice" states that in the late 1870s Church was "entrusted by the English foreign bondholders" with "full powers to negotiate the readjustment of the National Debt" of the state of Ecuador and another time in the mid-1890s in Costa Rica, "staying three months on behalf of the bondholders of that country."

6. Committee of the Holders of Spanish Bonds, *Spanish Bonds: A Full Report of the Public Meeting Held on August 19, 1851, Further to Consider Bravo Murillo's Confiscatory Scheme for the Arrangement of the Debt, &c.; Letters of J. Capel, Sir H. Bayly and Mr. Tasker; Memorial of Mr. Wadsworth, with Lord Palmerston's Reply; Criticisms of the English Press etc.* (London: John King, 1853?).

7. With Joanna Sławatyniec, we computed the centrality of the various leaders of the bondholder syndicate. In their current state our computations confirm that John Diston Powles was the undisputed boss of the industry. Gerstenberg (who would launch the Corporation of Bondholders with Hyde Clarke, and later Lubbock) is number three. Edward Haslewood, the author of *New Colonies on the Uplands of the Amazon*, is number two, having participated in no fewer than seven different bondholders' committees in 1854–66. Tasker, was a veteran of the earlier period of bondholding when Diston Powles's reign was initiated. See Marc Flandreau, Joanna Sławatyniec, and Alberto Gamboa, "The Dart and Elliott Associates of Victorian Britain."

8. See Quintin Quevedo, "The Madeira and Its Head-Waters," in *Explorations Made in the Valley of the River Madeira, from 1749 to 1868*, ed. George Earl Church (London: National Bolivian Navigation Company, 1875), 167–88.

9. On February 22, 1864, we again spot Gerstenberg participating in discussion of the lecture by William Bollaert of a paper by Raimondy (who was also the Lima correspondent of the Anthropological Society of London). Gerstenberg is said to be "knowledgeable on German settlements on the Eastern slopes of the Andes." At some point, he makes a statement about his concern with a navigation monopoly owned by a Brazilian company, ending in 1880: "He considered this a great hardship, and he hoped, wherever an arrangement of our present difficulties with Brazil took place, that the British Government would bring forward the question of the opening of the Amazon." *Proceedings of the Royal Geographical Society of London* (1864): 62. In 1867 the Amazon would be open to international navigation.

10. See Hecht, *Scramble for the Amazon*, 96–97.

11. When Church's companies were launched, the only existing way to Bolivia's mainland was via Peru. A road opened in 1870 connecting Mollendo and Puño in Peru then via steamer over Lake Titicaca to "a good and comparatively level wagon road" to La Paz. Craig, *Recollections*, 21. There had been plans for a more direct western route connecting Bolivia to the Pacific through the desert via the port of Antofogasta, at that time a Bolivian possession. The War of the Pacific begun in 1879 deprived Bolivia of its Pacific border to Chile and transformed it into a landlocked country. On foreign economic rivalries in the background of the War of the Pacific, see V. G. Kiernan, "Foreign Interests in the War of the Pacific," *Hispanic American Historical Review* 35 (1955): 14–36. On the western route, see Harold Blakemore, *From the Pacific to La Paz: The Antofagasta (Chili) and Bolivia Railway Company, 1888–1988* (London: Lester Crook Academic, 1990).

12. BT 31/1595/5315.

13. To courts of law exploring the matter later it was explained that after arriving at La Paz in August 1868, Church suggested to the government of Bolivia a "scheme for making a complete navigation from the interior of Bolivia down to Rio Mamoré and Rio Madeira to the River Amazon, and so to the Atlantic" (*Wilson v. Church* 13 Ch. D. 1 at 2). Craig (*Recollections*, 36, see also 42) claims that Church was an agent of the government of Bolivia, tipped by Quintin Quevedo with a ready-made project. The "Bolivian Government had been fortunate to secure for the work of organization, the services of a man fully prepared to meet all the exigencies of the situation."

14. J. Keller and F. Keller, *Exploración del Rio Madera en la Parte Comprendida Entre la Cachuela de San Antonio y la Embocadura del Mamoré* (La Paz: Imprenta de la Union Americana, 1870). See also J. Keller and F. Keller, "Relation of the Exploration of the River Madeira," in *Explorations Made in the Valley of the River Madeira, from 1749 to 1868*, ed. George Earl Church (London: National Bolivian Navigation Company, 1875), 1–76; and Craig, *Recollections*, 35.

15. *Report of Proceedings Connected with the Foreign Debt of the Republic of Ecuador, at a Meeting Held at the London Tavern, in the 28th of June, 1853* (London: Clay, 1853; Church collection, xfHT C81). Prominent bondholder Baron de Goldsmid, who must have been Isaac Lyon Goldsmid (1778–1859), was marked dead in the list when J. D. Powles, that is, John Diston Powles (1787–1867) was not. The inscription was therefore probably made between 1859 and 1867, long before the time when Markham tells us that Church was a formal envoy of bondholder committees. As indicated in the text, Church and Haslewood (assuming that he got the copy of the volume from Haslewood too) were connected by 1871. On the other hand, as I only count seven annotations when the summary list mentions eight persons dead, it might be that this was after the death of J. D. Powles and that Powles's death was such an evident thing to the person who did the counting that it was not worth scribbling. If so, it may be a later list, perhaps in preparation for Church's subsequent missions. All scenarios, however, in combination with the presence of Haslewood in the promoters of the Church railway and the presence of the book in Church's collection, point to Haslewood having initiated Church in the art of scavenging.

16. The rest of the discussion relies extensively on *Wilson v. Church*: 9 Ch. D. 552, 12 Ch. D. 454, and 13 Ch. D. 1. The House of Lords decision is *National Bolivian Navigation Company et al. v. William Millar Wilson et al.* 5 A.C. 176.

17. The Bolivian concession was dated December 7, 1869, and committed Church to create a National Bolivian Company with a capital of at least one million gold dollars. The Brazilian concession was dated April 20, 1870, and committed Church to launch a Madeira and Mamoré Railway Company.

18. Memorandum and Articles of Association, Madeira and Mamoré Railway Company, National Archive, BT/1595/5315, February 27, 1871; *Journal of the Ethnological Society of London* 1 (1869/70): 274–87. Other original promoters included one Frank H. Colins and one Ebenezer Westmoreland Pearson, a clerk with Morton C. Fisher's firm.

19. Both received $400,000 too. Together, Church, Reynolds, and Fisher owned 1.9 million of the 2 million shares of the Navigation Company.

20. W. Rubinstein, "Jewish Top Wealth-Holders in Britain, 1809–1909," *Jewish Historical Studies* 37 (2002): 133–61.

21. Craig, *Recollections*, 46, states that these "£12,500 seems to have been the only cash received by the company in return for the stock disposed of."

22. The other additions were Charles G. Pym, a young offspring from a political dynasty of Bedfordshire; J. B. Kendall; and Baron C. F. d'Escury, a Dutch noble. They all received 25 shares. Markbreit received 200 and de Campero 250. BT 31/1595/5315, May 14, 1872. George Earl Church, *The Route to Bolivia via the River Amazon: A Report to the Governments of Bolivia and Brazil* (London: Waterloo, 1877), 20, 22, 41. Each share was nominally worth £20. The amounts showered could be large, but were small compared to the bulk of the holdings, which remained with the Navigation Company. Note that the report to the Board of Trade was only made available with some delay, since it should have been filed in March. This may have been a tactical move to prevent the release of the information until the Bolivian loan was well established on the market.

23. BT 31/1595/5315, January 14, 1875; William Scully was the author of *Brazil: Its Provinces and Chief Cities: The Manners and Customs of the People etc.* (London: John Murray, 1866). On the Sociedade, see Alexandre Carlos Gugliotta, "Tavares Bastos (1839–1875) e a Sociedade Internacional de Imigraçao: Um Espaço a Favor da Modernidade" (manuscript presented at the "Usos de Passado" conference), http://www.rj.anpuh.org/resources /rj/Anais/2006/conferencias/Alexandre%20Carlos%20Gugliotta.pdf.

24. 13 Ch. D. 1 at 23–24.

25. 13 Ch. D. 1 at 28.

26. Craig, *Recollections*, 50.

27. *Edinburgh Gazette*, December 24, 1872. At that date, Sampson was already a consul general for the Republic of Argentina.

28. This appears to have been an integral part of the scheme, as evidenced by a December 22, 1869, agreement between Bolivian authorities and Church when the "bases of a loan" had been agreed upon, 13 Ch. D. 1.

29. *Journal of the Anthropological Society of London* meeting of June 16, 1868; Toshio Suzuki, *Japanese Government Loan Issues on the London Capital Market, 1870–1913* (London: Athlone, 1994), 52.

30. Craig, *Recollections*, 53. Joseph Conrad, *Lord Jim: A Tale* (London: Blackwood, 1900), 302.

31. From Wilson v. Church 13 Ch. D. 1, at 8. A copy of the prospectus is reprinted in

the *Examiner*, January 20, 1872, 77, the *Newcastle Journal*, January 22, 1872, 1, and so on. Note that unlike the prospectus, the general bond did not contain a single reference to the works contract, nor to the Madeira and Mamoré Railway Company, nor to the Public Works Construction Company. This was probably not coincidental because the general bond, unlike the prospectus, had contractual value.

32. *National Bolivian Navigation Company et al.* v. Wilson et al. 5 A.C. 176. According to Andrew St. George, Erlanger would have hosted Philip Rose in his offices at 43 Lothbury after Rose left Baxter, Rose, Norton and Co. in 1872, to "focus" on helping Disraeli; Andrew St. George, *A History of Norton Rose* (Cambridge: Norton Rose, 1995), 139.

33. *Wilson v. Church* 13 Ch. D. 1 at 28.

34. *Proceedings of the Royal Geographical Society of London* 16 (1872): 157.

35. Sir Clements Robert Markham to Colonel George Earl Church, May 10, 1872. Markham invites Church to a dinner of Royal Geographical Society, Church Collection Mss., Zt GM34 t.

36. Computations from information in *Wilson v. Church* 13 Ch. D. 1, especially p. 8. Of the £110,000 that came into the hands of the Navigation Company, £20,000 were acknowledged as having been paid to Church as the price for his Brazilian concession. The rest must have reached his pockets in more circuitous routes. Craig (*Recollections*, 64–65) has suggested a number of ways through which money leaked. Part would have gone in the few tiny boats such as the *Explorador*, purchased for the expeditions, but which "became worthless from want of care." He also alludes to "legal expenses" and "heavy expense to maintain necessary agencies at various points." Since the Madeira and Mamoré "squatted" at one shareholder's firm, it is unclear what is meant by this. Beyond the showering of stocks, there is no account of the bribes that must have been paid at various stages.

37. Not all this was appropriated by Church. He had had expenses, needed to shower supporters. This was typical of the contemporary schemer. For instance, Albert Grant felt bound to help Milan's construction of Galleria Vittorio Emanuele and present London's Metropolitan Board of Works with Leicester Square.

38. "Monetary and Commercial," *Freeman's Journal*, April 14, 1871; BT 31/1597/5329; St. George, *Norton Rose*, 139.

39. For a discussion of the links between the overall financial scheme and the requirement to execute the work before the first month of 1874, see *National Navigation et al. v. William Millar Wilson et al.* 5 A.C. 176 at 16.

40. *Proceedings of the Royal Geographical Society and Monthly Record of Geography* 18 (1873/74), special meeting of the Society to discuss "Across the Andes from Callao" by Thomas J. Hutchinson and "Railroad and Steam Communication in Southern Peru" by C. R. Markham, 203–20, 219. There was coverage in the contemporary press, which mentioned the participation of Senor Don Pedro Galvez, Peruvian minister in London.

41. See Kiernan, "Foreign Interests," 18.

42. *National Navigation et al. v. William Millar Wilson et al.* 5 A.C. 176 at 7. Compare with Craig, *Recollections*, who talks about an imperial decree in the official journal of the Brazilian government, dated November 24, 1877.

43. *National Navigation et al. v. William Millar Wilson et al.* 5 A.C. 176 at 7.

44. H. Spencer, "Morals of Trade," *Westminster Review* (April 1859), reprinted in H. Spencer, *Essays: Moral, Political and Aesthetic* (New York: Appleton, 1881), 107–48, 107.

45. *Wilson v. Church* 13 Ch. D. 1 at 31.

46. 13 Ch. D. 1 at 24–25.

47. 13 Ch. D. 1 at 27.

48. *National Navigation et al. v. William Millar Wilson et al.* 5 A.C. 176 at 8.

49. *National Navigation et al. v. William Millar Wilson et. al.* 5 A.C. 176 at 10.

50. Craig, *Recollections*, 449.

51. "Biographical Notice," xviii, in Church's *Aborigines*; Craig, *Recollections*.

52. "Biographical Notice," xviii, in Church's *Aborigines*. The *London News* hailed the project, claiming "it is well known that railways in the Argentine Republic give a higher return upon their capital than those of any other country, and that their shares and debentures command high premium" (March 3, 1889). The whole thing was to come to an end with the Baring Crisis (Markham insists that despite the crisis, Church was able to deliver). The *York Herald* of January 22, 1891, published a letter from Church saying the entire railway would be "opened by the end of this month [January]." See *Burdett's Stock Exchange Official Intelligence*, 1891, for a technical description of the line, which gives Church as contractor. It is of course interesting that Church was performing the job for which he had not been able to find any proper agent.

53. BT 31/246/807; Sir Robert Palmer Harding was Turquand's vice president in 1880–81 and 1881–82. On his role on committees and knighthood, see J. R. Edwards, *A History of Financial Accounting* (London: Routledge, 1989), 288.

54. Craig, *Recollections*, 41.

55. *New York Times*, January 6, 1910. This may have referred to "Argentine Geography and the Ancient Pampean Sea," Church's address to the geographical section at the British Association for the Advancement of Science, Bristol, 1898.

56. Isaiah Bowman, "Geographical Aspects of the New Madeira–Mamore Railroad," *Bulletin of the American Geographical Society* 45 (1913): 275–81.

57. Hecht, *Scramble for the Amazon*, 160.

58. 1839 (169) XIX. *First Report of the Commissioners Appointed to Inquire as to the Best Means of Establishing a Constabulary Force in the Counties of England and Wales*, 49.

CHAPTER SIX

1. *Select Committee*, p. 5 Q. 127 (de Zoete interview); other examples include *General Report*, p. xlvi; Report on Costa Rica Loans, p. 208; N. M. de Rothschild, p. 270 Q. 5837; Medley, p. 279, Q. 5987; and so on.

2. Letter by Hyde Clarke's supporter Brookes, *Athenaeum*, September 19, 1868, 369.

3. *Select Committee*; "A Bona Fide Traveller," *Times*, May 29, 1893; online research aid for the John Beddoe Papers, Bristol University; Isabel Burton, *The Life of Captain Sir Richard F. Burton*, vol 1., 20; Edward Haslewood, *New Colonies on the Uplands of the Amazon*; Augustus H. Pitt Rivers, *Primitive Warfare: Illustrated by Specimens from the Museum of the Institution; A Lecture at the Royal United Service Institution* (London: United Service Institution, 1867), 8.

4. T. Asad, ed., *Colonial Encounter*, 9.

5. *Times*, June 1, 1872.

6. *Select Committee*, p. 5, Q. 127–28.

7. This was part of part of a broader debate on whether industry or the City was the true driver of imperialism. See in particular, the first volume of their reference study: P. J. Cain and Anthony G. Hopkins, *British Imperialism*. According to Cain and Hopkins, one should look toward the City to understand where imperialism came from. The City would have acted initially ("until 1850") as the agent of the landed class to which it was closely tied. Over time, it became more of its own master, expanding the empire as it exported capital.

8. Some historians have disputed the validity of Cain and Hopkins's hypothesis. Most prominently, Martin Daunton has argued that there were complex internal divisions within the City, which makes it impossible to assume the existence of a singular interest. Daunton, "'Gentlemanly Capitalism' and British Industry, 1820–1914," in *State and Market in Victorian Britain: War, Welfare and Capitalism* (Woodbridge, UK: Boydell, 2008).

9. "Foreword: The Continuing Debate on Empire," in Cain and Hopkins, *British Imperialism*, 10.

10. Philip Augar, *Death of Gentlemanly Capitalism* (London: Penguin, 2000), xiii.

11. Steven Shapin, *A Social History of Truth: Civility and Science in Seventeenth-Century England* (Chicago: University of Chicago Press, 1994).

12. Hyde Clarke, "On the Geographical Distribution of Intellectual Qualities in England," *Journal of the Statistical Society of London* 34 (1871): 357–73; Francis Galton, *Hereditary Genius* (London: Macmillan, 1869). Clarke's argument was also related to the study of Alphonse de Candolle on the determinants of the spatial distribution of men of science. See Alphonse de Candolle's *Histoire des sciences et des savants* (Geneva: Georg, 1869).

13. Jim Endersby, *Imperial Nature: Joseph Hooker and the Practices of Victorian Science* (Chicago: University of Chicago Press, 2008). Other examples of the use of gentlemanly science include John C. Waller, "Gentlemanly Men of Science: Sir Francis Galton and the Professionalization of the British Life-Sciences," *Journal of the History of Biology* 34 (2001): 83–114; Pratik Chakrabarti, "The Travels of Sloane: A Colonial History of Gentlemanly Science," in *From Books to Bezoars: Sir Hans Sloane and His Collections*, ed. Alison Walker, Arthur MacGregor, and Michael Hunter (London: British Library Publishing, 2012).

14. See especially George Robb, *White-Collar Crime in Modern England: Financial Fraud and Business Morality, 1845–1929* (Cambridge, UK: Cambridge University Press, 1992). See also John Armstrong, "The Rise and Fall of the Company Promoter and the Financing of British Industry," in *Capitalism in a Mature Economy: Financial Institutions, Capital Exports and British Industry, 1870–1939*, ed. J. J. Van Helten and Y. Cassis (London: Edward Elgar, 1990), 115–38. There is a very recent burgeoning literature devoted to the subject of Victorian white-collar criminality and an accompanying renewal of interest for the evolution of ideas on corporate wrongdoing. See, for example, James Taylor, *Boardroom Scandal: The Criminalization of Company Fraud in Nineteenth Century Britain* (Oxford: Oxford University Press, 2013).

15. David Morier Evans, *City Men and City Manners* (London: Groombridge, 1852),

139ff.; Christopher Richardson, *Mr. John Diston Powles: or, the Antecedents, as a Promoter and Director of Foreign Mining Companies, of an Administrative Reformer* (London: Author, 1855); Eric Rosenthal, *On 'Change through the Years: A History of Share Dealing in South Africa* (Flesch Financial Publications, 1968), 162; Hannah Arendt, *The Origins of Totalitarianism* (New York: Harcourt Brace, 1951), 124ff. I am grateful to my student Mariusz Lukasiewicz for the Barnato anecdote and reference.

16. For Lubbock v. Lyell see the discussion in Patton, *Work of Sir John Lubbock*, 73–74. Lubbock started the accusation. Patton acknowledges that Lyell's defense, which hinted at the fact Lubbock drew from the Danes, "had a point." He adds that Lubbock had a large unacknowledged debt to Morlot and Henry Darwin Rogers, "whose three central questions relating to human antiquity were replicated almost verbatim and without reference in Lubbock's paper on the same subject"; For a thorough discussion of the dispute, see Leonard G. Wilson, "A Scientific Libel: John Lubbock's Attack upon Sir Charles Lyell," *Archives of Natural History* 29 (2002): 73–87; For Speke v. Burton see above, chapter 2; for *Pike v. Nicholas*, see "The Origins of the English: *Pike v. Nicholas*," *Anthropological Review* 7 (1869): 279–306; and Pike v. Nicholas (1868) P. 87.

17. The Livingstone or the Speke and Grant parties and the broader controversy on the sources of the Nile and Lake Nianza provide cases in point. It was their position within the Geographical Society that permitted explorers to discover the sources of Nile.

18. For a classic portrait of the links between gentlemen, dilettantes, and money market funds managers, see R. S. Sayers, *Gilletts in the London Money Market, 1867–1967* (London: Clarendon, 1968); or again, for the nostalgic take, P. Augar, *Death of Gentlemanly Capitalism* (London: Penguin, 2000).

19. William Blanchard Jerrold, "On the Manufacture of Public Opinion," *Nineteenth Century* 13 (1883): 1080–92, 1080. Jerrold was the son of the actor, dramatist, journalist, and professional puffer Douglas Jerrold.

20. For a background discussion of the rise of societies as voluntary associations, see R. J. Morris, "Voluntary Societies and British Urban Elites, 1780–1850: An Analysis," *Historical Journal* 26 (1983): 95–118; see also Morris, "Clubs, Societies and Associations," in *The Cambridge Social History of Britain, 1750–1950*, vol. 3, *Social Agencies and Institutions*, ed. F. M. L. Thompson (Cambridge, UK: Cambridge University Press, 1992), 395–444. Morris emphasizes the rise of voluntary associations as part of the definition of a new civil society emphasizing that it resulted in the import of new words of common usage such as *society, chairman, agenda, membership*, and *rules*.

21. "Membership List," *Memoirs Read before the Anthropological Society of London* 2 (1866), 14.

22. Endersby, *Imperial Nature*.

23. *Society of Accountants in Edinburgh v. Corporation of Accountants* (1893) 20 R. 750.

24. See Robb, *White-Collar Crime*, for modern discussion. For a contemporary perspective, see "The Way to Form a Directory," *Evening Post* (New Zealand), March 18, 1873; and Anthony Trollope, *The Way We Live Now* (London: Chapman & Hall, 1875), in which the Great Railway from Salt Lake City to Vera Cruz has such a venal noble director. See F. Braggion and L. Moore, "The Economic Benefits of Political Connections in Late Victorian Britain," *Journal of Economic History* 73 (2013): 142–76, for an empirical study of the value of political connectedness in late nineteenth-century Britain.

25. "Membership List," *Memoirs Read before the Anthropological Society of London* 2 (1866): 24.

26. *Select Committee*, p. 252, Q. 5482.

27. I have in mind, for instance, the volume edited by Aileen Fyfe and Bernard Lightman in which they describe the locations of Victorian science; Aileen Fyfe and Bernard Lightman, eds., *Science in the Marketplace: Nineteenth-Century Sites and Experiences* (Chicago: University of Chicago Press, 2007).

28. Henry Wheatly and Peter Cunningham, *London, Past and Present: Its History, Associations, and Traditions* (Cambridge, UK: Cambridge University Press, 2011), 189.

29. See chapter 7 in A. J. Arnold and S. McCartney, *George Hudson: The Rise and Fall of the Railway King* (London: Hambledon & London, 2004).

30. Jerrold, "Public Opinion," 1082.

31. *Journal of the Anthropological Society of London* (June 16, 1868).

32. Publication of the paper itself, to the extent that it took place, was often postponed. The Anthropological Society released the articles as part of the *Memoirs Read before the Anthropological Society of London.*

33. For instance, Landes, *Bankers and Pashas* at 65 has an ironical description of such a shareholder meeting.

34. This is precisely taken from the *Report* of the 1851 Spanish bondholders' meetings, given by Tasker to Haslewood, and that appears to have educated Church; Committee of the Holders of Spanish Bonds, *Spanish Bonds.*

35. Stocking, "What's in a Name?" 383–84.

36. "Mexican Bondholders," *Morning Post*, September 7, 1848.

37. *Athenaeum*, August 29, 1868, 271.

38. The minutes were later published as part of the report on Hyde Clarke and the various events that had led to the *Athenaeum* controversy. The contents of the Council's discussion of the amalgamation appeared in two places within the *Journal of the Anthropological Society.* First in discursive form as appendix to the special general meeting of September 2, 1868, vol. 6, under "Official Reports of the President and Director of the Anthropological Society of London Respecting the Failure of the Negotiation for Amalgamation," clxxxix ff., and a second time as appendix to the special general meeting of October 28, 1868, vol. vii, at p. vi and vii.

39. William James, *The Meaning of Truth* (a sequel to *Pragmatism*), Longman, Green, 1969.

40. "Professor Huxley on Celts and Teutons," *Anthropological Review* 8 (1870): 207–9.

41. Disraeli, "The Representation of the People," March 18, 1867 (HC Deb March 18, 1867, vol 186 cc6–94).

CHAPTER SEVEN

1. Stocking, "What's in a Name?" 369–90; Evelleen Richards, "The 'Moral Anatomy' of Robert Knox: The Interplay between Biological and Social Thought in Victorian Scientific Naturalism," *Journal of the History of Biology* 22 (1989): 373–436. On p. 426 Richards writes: "The Anthropological Society of London organized a public meeting in defense of Eyre, at which Captain Bedford Pim (*who had been hastily admitted to the society for*

*the purpose*) delivered a racist diatribe on 'The Negro and Jamaica' to the loud cheers of his audience and their unanimous vote of thanks" (my italics). Richards gives as source for this "Stocking, 'What's in a Name?' p. 379." However, on that page (and in others) all that Stocking writes is: "The response of the A.S.L. was a public meeting at which Captain Bedford Pim gave a paper on 'The Negro and Jamaica.'" I have found in many other places traces of what seems to be an urban legend. The only explanation I see is that someone may have come across the indication of Pim becoming an A.S.L. fellow "London (Hampstead)" on January 16, 1866, indeed, "just in time" to deliver the lecture. However, had previous scholars checked Pim out more rigorously, they would have found that he had been an Anthropological Society's local secretary for Nicaragua as early as June 1864. At the Anthropological Society, his arrival was preceded by his longtime associate and partner and Hooker's former protégé, the botanist Seemann. I believe that the desire to exclude the Anthropological Society from the origins of the Anthropological Institute is what explains the success, precisely among most serious scholars, of the myth of Pim's hasty admission.

2. Thomas Carlyle [calling himself Anonymous], "Occasional Discourse on the Negro Question," *Fraser's Magazine for Town and Country* 40 (1849): 670–79; as explained in chapter 1, the article had been written in the midst of mounting resentment and anxiety over the situation that had resulted from abolition. Catherine Hall, "The Economy of Intellectual Prestige. Hall describes Carlyle's willingness to put his "prestige" with the middle class to the service of the planters' cause as a piece of good fortune for the declining lobby.

3. Bedford Pim and Berthold Seemann, *Dottings*, 210, 215, 220. There are reasons to believe that Pim's part in this was written with the support of some hired pen. Pim, a sailor, would have "written badly enough." Pim and Seemann, *Dottings*, 434: "*Les marins écrivent mal, mais avec assez de candeur*" ("sailors write badly enough but with enough ingenuousness"). One is reminded of John Hanning Speke, another challenged writer. See David Finkelstein, *House of Blackwood*. Seemann was highly regarded as a writer. The *Athenaeum* did not find itself unfit to publish Seemann's letters from Nicaragua, even though this was a man who openly supported human zoos, another suggestion of the Liberal science lobby's tolerance of the horror.

4. C. H. Coote, "Pim, Bedford Clapperton Trevelyan (1826–1886)," in *Dictionary of National Biography*, vol. 45, ed. Sidney Lee (London: Smith, Elder, 1896), 306–7. I cannot reconcile the elements found in Pim's Navy file and some of the material in the *Dictionary*. Pim's Navy file has him a lieutenant on October 2, 1850, but according to the dictionaries, it would be on September 6, 1851. For precise chronology on Pim's career, see "Pim—Bedford—Clapperton—Trevelyan," Admiralty: Officers' Service Records (Series III), ADM 196/16/296, p. 296. I take the latter to be the authoritative source.

5. R. L. Stevenson, "Our City Men: No. 1—A Salt-Water Financier," *London* 1 (1877): 9–10. Another was the "Paris Bourse," *London* 4 (1877): 88. For a presentation of context, see Caroline A. Howitt, "Stevenson and Economic Scandal," *Journal of Stevenson Studies* 7 (2010): 143–56.

6. Berthold Seemann, *Narrative of the Voyage of HMS Herald*, 2 vols. (London: Reeve, 1853).

7. For a synthesis of official dictionaries' content (explaining the errors found in the article) but also original research of the early years of Pim's Arctic career, see W. Barr, "Franklin in Siberia? Lieutenant Bedford Pim's Proposal to Search the Arctic Coast of Siberia, 1851–52," *Arctic* 45 (1992): 36–46.

8. See Barr, "Franklin in Siberia?" 37.

9. "The Proposed Further Search for Sir John Franklin," *Times*, November 14, 1851, 3; "Sir J. Franklin's Expedition: Royal Geographical Society," *Times*, November 25, 5.

10. B. Pim, "Remarks on the Isthmus of Suez, with Special Reference to the Proposed Canal," *Proceedings of the Royal Geographical Society of London* 3 (1858/59): 177–206; See Pim's own discussion in Pim and Seemann, *Dottings*, 230.

11. Barr, "Franklin in Siberia?" 38. Elizabeth Matthews's book provides an imaginary description of the "encounter" between a Lieutenant Pim and his Captain Kellett aboard *HMS Resolute. From the Canadian Arctic to the President's Desk:* HMS Resolute *and How She Prevented a War* (London: Alto Press Book, 2007), 2.

12. Pim and Seemann, *Dottings*, 268–69.

13. Ephraim George Squier, *Waikna; or, Adventures on the Mosquito Shore* (New York: Harper & Bros., 1855). Pim and Seemann, *Dottings*, 268–69. In his earlier *Gate of the Pacific* (London: Lovel, Reeve, 1863), 75–76, Pim wrote: "In appearances, the aborigines are about the middle height, very swarthy complexion, long coarse black hair, good eyes, and thin lips; the most remarkable feature is the nose which is sharp, thin, and small and looks more so from the cheek-bones being high. They do not think of themselves as is evidenced by their native appellation, 'Waikna,' man; but this conceit is not altogether unjustifiable, for they are brave, warlike, and about the best canoemen in the world."

14. Search for the phrase "Miskito Indians" in "full text," limited to the 106 Anthropology titles carried by JSTOR (http://www.jstor.org/) at the time of the search (January 25, 2014): 333 results; for an early bibliography of Miskito science, see Courtenay de Kalb, "A Bibliography of the Mosquito Coast of Nicaragua," *Journal of the American Geographical Society of New York* 26 (1894): 241–48.

15. Michael D. Olien, "Were the Miskito Indians Black? Ethnicity, Politics and Plagiarism in the Mid-Nineteenth Century," *New West Indian Guide* 62 (1988): 27–50. See also Olien, "E. G. Squier and the Miskito: Anthropological Scholarship and Political Propaganda," *Ethnohistory* 32 (1985): 111–33.

16. Richard White, *The Middle Ground: Indians, Empires, and Republics in the Great Lakes Region, 1650–1815* (New York: Cambridge University Press, 1991), xi. For remarks on the spatial implications of this "imperialism of the margin" for the world system, which my book very much espouses, see the discussion of White in Martin W. Lewis and Kären Wigen, *The Myth of Continents: A Critique of Metageography* (Berkeley: University of California Press, 1997), 151. I am grateful to Carolyn Biltoft for having pointed out the relevance of this remarkable work to my argument.

17. Whether Miskito kings had solely a role for external relations or instead had a local importance is discussed in Philip A. Dennis and Michael D. Olien, "Kingship among the Miskito," *American Ethnologist* 11 (1984): 718–37.

18. Naylor, *Penny Ante Imperialism*. For an anthropological discussion of the issue of kingship, see for instance Dennis and Olien, "Kingship."

19. Consider the following, from the *Chester Chronicle*, December 16, 1825: "Negro Joke—A young Scotsman, who had just gone out to take possession of an estate in Jamaica, while wandering over the grounds with an old Negro, observed some pompions [pumpkins] growing on a rocky piece of waste ground, and inquired of his guide what they were. "Dem Cotchman Massa," said the negro. "And why are they called Scotchmen?" said the other. "Because," replied Sambo, "dem grow 'mong de rock and de stone, dem grow ebery where.""

20. For a discussion of the history of the term in American English, see Joseph Boskin, *Sambo: The Rise and Demise of an American Jester* (New York: Oxford University Press, 1988).

21. James Stuart Olson, *The Indians of the Central and South America: An Ethno-Historical Dictionary* (Westport, CT: Greenwood, 1991), 289.

22. For discussion of the McGregor operations in relation to the Mosquito Coast see Naylor, *Penny Ante Imperialism*, 79ff.

23. David Sinclair, *The Land That Never Was*; Frank G. Dawson, *The First Latin American Debt Crisis*; and Dawson, "MacGregor." For a critical perspective, see Damian Clavel, "Financial Fraud or Failure?"

24. Naylor, *Penny Ante Imperialism*, 152.

25. Naylor, *Penny Ante Imperialism*, 137, 198–99, 245n20; N. Draper, *The Price of Emancipation*.

26. Bard [Squier], *Waikna*, 33; Charles L. Stansifer, "The Central American Career of E. George Squier" (diss., Tulane University, 1959), 149.

27. Naylor, *Penny Ante Imperialism*, 187.

28. FO 94/537; FO 881/4013; PRO 30/22/93.

29. Sampson, *Central America*.

30. A later prospectus for the Honduras Inter-Oceanic Railway put down distances for the isthmus route from Liverpool to San Francisco (touching at Jamaica) as follows: by way of Panama 7,980 miles; Nicaragua (Pim's project) 7,720 miles; Honduras (Squier's Project) 7,320 miles. The New York to San Francisco route was: by Panama 5,224 miles; by Nicaragua 4,700 miles, Honduras 4,121 miles

31. Quoted in *Morning Post*, February 5, 1855.

32. On the launching of the Central American Transit Company, see ADM 128/55 "Declaration by the Central American Transit Company," as well as the "Tariff of Rates," which set the price of passage for an adult at $25 (both documents circa 1862). According to figures reported by Collinson, in the mid-1860s the Panama Railroad Company charged £5 s4 d2 per adult and thus the overall charges for transit were comparable (in 2013 pounds, this is £434 using the retail price index).

33. FO 925/1208, *Map of Honduras and San Salvador, Central America, showing the line of the proposed Honduras, Inter-Oceanic Railway, territories of Indian Tribes and mines etc. with notes*, by E. G. Squier, 1854. *London Standard*, November 18, 1871.

34. BT 31/246/807 and BT 41/301/1750. The other founding signatures were Abraham Darby, heir to a dynasty of industrialists and now involved in foreign investments; Joseph Robinson, a common name but from the address (Pountney Hill), which was identical to the previous gentleman, probably that Robinson who was one of Darby's regular business associates, along with members of the Wilkinson family; Major-General George Borlase

Tremenhere, Royal Engineers; and William Wheelwright, an American businessman who was involved in steamboat and train transportation in various parts of South America. The first board of directors included the same plus John Pemberton Heywood, a merchant banker, Charles Holland, John Lewis Ricardo, Edward William Watkin, Howell L. Williams, and George Wilson; American directors included Shepherd Knapp, George W. Riggs, Paul Spofford, and Fletcher Westray. I also find one John Ingram Travers among the main stockholders when the company was created, a merchant and the author of a brief article, "Memorandum on the Present Statistics of the Currant Trade," *Journal of the Statistical Society of London* 20 (1857), 313.

35. *Morning Chronicle*, January 12, 1859.

36. When he was an exile in London in the 1840s, the future French emperor published a pamphlet on a canal through Nicaragua. *Le Canal de Nicaragua*, printed in 1847, was reprinted in vol. 2 of *Oeuvres de Napoléon III* (Paris: Plon, 1856). On the simultaneous visits of the promoters of the company to French and British authorities in June 1860, see *Sheffield Independent*, June 2, 1860 (Napoleon III); *Morning Post*, June 15, 1860 (John Russell). At about the same time, Burton also sought French support for his expeditions; see Brodie, *Devil Drives*.

37. Pim and Seemann, *Dottings*, 332, 350.

38. Pim to Admiral Alexander Milne, April 2, 1860, ADM 128/55.

39. Tulane University Library, Latin American Library Manuscripts, Collection 50, Mosquito Territory, Nicaragua Documents, 1859–1867, Folder 1 and 3. Deed issued in Bluefield, Nicaragua. Bedford Pim, "Extract from a Paper by Commander Bedford Pim, R. N., on a New Transit-Route through Central America," *Proceedings of the Royal Geographical Society of London* 6 (1861): 113, argues that he would have initially considered the name "Gorgon's Bay."

40. Mosquito Territory, Nicaragua Documents, 1859–1867, Folder 4. This document is also reproduced in the appendixes of *Dottings*, 436.

41. Mosquito Territory, Nicaragua Documents, 1859–1867, Folder 2 and 5. Folder 5 has the contract (May 1864) whereby William Cope Devereux transferred his rights to the two cays to Pim. The lawyer was George Salmon, who is associated as director or shareholder in all Pim's promotions of the time. Devereux, *A Cruise in the "Gorgon,"* cover page.

42. See, for example, the answer sent by the Colonial Office to Mrs. Hedgcock, widow of a speculator in Mosquito Land rights, following an inquiry about rights of hers in the area of the Black River: "I am at the same time desired to add that this Dept. has always declined to take notice of any alleged grants from the King of the Mosquito Territory, inasmuch as that Country is not a British Colony" (November 4, 1861); CO 123/108. For a discussion of the activities of Mrs. Hedgcock's husband, Thomas Hedgcock, during the 1840s, see Naylor, *Penny Ante Imperialism*, 122, 125.

43. Bedford Pim to H.M. Minister Plenipotentiary, Central America, Charles Lennox Wyke, January 22, 1860; ADM 128/55.

44. Bedford Pim to Charles Lennox Wyke, January 13 and 22, 1860; ADM 128/55.

45. Carl Sandburg, *Storm Over the Land: A Profile of the Civil War Taken Mainly from Abraham Lincoln: The War Years* (Harcourt, Brace, 1942), 88.

46. For details on Cauty's operations against Walker, see William O. Scroggs, *Filibusters*

*and Financiers: The Story of William Walker and His Associates* (New York: Macmillan, 1916), esp. chap. 18: "The Vengeance of Vanderbilt." Cauty would be arrested in 1863 as he was on his way to Nicaragua with arms that he claimed were for San Salvador but which Union military authorities suspected were for the Confederate army. See J. B. Moore, *History and Digest of the International Arbitrations to which the United States Has Been a Party*, vol. 4 (Washington, DC: Government Printing Office, 1898), 3309–10.

47. "News from Central America, S. J. del Norte Correspondence, 30 January 1860," *New York Herald*, February 13, 1860; "News from Nicaragua: Our Greytown Naval Correspondence, 31 January 1860," *New York Herald*, February 14, 1860.

48. Vice Admiral and Commander in Chief Houston Stewart to Secretary of the Admiralty in London, March 6, 1860; ADM 128/55, Admiralty: North America and West Indies Station: Correspondence, Reports and Memoranda. Jamaica Division. Central America and Greytown (Nicaragua), "Reported proceedings of Commander Pim of the 'Gorgon' in reference to a new transit route." Vice Admiral and Commander in Chief Houston Stewart to Commodore Kellett, March 6 1860.

49. ADM 128/55; Houston Stewart to Commander Pim, March 6, 1860.

50. ADM 128/55; Bedford Pim to Admiral Alexander Milne, April 2, 1860.

51. ADM 128/55; Bedford Pim to Admiral Alexander Milne, April 2, 1860.

52. ADM 128/55; Alexander Milne to Bedford Pim, April 16, 1860; Pim to Alexander Milne, April 18, 1860.

53. ADM 128/55; Alexander Milne to Secretary of the Admiralty, April 20, 1860.

54. ADM 128/55; Milne report, May 2, 1860.

55. Special Resolution of the Honduras Inter-Oceanic Railway Company, October 11, 1859, BT 31/246/807. The initial capital was £2,050,000. The extra outlay was £1,025,000. Tenders for the construction set the project around £2.5 million sterling.

56. ADM 128/55. W. G. Romaine to Alexander Milne, May 29, 1860.

57. In early October 1860 the *Gorgon* log makes reference to Pim and his gig being "with *H.M.S. Victory*" in relation to a court martial. The *Victory*, anchored at Portsmouth, was where a number of courts martial were held. I had initially wondered whether this might have been Pim's own, although I realize now that such would have been a dangerous proceeding against a hero who enjoyed powerful support. Indeed, the index of "Officers of the Royal Navy tried by Court Martials 1856–1879" (AMD 194/47) returned no trace of Pim being court-martialed. A more likely interpretation may be that Pim was assigned various administrative tasks, as a reference to Navy exams suggests. Could this visit to the *Victory* have also been to ask Pim for further particulars?

58. W. Cope Devereux (*A Cruise in the "Gorgon,"* 28–29).

59. ADM 128/55; Under Secretary for Foreign Affairs, to Rear Admiral Alexander Milne, December 10, 1860; Consul Green to Milne, January 10, 1861.

60. Excerpts of W. Cope Devereux's account of the final days of Pim with the *Gorgon* in March 1861 (*A Cruise in the "Gorgon,"* 45–48): "For some time past Captain P—has talked of returning to England, although one cannot realize such a calamity still the thoughts of it will goblin-like dance before us." "Selfish mortals, we hope that something will turn up to dissuade our gallant captain from leaving us." "To me he has been so considerate, courteous, indulgent and so kind that with him I shall lose more than words can express."

21st March: "*Fury* arrived from China. On board is an acting commander. I hope he likes his ship too well to leave her!" March 31: "Captain P—takes with him our sincere love and respect, and leaves an impression which time cannot efface."

61. To "18 s. a day" and "not more than 20 s a day on attaining rank of rear admal"; See "Pim—Bedford—Clapperton—Trevelyan," Admiralty: Officers' Service Records (Series III), ADM 196/16/296, p. 296. ADM 12/695 has mention of "half pay." Pim and Seemann, *Dottings*, 363.

62. Coote, "Pim, Bedford."

63. J. Collinson and B. Pim, *Descriptive Account of Captain Bedford Pim's Project for an International Atlantic and Pacific Junction Railway across Nicaragua* (London: J. E. Taylor, 1866), 15.

CHAPTER EIGHT

1. Sophia Soltau-Pim, *Job and Fugitive Pieces* (London: Gee, 1885), 37; full text of the acrostic after the conclusion.

2. Pim and Seemann, *Dottings*, 361. Underscoring continued French interest in the region, we have Napoleon's public relations and commercial propaganda hit man, Michel Chevalier, signing a treaty with Nicaragua in 1869 for the building of a canal a few years later. See Michel Chevalier and Thomas Ayon, *Convention pour l'exécution d'un canal maritime interocéanique sur le territoire de la République de Nicaragua* (Paris: P. Dupont, 1869).

3. Pim and Seemann, *Dottings*, 361ff. A copy of the Nicaraguan concession, dated March 5, 1864, is found in Charles Toll Bidwell, *The Isthmus of Panama* (London: Chapman & Hall, 1865), 393–96.

4. BT 31/14380/2935; BT 31/1304/3339.

5. *London Standard*, September 21, 1869.

6. See *London Daily News* of February 8, 1868: "The prospectus of the Javali Company . . . is now published. This mine, situated in the Chontales district of Nicaragua, has been purchased and opened by the Central American Association (Limited), whose operations consist in acquiring and re-selling properties of this description." The capital of the Javali Company was £100,000.

7. Marc Flandreau, "New Facts and Old Fictions."

8. For a discussion of the question of Poyaisian land grants, see Naylor, *Penny Ante Imperialism*. For Pim's announcements to holders of "Mosquitian" and "Poyaisian" land securities, see for instance, *Morning Post*, May 6, 1868; *London Standard*, May 7, 1868.

9. For details on this see Samuel Pasfield Oliver, *Off Duty: Rambles of a Gunner through Nicaragua, January to June 1867* (London: Taylor & Francis, 1879), xviii.

10. Pim and Seemann, *Dottings*, 367.

11. Pim and Seemann, *Dottings*, 357ff., 366. From the initial application (BT 31/1304/3339) it is unclear who the said "gentleman" could have been; See also *Acts of the Legislature of the State of New Jersey*, 1868, p. 816;

12. Central American Association, list of shareholders 1867, BT 31/14380/ 2935; Pim and Seemann, *Dottings*, 59.

13. From prospectus, *London Standard*, December 30, 1869.

14. Matthew Fontaine Maury, *Letter on the Physical Geography of the Nicaraguan Railway Route, from M. F. Maury to Bedford Pim* (London: J. E. Taylor, 1866).

15. John Henry Murchison, *British Mines Considered as Means of Investment* (London: Mann, 1854). On his association with Burton, see James L. Newman, *Paths without Glory: Richard Francis Burton in Africa* (Dulles, VA: Potomac Books, 2009). See also Nicaragua Railway Company Ltd., Company Records, National Archive, BT 31/1304/3339, application to the Registrar, November 23, 1866. Burton's mining interest coincided with his residency in Brazil. See Brodie, *Devil Drives*, 238–39. Burton's book on Brazil, published in 1869, is called *Explorations of the Highlands of Brazil, with a Full Account of the Gold and Diamond Mines* (London: Tinsley Bros., 1869).

16. George Earl Church, *Route to Bolivia*, 201–5.

17. Other names in the group of interlocking directors included Alfred A. Pollock, George Henry Walker, Parke Pittar of Pittar & Alcok (Anglo-Indian merchants who became insolvent in 1872), John Ralph Grimes, George Noakes, and Thomas Staunton. Since the records of most companies in the group are kept at the National Archive in Kew Gardens, and many were quoted on the stock exchange, a complete study of the mining speculation that occurred then is possible from existing sources and would yield important insights.

18. *London Standard*, August 24, 1867.

19. Bedford Pim, "Extract from a Paper"; Pim, "Proposed Interoceanic and International Transit Routes through Central America," in *Report of the Thirty-Third Meeting of the British Association for the Advancement of Science, Held at Newcastle upon Tyne in August and September 1863* (London: John Murray, 1864), 143–45.

20. *Journal of the Anthropological Society of London* 2 (1864). The exploration party comprised one Mr. Paul, possibly B. H. Paul, who shall join the Anthropological in November 1865; Pim and Seemann, *Dottings*, 84ff. Pim's party departed from Southampton to Greytown, Nicaragua, on November 17, 1864.

21. *Journal of the Anthropological Society of London* 3 (1865): ii. The rules of the British Association stated that the general committee was composed of, in addition to the Council, "any member" who had communicated a paper to any society, had it printed by that society, and related it to "such subjects as are taken into consideration at the sectional meetings of the Association."

22. Pim made sure John Collinson put out a paper, "Explorations in Central America Accompanied by Survey and Levels from Lake Nicaragua to the Atlantic Ocean." *Proceedings of the Royal Geographical Society of London* 12 (1868): 25–46.

23. *Journal of the Anthropological Society of London* (January 16, 1866): lxxxviii. The W. T. Pritchard appointment occurred on February 6; *Journal of the Anthropological Society of London* 4 (1866): cix, cx; and *Popular Magazine of Anthropology* 1 (1866): 85, referring to an article (n.d.) from the *Court Circular*, praising the sending of Pritchard, who went with Seemann. When the spring sessions of 1866 ended, the president (Hunt) "much regretted" the delay with the report (*Journal of the Anthropological Society of London* 4, ccxii). It was never subsequently produced or, at least, discussed.

24. Pim and Seemann, *Dottings*, 367.

25. Pim, *Gate of the Pacific*.

26. John Collinson, *Descriptive Account.*

27. Collinson, *Descriptive Account,* 11 and 13; this continues: "Moravian missionaries, all Europeans, established there sixteen years, have not lost a man. . . . Not one has been invalided. . . . Many a place has been set down as sickly, simply from the rapid deaths of the disorderly and dissipated members of its resident society, and from no radical fault in the locality itself." This too: "Never missed a single face out of the small population of the town."

28. Pim and Seemann, *Dottings,* 274. Pim construes this elsewhere as subaltern belief: "My officers, although better young men never set foot on a ship's deck, firmly believing the reports in circulation, that both country and climate were execrable, and that they would be fortunate to escape with life, at the cost of shattered health" (357–58). The puffing of Nicaragua is all over the place in Pim's narrative, such as when he sets out to refute the claim that had been made of cases of anthropophagy in the area (an embarrassing objection for a promoter of colonization). The confusion, Pim explains, had arisen from a certain local delicacy, a "monkey eaten with roasted plantain and plenty of cacao to wash down the solids," which he was "not ashamed to say" he had himself "eaten with very considerable relish" (*Dottings,* 251).

29. Pim and Seemann, *Dottings,* 239. Anonymous, "Berthold Seemann," *Journal of Botany, British and Foreign,* n.s. 1 (1872): 1–7. Seemann's estate was reluctant to admit tropical disease, and the *Journal of Botany, British and Foreign* thus asserted "there were circumstances which pointed towards some cardiac complication."

30. The 1867 voyage is described in a pamphlet by S. P. Oliver, an English artillery officer and a fellow of the Royal Geographical Society, who had accompanied Pim and would make a career of his own as an explorer and antiquarian; Oliver, *Description of Two Routes through Nicaragua* (Gosport, UK: J. P. Legg, 1867). The pamphlet is found at Tulane University Library, Mosquito Territory, Nicaragua Documents, 1859–1867, Folder 50. The material would again be narrated in a volume that mixed up exploration and financial promotion, Oliver, *Off Duty.*

31. John Collinson, "On the Indians of the Mosquito Territory," *Journal of the Anthropological Society of London* 6 (1868): xiii–xvii (summary plus discussion on November 5, 1867); the paper was published later as John Collinson, "The Indians of the Mosquito Territory," *Memoirs Read before the Anthropological Society of London* 3 (1870): 148–56. His *Royal Geographical* paper is Collinson, "Explorations in Central America."

32. The article would eventually be published by R. S. Charnock and C. Carter Blake as "Notes on the Wolwa and Mosquito Vocabularies," *Transactions of the Philological Society* 15 (1874): 350–53. The two-page list of words was also reproduced both in Collinson's "The Indians" and in his "Explorations in Central America." For a discussion of the events surrounding the Exeter meetings, see above, chapter 2. Evidence and discussion of the article ending in the "Poster sessions" is found in "Anthropology at the British Association," *Anthropological Review* 7 (1869): 414–32, 428, and in the *Report of the Thirty-Ninth Meeting of the British Association for the Advancement of Science, held in Exeter in August 1869* (London: John Murray, 1870), 129.

33. *Journal of the Anthropological Society of London* 6 (1868): cviii.

34. "Discussion of Collinson," *Journal of the Anthropological Society of London* 6 (1868): xv.

35. One curious element that both Squier and Pim's anthropologies have in common is their ignorance of the local ruling class and the creole population, widespread in the Bluefield area.

36. Pim and Seemann, *Dottings*, 325ff.: "a rainy fact" and "cheap philanthropy."

37. Pim and Seemann, *Dottings*, 327ff. ("cheap philanthropy"): the "emancipated gentlemen had one and all declined to work, and had never wavered in that resolution from the time of their manumission to the present day, thus turning the tables with a vengeance; they had been allowed to squat on the land of their masters, but only cultivated sufficient to keep body and soul together; living in a squalid, half-starved condition the best part of the time."

38. Pim and Seemann, *Dottings*, 327: list of former slave owners looking for compensation that includes names such as Bowden, Quin, Brown, and Hooker, who had also been land promoters. One George Augustus Brown had been associated in the early 1840s with Hedgcock in the British Central American Land Company (Naylor, *Penny Ante Imperialism*, 260–62). They claimed indemnity under the British Slave Compensation Commission that had operated between 1834 and 1845. They were turned down since the territory was not formally a British possession.

39. Pim and Seemann, *Dottings*, 368, 450: "Proceedings of the meeting held in Mosquito Territory in May 1867."

40. Naylor, *Penny Ante Imperialism*, 219. Naylor provides what is to date the best available perspective on the long-running history of the politics of property rights along the Mosquito Coast.

41. *London Standard*, October 11, 1867.

42. Pim and Seemann, *Dottings*, 292, 433. Or on p. 356, "one who must ever take a warm interest in their welfare."

43. Pim and Seemann, *Dottings*, 82.

44. *London Standard*, October 11, 1867.

45. Pim and Seemann, *Dottings*, 433.

46. Bard, *Waikna*, 243.

47. Their evocation is prominent in *Waikna* and occupies the very first pages of the novel with a "dialogue" that "took place, or might have taken place" on the Mosquito Coast: "[Scene: A lonely shore. Enter Yankee and Mosquito Man] Well, my dark friend, who are you? 'Waikna' A man! And what is your nation? 'Waikna!' A nation of men! Pretty good for you, my dark friend! There was once a great nation—a few old bricks are about all that remains of it now—but then what do you know about the Romans? 'Him good for drink—him grog?' Bah! No! 'Den no good! Ba, too!'" Bard, *Waikna*, v.

48. W. Bollaert, "Observations on the Past and Present Populations of the New World," *Journal of the Anthropological Society of London* 1 (1863): iii–x. Bollaert and Carter Blake referred there to the broader concept of Sambo as a West Indies mixture of black and Indian.

49. Collinson, *Descriptive Account*, 11 and 13.

50. Pim, *Gate of the Pacific*, 75–76.

51. Expressions found verbatim in a summary of Collinson's 1867 paper described Miskitos as follows: "Complexion dark; with finely-marked features; small noses; high cheek-bones; and long, coarse, black hair." John Collinson, "On the Indians," xiii.

52. Pim and Seemann, *Dottings*, 404. Collinson's description also added that "the chief

of the entire territory must be of the Mosquito tribe, and reigns by direct descent through the male line, etc.," a curious emphasis for an Englishman, in which the not-so-cynical may read the competition that existed at the time between the king and his sister who was reported to be closer to the authorities in Nicaragua. Collinson, "On the Indians."

53. Pim and Seemann, *Dottings*, 309.

54. "Discussion of Collinson."

55. Pim, *Negro and Jamaica*, vii.

56. Pim, *Negro and Jamaica*, 32–33. Thus Pim is an expert on the region, and this is what enables him to get into details and dispute the figures of household wealth computed by Dr. Underhill of the Baptist Missionary Society and used at the time as authoritative. On Dr. Underhill's surveying of Jamaica, see T. C. Holt, *The Problem of Freedom*, 270–72.

57. Pim and Seemann, *Dottings*, 432.

58. Pim and Seemann, *Dottings*, 432.

59. Pim and Seemann, *Dottings*, 222.

CHAPTER NINE

1. *London Standard*, May 22, 1868.

2. Charles Tilston Beke, *The British Captives in Abyssinia* (London: Longmans, 1867). This was the second and considerably expanded (398 pages versus 61) edition of a same titled pamphlet published by Longmans in 1865.

3. The hostage crisis was the subject of an abundant contemporary literature by press correspondents and military participants: G. A. Henty, *The March to Magdala* (London: Tinsley Bros., 1868); Captain Henry Hozier, *The British Expedition to Abyssinia, Compiled from Authentic Documents* (London: Macmillan, 1869). Hozier had been the assistant military secretary to Lord Napier. For slightly more recent references, see, for example, Darrell Bates, *The Abyssinia Difficulty: The Emperor Theodorus and the Magdala Campaign, 1867–68* (Oxford: Oxford University Press, 1979); F. Myatt, *The March to Magdala: The Abyssinian War of 1868* (London: Leo Cooper, 1970); S. Rubenson, *King of Kings: Tewodros of Ethiopia* (Addis Ababa, 1966).

4. Hozier, *British Expedition*, 226 (quoting, no source indicated).

5. T. E. Kebbel, ed., *Selected Speeches of the Earl of Beaconsfield*, vol. 2 (London: Longmans Green, 1882), 529–34.

6. House of Commons, Abyssinian Expedition, Vote of thanks to Sir Charles Napier, HC Deb July 2, 1868, vol. 193 cc522–9.

7. Harcourt, "Disraeli's Imperialism," 87–109. For the sake of simplicity, the discussion omits a critical part of Harcourt's argument about the fact that promotion of empire at this particular juncture also served to uphold "domestic peace which was threatened by class divisions."

8. Russell had been foreign affairs secretary since 1859; he became prime minister in October 1865.

9. *Journal of the Anthropological Society of London* (May 12, 1863); *Transactions of the Ethnological Society of London* 2 (1863) back matter.

10. *Journal of the Anthropological Society of London* (December 5, 1865).

11. Platt, *Cinderella Service*, 50.

12. "Consular Jurisdiction in the Levant," letter by Hyde Clarke, *Times*, August 10, 1868.

13. In report from Plowden to the Earl of Clarendon, dated June 25, 1855, P.P. 1865 (3536) LVII, *Abyssinia. Papers Relating to the Imprisonment of British Subjects in Abyssinia*, 41; See also P.P. 1866 (3748) LXXV, *Abyssinia. Further Correspondence Respecting the British Captives in Abyssinia*

14. P.P. 1866 (3748) LXXV, *Abyssinia. Further Correspondence Respecting the British Captives in Abyssinia. Parliamentary Papers*, 6–47.

15. P.P. 1865 (3536) LVII, *Papers Relating to the Imprisonment of British Subjects in Abyssinia*, 2, Document 3, King of Abyssinia to the Queen of England.

16. P.P. 1865 (3536) LVII, *Papers Relating to the Imprisonment of British Subjects in Abyssinia*, 1, instructions to Cameron.

17. Writing about Plowden and his dispatches, historian J. R. Hooker claimed that "as a political agent, Plowden was valuable; as a writer of travel literature he was engaging and intelligent; but as a consul he was useless, his commercial reports being limited to three in 1852. He was never at his post after 1855"; Hooker, "The Foreign Office and the 'Abyssinian Captives,'" *Journal of African History* 2 (1961): 245–58. Freda Harcourt has argued that Hooker's piece was "superficial and inaccurate" (Harcourt, "Disraeli's Imperialism," 99). Judged on the previous quote, Hooker also failed to understand what informal empire was about. In the 1850s, there was not yet any trade to speak of (so why report), and it was part of the tacit understanding of the consular service that for some destinations, consuls were not expected to stick to their post but could very well justify their limited cost by acting as political agent and writer of travel literature that aroused an interest conducive of imperial expansion.

18. P.P. 1866 (3748) LXXV, *Abyssinia. Further Correspondence Respecting the British Captives in Abyssinia.*

19. Beke, *British Captives*, xvii–xviii, 86–87.

20. P.P. 1865 (3536) LVII, *Papers Relating to the Imprisonment of British Subjects in Abyssinia*, 11–12.

21. Beke, *British Captives*, iv, v, xiv, xvi.

22. Ibid., 113–114.

23. *Journal of the Anthropological Society of London* 4 (1866): lxxix.

24. *Journal of the Anthropological Society of London* 4 (1866): lxxviii. This is a reference to William Wilberforce (1759–1833), the English philanthropist who had been the leader of the movement to abolish the slave trade. Bishop Samuel Wilberforce (1805–1873), who fought Huxley on the subject of the origins of man during the British Association meetings in Oxford in 1860, was William's third son.

25. *London Standard*, May 22, 1868, 4.

26. Albert H. Markham, *The Life of Sir Clements R. Markham* (London: John Murray, 1917); Elizabeth Baigent, "Markham." Roderick Murchison had advised Russell on the routes to Magdala.

27. A discussion of the conventional view of a blunder causing aging Russell is Nini

Rodgers "The Abyssinian Expedition of 1867–1868: Disraeli's Imperialism or James Murray's War," *Historical Journal*, 27 [1984]: 129–49). Rodgers exonerates Russell and blames his staff. The story of the long-to-come rewarding of Beke offers a symmetrical episode to Church's and Pim's long-to-come sanctions: while Conservatives had been happy to exploit Beke's resentment, it is understandable that there was some political reluctance within the executive to reward scientists for their predatory behavior toward politicians. And indeed, with the return of a Liberal government at the general election of 1868, resistance to reward Beke generously continued. The *Dictionary of National Biography*, unreliable to tell us what happened really but a very reliable ethnographical recording of the Victorian masks, reads: "During the Abyssinian difficulty Beke furnished maps, materials, and other information to the British government, and to the army, by which many of the dangers of the expedition were averted, and in all probability many lives saved. Beke received a grant of 500 £ from the Secretary of State for India, but his family and friends regarded this remuneration as very inadequate for public services extending over a period of thirty or forty years, and culminating in his aid and advice in connection with the Abyssinian campaign. In June 1868, Professor E. W. Brayley, F.R.S., drew up a memorandum of the public services of Beke in respect of the Abyssinian expedition. Two years later the queen granted Beke a civil-list pension of 100 £ per annum in consideration of his geographical researches, and especially of the value of his explorations in Abyssinia." See G. Barnett Smith, "Beke, Charles Tilstone, 1800–1874, " in *Dictionary of National Biography*, vol. 4, ed. Leslie Stephen (London: Smith Elder, 1885), 138–40.

28. Abyssinian expedition, vote of thanks to Her Majesty's forces; HC Deb July 2, 1868, vol. 193 cc522–9.

29. For a clarifying contribution, which provides elements of criticism on the humanitarian thesis, see Nora Kelly Hoover, "Victorian War Correspondents G. A. Henty and H. M. Stanley : The 'Abyssinian' Campaign 1867–1868," (unpub. Ph.D. diss., Florida State University, 2005).

30. Bronislaw Malinowski, "Practical Anthropology," *Africa: Journal of the International African Institute* 2 (1929): 22–38.

31. For a similar point, see Conrad C. Reining, "A Lost Period of Applied Anthropology," *American Anthropologist* 64 (1962): 593–600.

32. *Popular Magazine of Anthropology* 1 (1866).

33. Sir B. C. Brodie, "Address to the Ethnological Society of London, Delivered at the Anniversary Meeting on the 27th May 1853," *Journal of the Ethnological Society of London* 4 (1856): 98–103.

34. *Anthropological Review* 8 (1870): 187–216.

35. *Journal of the Anthropological Society of London* (March 14, 1865): clxxxiii (for the quote by Winwood Reade), clxix (for Burton's).

36. Colenso may have felt that this was the danger, for his rejection of Reade's theses covertly disputed that the Anthropological Society was the right place for such a discussion. He was joining in the conversation, he said, because while he "differed on some points from the opinions expressed in the paper" by Reade, he thought that the "subject was one that required much consideration" and felt also that the "opinion on the subject of [missionaries'] labours of *laymen who had resided*" in Africa was important (my italics);

*Journal of the Anthropological Society of London* 3 (1865): clxix. It is interesting to note Colenso's skillful attempt at disqualifying anthropologists by construing them as "laymen," thus construing missionaries as experts.

37. As Hunt reminded participants in the ensuing debate: "It should be borne in mind that that was a scientific society, in which all facts bearing on the subject under discussion should be brought forward. The statements made were open to the strictest scrutiny, and however much any members might differ from the opinions expressed, it was right that they should be received with consideration and replied to if thought erroneous." *Journal of the Anthropological Society of London* (March 14, 1865): clxviii.

38. B. Pim, *Negro and Jamaica*, 50.

39. R. W. Stewart, *Hand List of the Hughenden Papers* (London: The National Trust, 1961; revised 1968), Bodleian Library, Oxford.

40. E. J. Feuchtwanger, "Spofforth, Markham (1825–1907)," in *Oxford Dictionary of National Biography* (Oxford: Oxford University Press, 2004).

41. The Disraeli correspondence includes many leaders of the foreign debt-scavenging industry, including John Diston Powles, Sir Henry Drummond Wolff, or George Augustus Frederick Cavendish-Bentinck.

42. Robert Blake, *Disraeli* (New York: St. Martin's, 1967), 28.

43. Michael P. Costeloe, *Bonds and Bondholders*, 4, 272, 288.

44. John Stephen Flynn, *Sir Robert N. Fowler, Bart., M.P.: A Memoir* (London: Hodder & Stoughton, 1893); Howard L. Malchow, "God and the City: Robert Fowler (1828–91)," in *Gentlemen Capitalists: The Social and Political World of the Victorian Businessman* (Palo Alto, CA: Stanford University Press, 1992).

45. Arthur Scratchley, *On Average Investment Trusts* (London: Shaw & Sons, 1875); David Chambers and Rui Pedro Esteves, "The First Global Emerging Markets Investor: Foreign & Colonial Investment Trust 1880–1913," *Explorations in Economic History* 52 (2013): 1–21.

46. Too many finds in the story told in these pages took me to the *London Standard* for this to be coincidental. The blind justice of the Internet search was only reminding me that this Conservative paper, which James Johnstone had started redressing out of oblivion in the 1850s was becoming a success in the 1860s (30,000–46,000 copies in 1860 rising to 160,000–170,000 in 1874 at the time of Disraeli's political triumph) as it established itself as a major provider of City information regarding foreign debts, exotic initial public offerings, and information relevant to foreign bondholders. In the early 1860s, James Johnstone "had accepted funds from the Tory party to assist his debts," one contributor in the 1860s being the future Lord Salisbury. There is evidence he approached Disraeli for support in 1862. This was a newspaper of which Austen Henry Layard, the Liberal under-secretary of state for foreign affairs under John Russell who had so much to suffer from the attacks of his "mismanagement" of the Abyssinia hostages crisis, asked "whether anybody read it" (something which the *Times* gladly printed on July 1, 1865). When Disraeli became prime minister in February 1868, the *Standard* wrote with delight that "no member of the Derby Ministry of the Conservative Party has earned so good a right to the place of Premier as Mr. Disraeli" (*London Standard*, February 26, 1868). See Dennis Griffiths, *Encyclopedia of the British Press, 1422–1992* (London: Macmillan, 1992); see also Laurel Blake and Marysa Demoor, *Dictionary of Nineteenth-Century Journalism in Great Britain and Ireland* (Gent:

Academia Press, London: The British Library, 2009), esp. 596–97 on the *London Standard*. More specific perspectives include the history of this journal by Dennis Griffiths, *Plant Here the "Standard"* (London: Palgrave Macmillan, 1995), 103–5. Martin Hewitt, *The Dawn of the Cheap Press in Victorian Britain: The End of the Taxes on Knowledge, 1849–1869* (London: Bloomsbury, 2014) provides an interesting perspective recasting the rebirth of the *Standard* from its ashes in the late 1850s in the context of the rise of the "penny press."

47. Spanish bonds, *Select Committee*, p. 210, Q 4780: Lewis reports being sorry to say that he was "also an inefficient member of the Council of the committee of foreign bondholders on the Spanish loan." American railways bonds: "Atlantic and Great Western Railway Bondholders meetings, Charles Lewis M.P. in the Chair," *London Standard*, July 1, 1875; joining the committee of the Foreign and Colonial Government Trust *Morning Post*, October 2, 1871; on his joint regulatory proposal with former Anthropological Society of London's member and Mexican bondholders' leader H. B. Sheridan, Hansard, HC Deb March 11, 1875, vol 222 c1680.

48. Disraeli Papers, 53/1 fol. 125–26; December 16, 1878.

49. Disraeli Papers, 53/1 fol. 131–32; December 26, 1878.

50. The letter continued: "I wish for justice, and nothing more, and I am at a loss to understand to whom I ought to appeal if not to your Lordship, whom I have loyally accepted as the leader of my Party ever since I entered on my political life" (April 2, 1879, Disraeli Papers, 53/1 fol. 137–38). This led Algernon Turner to reiterate Lord Beaconsfield's regrets that Pim should be "dissatisfied." Lord Beaconsfield feared, the letter continued, that Pim misapprehended the position in which he stood: "Questions of promotion in the Navy or decoration for naval service are matters of grave administrative importance, but wholly beyond his province as Prime Minister." Disraeli Papers, 53/1 fol. 139–40.

51. Walford, "Captain Bedford Pim," 14, pamphlet found in Disraeli Papers, 53/1 fol. 139–40; On the 1866 motion by E. P. Bouverie for the creation of a commission on bribery and corruption at the elections for the borough of Totnes and subsequent report, HC Deb 01 May 1866 vol 183 cc260–9 and HC Deb 09 April 1867 vol 186 cc1353–73; P.P. 1867 (3774) XXVIII, *Report of the Commissioners Appointed to Inquire into the Existence of Corrupt Practices at the Last Election, and on Former Similar Occasions, for the Borough of Totnes; Together with the Minutes of Evidence and Appendix*; Andrew St. George, *Norton Rose*, 61; H. J. Hanham, *Elections and Party Management: Politics in the Time of Disraeli and Gladstone* (London: Longmans, 1959), 277.

52. "Pim—Bedford—Clapperton—Trevelyan," Admiralty: Officers' Service Records (Series III), ADM 196/16/296, p. 296; Pim and Seemann, *Dottings*, cover page.

53. See Pim's foreword to his mother Sophia Soltau-Pim's (Mrs. Edward Bedford Pim) *Job and Fugitive Pieces* (London: Gee, 1885), 34, where he writes that he was contented she had at least had the happiness of seeing "her son rise to the top of his father's profession against obstacles which her patience and her faith did more to overcome than any other human assistance."

54. Edward John Eyre, *Journals of Expeditions of Discovery into Central Australia and Overland from Adelaide to King George Sound*, 2 vols. (London: Boone, 1845); for evidence of Eyre as a "scientific authority," see for instance Charles Staniland Wake, "Tribal Affinities among the Aborigines of Australia," *Journal of the Anthropological Society of*

*London* 8 (1870/71): xiii–xxxii, which has several references to Eyre's *Journals of Expeditions*; Wake became after the death of Hunt a director of the Anthropological Society, and editor of the new periodical.

55. Charles Ewald, *The Right Hon. Benjamin Disraeli, Earl of Beaconsfield KG and His Times*, vol. 2 (Edinburgh, UK: William McKenzie, 1881), 45.

56. I suspect that under the same heading, scholars might discuss other aspects of the emergence of colonial collective action institutions at this very juncture. An excellent example is provided by the Colonial Institute, launched in June 1868. This has been recently studied by Edward Beasley, *Triumph of Theory*. The initial blueprint included an academic dimension. It was spoken of "establishing a society, which shall occupy as regards colonies, the position filled by the Royal Society with respect to science, or the Royal Geographical Society, with respect to Geography." This society would be for the "encouragement of all objects likely to create a better knowledge and understanding of the colonies," promote "good feelings" between them and the mother country and also promote "a closer intercourse between the colonies themselves" (*Morning Post*, June 29, 1868). In the next few months (the last months of Disraeli's rule, the matter being finalized a few weeks after the fall of Disraeli), the *Morning Post* reported on the "first ordinary general meeting" being held at the Institution of Civil Engineers ("The Colonial Society," *Morning Post*, December 18, 1868). The politics of the Colonial Institute and its competition with other existing institutions deserve interest.

CHAPTER TEN

1. Census of 1881.

2. On the *Athenaeum* and its place in British intellectual life, see Walter James Graham, "*The Athenaeum*" in *English Literary Periodicals* (New York: Nelson, 1930), 317–21; L. A. Marchand, *The Athenaeum: A Mirror of Victorian Culture* (Chapel Hill: University of North Carolina Press, 1941); Alvin Sullivan, "*The Athenaeum*," in *British Literary Magazines: The Romantic Age, 1789–1836*, ed. A. Sullivan (Westport, CT: Greenwood, 1983), 21–24.

3. *Athenaeum* 2129 (1868): 210.

4. Ibid.

5. Articles and letters in the *Pall Mall Gazette* covering the tensions between the two societies: June 3, 1868, June 5, October 13, 15, and 16, 1868. The *Pall Mall Gazette* had been created in February 1865 (*Handlist of English and Welsh Newspapers*, 100) and is generally described as a pro-Disraeli sheet.

6. See Marchand, *Athenaeum*, esp. 359–68, 369–71. For reactions to Clarke's letter, see Minutes of the Anthropological Society of London, September 2, 1868, p. 15, RAI Archive, A-4.

7. *Engineer*, March 15, 1895, 217.

8. For instance, http://www.gnomesoflasallestreet.com/panicscrashes.htm.

9. H. Clarke, "On the Mathematical Law of the Cycle," *Herapath's Railway Magazine and the Annals of Science* 5 (1838): 378–80, 379.

10. H. Clarke, "Physical Economy: A Preliminary Inquiry into the Physical Laws Gov-

erning the Periods of Famine and Panic," initially published in *Railway Register and Record of Public Enterprise for Railways* (London, 1847); as a pamphlet, quoted in W. S. Jevons, *Investigation in Currency and Finance* (London: Macmillan, 1884), 222-23; via Jevons, in J. A. Schumpeter's *History of Economic Analysis* (Oxford: Oxford University Press, 1954), 743, although Schumpeter confessed that he knew Clarke's paper only from Jevons's report. The origin of Clarke's paper is conventionally traced to an 1838 article "On the Mathematical Law of the Cycle" published in *Herapath's Railway Magazine and the Annals of Science* 5 (1838): 378-80, but this paper is mostly concerned with conjecturing that all "cycles" dampen over time and suggesting that solutions of this law would "afford another elucidation in the history of creation" (380). Hyde Clarke and Jevons had other connections as well—Jevons would himself be among the life-subscribers and members of the Corporation of Foreign Bondholders when its certificates were sold by Hyde Clarke, Lubbock, and Gerstenberg in 1872. Jevons's membership in the Corporation of Foreign Bondholders is evidenced from the registries in the Corporation of Foreign Bondholders Archive, certificates of membership (CLC/B/060/MS34592/001 and 002, 1873).

11. H. Clarke, *Theory of Investment in Railway Securities* (London: Weale, 1846).

12. Clarke, "Preliminary Inquiry," 3, 5, 8.

13. Clarke's series of articles published between 1846 and 1848 in the *Civil Engineer's and Architect's Journal* were later reprinted as *Contributions to Railway Statistics in 1846, 1847, & 1848* (London: John Weale, 1849); see A. Odlyzko, "Collapse of the Railway Mania."

14. This is according to *The Railway Directory for 1844, Containing the Names of Directors and Principal Officers of the Railways in Great Britain and Ireland* (London: Railway Times, 1845) in John Palmer's papers, Brunel University Special Collections, which holds a photocopy bearing the indication that it came from the Bodleian. An indication on the photocopy attributes the marked comments to Hyde Clarke. A copy of this via document was shown to me by Gabriel Geisler Mesevage.

15. Dane Keith Kennedy, *The Magic Mountains: Hill Stations and the British Raj* (Berkeley: University of California Press, 1996), 152. Kennedy also relates Hyde Clarke to B. H. Hodgson, British envoy to Nepal and the author of a report "On the Colonization of the Himalaya by Europeans," republished in Hodgson's *Essays on the Languages, Literature, and Religion of Nepal and Tibet: Together with Further Papers on the Geography, Ethnology, and Commerce of Those Countries* (London: Trübner, 1874). Daniel Thorner, *Investment in Empire: British Railway and Steam Shipping Enterprise in India, 1825-1849* (Philadelphia: University of Pennsylvania Press, 1950), 12; and Zaheer Baber, *The Science of Empire: Scientific Knowledge, Civilization and Colonial Rule in India* (New York: State University of New York Press, 1996), 208, mention the role of Hyde Clarke in relation to the fostering of communication technologies in India.

16. Hyde Clarke, *Colonization, Defence, and Railways in Our Indian Empire* (London: John Weale, 1857). For discussion, see Kennedy, *Magic Mountains*, 151-53.

17. Hyde Clarke, "On the Organization of the Army of India with Especial Reference to the Hill Regions," *Journal of the Royal United Service Institution* 3 (1860): 18-27. Hyde Clarke's later correspondence with the Reverend George Percy Badger (1815-88), chaplain and oriental scholar, illustrates Hyde Clarke's continuing role in collecting ethnological

information that might be relevant for imperial defense. An 1873 letter by Badger comments on the need to persuade Muslims in India that their position was as good as that of their co-religionists in Turkey in order to cement their wavering loyalty to British rule (British Library, George Percy Badger papers; India Office Records and Private Papers; Mss Eur C832).

18. Henry Trueman Wood, *A History of the Royal Society of the Arts* (London: John Murray, 1913), 454. The proposal resulted in a committee that did not decide anything. A reduced version of the project—the creation of an India section—would be implemented upon Clarke's return in 1868 and under his stewardship.

19. Clarke, *Colonization in Our Indian Empire*, 8.

20. My sources are one unpublished master's dissertation that covers the history of the railway up to the eve of the *Athenaeum* controversy, material from British consuls in Smyrna, and court cases. Javier Valenzuela, "The Construction of the Smyrna–Aidan railway in Southwestern Anatolia 1856–1866: A Discussion" (MA thesis, University of Texas at El Paso, 1975). See also National Archive, Kew Gardens, *Report of the Committee of Investigation* FO 78-2255–86.

21. D. Landes, *Bankers and Pashas*. Note that neither the Smyrna–Aidin railway nor Hyde Clarke are part of the Landes account.

22. Hyde Clarke, *The Imperial Ottoman Smyrna & Aidin Railway: Its Position and Prospects* (London: Koehler Bros., 1861), 42.

23. *London Standard*, October 1, 1868, 2. There were inferences of insider trading and nepotism.

24. See, for example, Charles R. Salit, "Anglo-American Rivalry in Mexico, 1823–1830," *Revista de Historia de América* 16 (1943): 65–84. For a more recent discussion of Freemasonry and imperialism, which argues that imperialism always found ready supporters in British overseas networks of Freemasons, see Jessica L. Harland-Jacobs, *Builders of Empire: Freemasons and British Imperialism, 1717–1927* (Chapel Hill: University of North Carolina Press, 2007).

25. Indications provided by Mr. Sakmar. See also Robert Morris, *Freemasonry in the Holy Land, or Handmarks of Hiram's Builders* (Chicago: Knight and Leonard, 1876).

26. This Pike has no connection with the Owen Pike of the Anthropological Society. See Charles S. Lobingier, *Supreme Council 33rd Degree, Part 1 or Mother Council of the World of the Ancient and Accepted Scottish Rite of Freemasonry, Southern Jurisdiction, United States of America* (Louisville, KY: Standard Printing, 1931), 872. Albert Pike, like many Freemasons from Charleston, had been an outspoken pro-slavery activist. He was also a man who, as lawyer, was known to have defended Indian rights. For instance, Sigurd Anderson, "Lawyers in the Civil War," *American Bar Association Journal* 48 (1962): 457–60.

27. Robert Morris, *Freemasonry in the Holy Land*, 52. See also the mention he makes of founders of Freemasonry in Smyrna, p. 59. Morris insists on spelling Clark in the text, but has Clarke in the legend of the etching showing Hyde Clarke (see text). Documents shown to me by Mr. Sakmar mention the existence of a Masonic lodge in Alexandria in 1865–69, No. 1082, operating under the United Grand Lodge of England and named after Hyde Clarke.

28. Kenneth R. H. Mackenzie, *Royal Masonic Cyclopaedia*, vol. 1 (London: Kenneth R. H. Mackenzie, 1877), 103. An illustration of the rivalry that existed between Masonic groups is provided by the way one leading Masonic newspaper, the *Freemasons' Magazine*, came down against Hunt at his death in 1869: "Dr. James Hunt, who kept an establishment for the cure of stammering, but was not so well-known as President of a society called the Anthropological Society, died lately. He had taken his usual part in agitating in the Biological Section at the meetings of the British Association at Exeter, but on this occasion with such ill-success as was attended with great excitement on his part. On going into the street he fell down—as some supposed from the effects of sunstroke—and became delirious. He was with difficulty removed to his home where he died within a few days." *Freemasons' Magazine and Masonic Mirror*, October 2, 1869, 279, under "Literature, science, music, drama and the fine arts." In this same newspaper and year, Hyde Clarke was featured twenty-one times, always with eulogy.

29. Stocking, *Victorian Anthropology*, 255.

30. A letter by Clarke kept in the archive of the Anthropological Society (A5) mentions a picture by photographer Alexander Svoboda, known as the author of *The Seven Churches of Asia: With Twenty Full-Page Photographs, Taken on the Spot, Historical Notes, and Itinerary* (London: Sampson Low, Son and Marston, 1869). In 1866 Bollaert read a paper sent by Hyde Clarke, "On Anthropological Investigations in Smyrna," *Journal of the Anthropological Society of London* 4 (1866): xcviii–cii. Abigail Green spots him in the correspondence of philanthropist Moses Montefiore, stating in August 1865 his concern that the money sent by Montefiore be applied "irrespective of creed or opinion." Green seems to imply that Clarke was at this point a British consul in Smyrna, although the language is not clear; Green, *Moses Montefiore: Jewish Liberator, Imperial Hero* (Cambridge, MA: Belknap, 2010), 325. According to the *Morning Post* (September 2, 1865), Montefiore was responding to the appeal made by Hyde Clarke. Other donors included Sir F. Goldsmid and the Ottoman Bank. The same letter shows one Rob W. Cumberbatch to have been the British consul in Smyrna.

31. *Journal of the Anthropological Society of London* (May 14, 1867): cxci. Clarke would be succeeded by a surgeon named James McGraith; *Journal of the Anthropological Society of London* 5 (1867): ccviii.

32. This position however, he would describe during the *Athenaeum* controversy as void of substance. "Hyde Clarke Esq. L.L.D." appears as a new member in January 31, 1865 as local Secretary for Smyrna, *Journal of the Anthropological Society of London*, 1865, vol. 3, p. cxxiv. He was signed in as a member on November 14, 1865 and in later discussion during the *Athenaeum* controversy, claimed that he was a permanent member having paid his twenty-guinea subscription, something that is confirmed by the Archive of the Anthropological Society, subscription ledgers, A-6. The appointment as Local Secretary for Asia took place on March 19, 1867. *Journal of the Anthropological Society* 5 (1867): cxxxiii.

33. See *Journal of the Anthropological Society of London* 6 (1868): clxxxviii. His statements at the 1868 anniversary meeting were summarized there as follows: "Dr. Hyde Clarke rose and made a few general remarks . . . to which Major Owen replied on behalf of the Council" (lxv). See letters about the *Athenaeum* controversy and the summary of debates in the *Journal of the Anthropological Society of London* for more details on this particular point.

34. As stated, Hyde Clarke and Robert des Ruffières stand out as rare instances of joint membership in the two societies. In 1879 the two men would be eager discussants of a paper by F. Galton, "On Composite Portraits," *Journal of the Anthropological Institute* 8 (1879): 132–44, 142–43. There Galton described a method, which he had designed with Herbert Spencer, of "superimposing optically" various drawings of human faces, and "accept the aggregate result" as a mean; records show that Clarke had been a fellow of the Statistical Society since 1856; records of the Royal Statistical Society, London; see "Special General Meeting, 2nd September," *Journal of the Anthropological Society of London* 6 (1868): cxcii.

35. The Norwich Meetings of the British Association for the Advancement of Science took place on August 19–26, 1868.

36. *Anthropological Review* 6 (1868): 323, 324, 327. The Murchison letter to the *Pall Mall Gazette* was published on June 5, 1868 and responded to a note published in this journal on June 3, 1868. The issue at stake was the role of Murchison in the failed amalgamation and his animadversions toward what Cannibals called the "scientifically exact" name of "anthropology." In a letter published later in the *Pall Mall Gazette*, Huxley claimed that the title "Society for the Promotion of the Science of Man" had been suggested by Murchison ("The Ethnological Society," *Pall Mall Gazette*, October 15, 1868).

37. *Athenaeum* (1868): 210.

38. My explorations in the archive of the printer and publisher Trübner have not enabled me to ascertain this claim. The detail given in the account books appears to concern the copies of the *Review* that were sold above and beyond subscriptions. This seems confirmed by the fact that surviving Trübner's accounts bear no mention whatsoever of the *Journal*, which I understand to have been more narrowly tied to membership than the *Review*.

39. Minutes of Anthropological Society of London, September 2, 1868, p. 15 RAI Archive, A-4.

40. For Hunt, see *Athenaeum* (August 22, 1868): 239–40; for Clarke, see *Athenaeum* (August 29, 1868): 271–72.

41. Marchand, *Athenaeum*, 97–165; Francis Bacon's *Great Instauration* (1620) is available as *Novum Organum: With Other Parts of the Great Instauration*; translated and edited by Peter Urbach and John Gibson (Chicago: Open Court, 1994).

42. *Athenaeum* (September 12, 1868): 334 (letter by Charles Harding).

43. Following a resolution carried unanimously by the Anthropological Society Council on August 18, 1868, a special general meeting was convened on September 2, 1868, to vote on the Council's motion to expel Clarke. Of the 42 members attending the special session, 26 supported expulsion and 16 opposed it, the vote thus failing to secure Clarke's exclusion. During that meeting, P. Martin Duncan proposed "a committee of five fellows of the Anthropological Society of London who are neither members of the Council nor friends of Mr. Clarke." Despite Hunt's objections, the resolution was carried. On October 28, another meeting took place to discuss the report that had resulted from the Duncan motion. In both cases, Hyde Clarke participated and much acrimony was spilled. On October 28 the committee took the "opportunity of expressing their regret at the unwarrantable statements made by Mr. Hyde Clarke, and their hope that he will publicly retract the

same at the earliest opportunity." A statement requiring Clarke's expulsion was considered but removed. The president concluded the meeting by stating that "if Mr. Clarke and Mr. Brookes came forward to apologize for the injury they had tried to do to the Society, he should be most ready to let bygones be bygones." They did not. Despite the legend, therefore, Clarke was never expelled from the Anthropological Society of London, and his name indeed would later be displayed in the society's membership lists. Source: Printed minutes and reports on the matter in "2nd September, 1868, Special General Meeting," *Journal of the Anthropological Society of London* 6 (1868): clxxxii–cxcvii; "October 28th, 1868, Special General Meeting," *Journal of the Anthropological Society of London* 7 (1869): i–xxi; RAI Archive, A-4, Minutes of the Anthropological Society of London, September 2 and October 28, 1868; Minutes of Council, A-3: 1, August 18, 1868.

44. *Journal of the Anthropological Society of London* 7 (1869): xii.

45. http://www.britishnewspaperarchive.co.uk/, consulted several times between January and March 2013.

46. "Ottoman (from Smyrna to Aidin)," *London Standard*, October 1, 1868 (section "railway intelligence"), on a meeting of bondholders of the Imperial Ottoman Smyrna and Aidin Railway. This article was not retrieved when I did my sampling of Hyde Clarke in the press, suggesting that the Internet search has not ended the ability that the world has to surprise us.

47. See Valenzuela, *Smyrna–Aidan Railway*, 29, 38; see also *Report of the Committee of Investigation*, FO 78-2255–88. The report ended with a recommendation for ending the era of poor local supervision and sending "a competent person" to Turkey.

48. I have assumed from cross-referencing the information on the matter that we are dealing with a member of the House of Ellissen, and Philipp is the most likely choice; C. L. Holtfrerich, *Frankfurt as an International Financial Center: From Medieval Trade Fair to European Trading Center* (Munich: Beck, 1999). The involvement of the Frankfurt House of Ellissen in the delights of Middle East mid-century finance is attested by Landes (*Bankers of Pashas*, 348).

49. *Pickering v. Stephenson* [1871] P. 163. In 1868, the chairman and directors countered by suing Ellissen for libel; see *Liverpool Daily Post*, November 16, 1868.

50. Sir John Wickens V.C. in *Pickering v. Stephenson* [1871] P. 163 at 8.

51. Hyde Clarke, "The Ottoman, Smyrna and Aidin Railway" (letter), *London Daily News*, September 4, 1868.

52. The decline in the share price was a central grievance of the directors. According to the article from the *London Standard*, October 1, 1868, Sir M. Stephenson, chairman of the company's board of directors and of the September 30 meeting, stated about the Ellissen letters that it was "something unparalleled to see any shareholder setting himself up in opposition to the interests of the whole body" and added that "the consequences were already apparent." In the subsequent court report it was stated that "their shares were consequently depreciated" (*Pickering v. Stephenson* [1871] P. 163).

53. *London Daily News*, September 4, 1868, 6.

54. For the entire paragraph, "28 October 1868, Report of the Committee of Investigation," *Journal of the Anthropological Society of London* 7 (1869): xii, xv.

55. *London Standard*, December 18, 1868.

56. Odlyzko, "Collapse of the Railway Mania." This could also be put in relation with recent research on Masonic activity in the Victorian era. It underscores the dispute that developed in the late 1860s on the transparency of governance in Masonic lodges—revolving on the problem of insider cliques capturing societies—the sort of topic Clarke relished. In 1870 the *Freemasons' Magazine* wrote: "Again, we call attention to the meagre accounts of business done, and the paucity of information which is doled out to the brethren concerning the doings at head-quarters. While possessing a Grand Secretary who is indefatigable in the performance of his duties, with a reputation not confined to this country for business qualifications, it is perfectly clear that somewhere great obstructionism prevails" (quoted by Aubrey Newman, "Masonic Journals in Mid-Victorian-Britain," 2012, article posted by the Masonic Press Project in partnership with King's College, http://www.masonicperiodicals .org/. On the same website, see also Rebecca Coombes, "Fraternal Communications: The Rise of the English Masonic Periodical," 2012).

57. *Athenaeum* (September 5, 1868): 302. One H. Brookes joined the Anthropological Society in November 1864.

58. "Scientific Societies," *Pall Mall Gazette*, October 13, 1868, 3.

59. Specifically, the *Pall Mall* article noted that the Ethnological Society "was supported by a well-known literary and scientific journal."

60. "28 October 1868, Report of the Committee of Investigation," *Journal of the Anthropological Society of London* 7 (1869): xii.

61. Pim and Seemann, *Dottings*, 143.

62. Special Notice Relating to List of Defaulters, July 30, 1868, A-8 Suppl. 2/27: "Regretting that you have not done me the favour of replying to my circular letter, sent to you some time ago, it is now my duty to inform you that our Finance Committee have directed that the names of those gentleman who do not, on or before September 1st next, discharge the arrears of contribution due from them, shall be placed in a separate list in the new list of Fellows of the Society. It would be so disagreeable to have to include your name among them, that I hope you will save me the necessity of doing so by forwarding the amount due."

63. "Your Council, having resolved during the present year to take strict measures for the recovery of subscriptions in arrear, they regret to have to report the resignation, which they attribute principally to this cause, of fifty-nine fellows." *Journal of the Anthropological Society of London* 7 (1869): lxxvii.

64. "Anniversary Meeting, January 18, 1870," *Journal of the Anthropological Society of London* 8 (1870): lxxvi.

65. *Journal of the Anthropological Society of London* 7 (1869): lxxiv ff. The policy of enforcing late subscriptions is mentioned on lxxvii: the Council had "resolved to take strict measures for the recovery of subscriptions in arrear"; details for year 1869 in *Journal of the Anthropological Society of London* 8 (1870): lxxv ff. Figures for 1868 and 1869 are (respectively): annual fees: 976 and 731; life membership: 84 and 92; arrears 152 and 77. The last numbers suggest that efforts had been made to have the money roll in. But total revenues from subscription declined by about 240£.

66. British Association for the Advancement of Science, *Notes and Queries on Anthropology for the Use of Travellers and Residents in Uncivilized Lands* (Charing Cross, UK: Stanford, 1874). On the creation and legacy of the *Notes and Queries*, see J. Urry, *Before Social Anthropology: Essays on the History of British Anthropology* (New York: Routledge,

1993), chap. 1: "Notes & Queries on Anthropology and the Development of Field Methods in British Anthropology, 1870–1920."

67. See chapters 1 and 4; in the Ecuador Land Company, Gerstenberg had partnered with anthropologist Bollaert, whose "Past and Present Populations" (*Journal of the Anthropological Society of London* 1 [1863]: iii–x), was one of the very first papers discussed at the Anthropological Society in 1863.

68. Silva Ferro, *Historical Account*, 6; Stansifer, "E. George Squier," 1–27.

69. *Pall Mall Gazette*, May 23, 1872.

70. Silva Ferro, *Historical Account*, 42.

71. FO 566/30, Index of the Diplomatic Correspondence related to Central America, November 10, 1862, November 30, 1862, where mentions are made as well of an Haslewood's inquiry regarding British government protection for investors in a Honduras railway.

72. The source for this section is principally the *London Standard* as well as the material in Clarke's interview in the select committee (where the open letters between Guttierez and Clarke are reproduced).

73. On May 7, 2013, the website www.abebooks.com was selling the copy of an autograph letter signed "Bedford Pim" to Don Carlos Gutierrez (1818–1882), minister plenipotentiary, Honduras government. The document seen online seems authentic, judging from the signature of Pim. The letter reads: [letter head: Crown Office Row, Temple, E.C.] "15 July 1872. Excellency, Referring to the terms upon which I accepted the appointment of Special Commissioner of Honduras on 23rd May Last, I have to request your Excellency to give the necessary instructions that my salary may be paid on the quarter days usual in this country, viz. 1 Jan. 1 April 1 July: and 1 October instead of as the broken periods counting from 23rd May last the date mentioned above when my salary commenced. If your Excellency sees no objection to this arrangement my account will stand as follows with your Government.—[details follow and a total of £550 is given, including salary for three months in advance for £ 530 and incidental expenses including journey to Paris and back for £ 20]. I think your Excellency will find this account quite correct and I shall feel obliged if you will kindly give the necessary orders for the [illegible] payment of the same. Your Excellency's obedient servant. [Signed]: Bedford Pim." The letter is endorsed "in the name & on behalf of the Honduras Government & as Minister Plenipotentiary" and signed "Carlos Gutierrez." Countersigned by Pim in receipt of £550 over a penny Inland Revenue stamp, and dated 23 July 1872. The letter is stamped for authentication of receipt.

74. *London Standard*, July 22, 1872.

75. Lewis Harrop Haslewood (the father of Edward) having died in 1869, it is a little unclear who this relative might be. It might have been Lewis Bernard Haslewood, who spent time in Jamaica in the late 1860s. In 1859 Digby Seymour had been (as the *Dictionary of National Biography* puts it euphemistically) "called before the benchers of the Middle Temple to answer charges affecting his character as a barrister in connection with some commercial transactions, and was [. . . eventually] censured by the benchers"; *Dictionary of National Biography*, vol. 51, ed. Sidney Lee (London: Smith, Elder, 1897, 335–36). His deeds were splashed out in a pamphlet published soon after by some vengeful Anonymous: *William Digby Seymour: Proceedings before the Middle Temple Benchers, and the Northern Circuit Committee* (London: W. H. Cox, 1862). Digby Seymour appears in 1872 in the *London Standard* in connection with the production of an "expert" on Liberia. The

"expert" in question was David Chinery (a F.R.G.S. as it should) and the dinner mentioned reminds one of the description from Drummond Wolff given in chapter 4; See "'The Republic of Liberia," *London Standard*, January 1, 1872. Digby Seymour's name would later appear among the individuals who supported a debt-equity swap whereby the securities of the Honduras railway debt were converted in stocks of a Honduras Inter-Oceanic Railway Company in 1873, after the failure of the 1872 loan. Other promoters and investors whose names can be deciphered include Sampson Copestake, Sir Thomas White, William George Fitzgerald, William Peel, John William Jones, Charles Frederic Denny, Henry Maxwell, and Egan Desmond; BT 31/1922/7919.

76. See "Proceedings of the Anthropological and Ethnological Societies of London prior to the Date of Amalgamation," appendix to the *Journal of the Anthropological Institute of Great Britain and Ireland* 1 (1872).

77. *London Standard*, July 22, 1872.

78. *London Standard*, July 27, 1872. The same day, the *Standard* was publishing a further letter by Hyde Clarke preceded by a "spirited" header where we recognize familiar names: "The following letter has been received from Mr. Hyde Clarke. It was not to be supposed that he would be received courteously by the meeting of Honduras bondholders, seeing that his communication to the Minister had an adverse effect upon the stock. Captain Bedford Pim and Mr. Digby Seymour were no doubt guns of too great caliber for the late representative of the Ottoman and Smyrna Railway and other unsuccessful companies. We should have liked to have seen a struggle between Mr. Lewis Haslewood and Mr. Hyde Clarke on their own respective grounds, and if poor John Diston Powles had been alive he would have witnessed the encounter with some degree of satisfaction."

79. See "Proceedings of the Anthropological and Ethnological Societies of London," xxxviii.

80. Blanchard Jerrold, "On the Manufacture of Public Opinion," *Nineteenth Century* (1883): 1080–92.

81. "The Funeral of Mr. Darwin," *Times*, April 27, 1882. I am grateful to Mr. Umit Sakmar for having drawn my attention to this article.

82. On the role of Clarke and the Corporation of Foreign Bondholders in the acquisition of Egypt, see Paul Frank Meszaros, "The Corporation of Foreign Bondholders."

83. *Morning Post*, March 4, 1895.

84. John Gross, *The Rise and Fall of the Man of Letters: A Study of the Idiosyncratic and the Humane in Modern Literature* (New York: Macmillan, 1970), 25.

## CHAPTER ELEVEN

1. For a more recent illustration, see Richard H. Grove's *Green Imperialism*, which provides a comparative study of patterns of emerging environmentalism within three early modern empires (Dutch, French, and British); *Green Imperialism: Colonial Expansion, Tropical Island Edens and the Origins of Environmentalism, 1600–1860* (Cambridge, UK: Cambridge University Press, 1995).

2. David Keith Van Keuren, "Museums and Ideology: Augustus Pitt Rivers, Anthropological Museums, and Social Change in Later Victorian Britain," *Victorian Studies* 28 (1984): 171–89.

3. Augustus Pitt Rivers, *An Address Delivered at the Opening of the Dorset County Museum* (Dorchester, UK: James Foster, 1884).

4. It is striking how much conventional narratives of the birth of the Pitt Rivers Museum underplay the Anthropological Society's connection. They discuss its descent from the sole vantage point of Pitt Rivers's "original collection" of guns—as was done initially by an early curator of the museum, Henry Balfour, in his introduction to A. H. Pitt Rivers [Lane Fox], *The Evolution of Culture, and Other Essays* (Oxford: Clarendon, 1906), v–xx—and gloss over the fact that the Anthropological Society, unlike the Ethnological Society, had a significant museum to which local secretaries such as R. B. N. Walker provided substantial loot. In a somewhat contradictory way, individuals such as Walker are acclaimed on the museum's website as one of Pitt Rivers's "field collectors." The museum's research aid also concedes the point, mentioning for instance, "it seems likely that all the R. B. N. Walker objects in the Pitt Rivers collection were received via the Anthropological Society."

5. For an exhaustive list of Parliamentary Papers on the subject, see Platt, *Cinderella Service*, 243–44.

6. See, for example, *Brisbane Courier*, February 1, 1887, for a mention of a project by the United Kingdom Pilots' Associations to erect a memorial to the memory of the late Admiral Bedford Pim. Reference to the London Seaman's Society is provided in the Disraeli Papers, as well as mention of Pim's journal, the *Navy*, Ms Dep. Hughenden 53/1. On Pim's harassment of the Navy, see ADM 6/102, ADM 1/6155.

7. Platt, *Cinderella Service*, 146.

8. "Our Consulates in Asia and Africa," *Economist*, November 30, 1867, 1351–52. On the significance of the Plowden and Abyssinia affair in revealing a broader issue, see Platt, *Cinderella Service*, 136.

9. "Our Consulates in Asia and Africa," *Economist*, November 30, 1867.

10. *Economist*, November 14, 1868, 1300–1301.

11. The "splendid specimen" is from Arthur D. Elliot, *Life of George Joachim Goschen, First Viscount Goschen, 1831–1907*, 2 vols. (London: Longmans, 1911), 47.

12. *Economist*, November 14, 1868, 1301.

13. Ibid.

14. Thomas J. Spinner, *George Joachim Goschen: The Transformation of a Victorian Liberal* (Cambridge, UK: Cambridge University Press, 1973), 9.

15. Earl of Cromer [Evelyn Baring], "Government of Subject Races."

16. Ibid.

17. The Pim v. Clarke dispute over Honduras operated along similar lines. We have seen in the previous chapter that there was a link between Pim and Disraeli. Supporting the same interpretation, we remark that in 1868, Hyde Clarke had taken issue with the consular jurisdiction in the Ottoman Empire, siding with Layard against Conservative Foreign Secretary (and Anthropological Society permanent fellow) Lord Stanley in his article in the *Times*.

18. See the "List of Members of the Ethnological Society of London" put out in the society's *Proceedings* (1863): 3.

19. His official biography has an entire page vaunting the high credit in which Liberals held Fowler. The occasion given, suggestively, is the case of Governor Eyre, in support of whom he spoke in Parliament. This was natural, since he was Disraeli's designated agent

for foreign and colonial subjects and informs on the subtle contours of humanitarianism. Gladstone, Flynn declares, felt that Fowler's speech "did credit to his feelings, and showed what difficulties the case presented to the candid mind"; John Stephen Flynn, *Sir Robert N. Fowler, Bart., M.P.: A Memoir* (London: Hodder & Stoughton, 1893). H. L. Malchow, "God and the City: Robert Fowler (1828–91)," in *Gentlemen Capitalists: The Social and Political World of the Victorian Businessman* (Palo Alto, CA: Stanford University Press, 1992), expands on Fowler's bipartisan inclinations, particularly when he speaks of a politician "jealous of bipartisan and independent activity" (179ff.); There is also interesting material on Fowler's role at the Aborigines' Protection Society in James Heartfield, *Aborigines Protection Society*. An indication of the standing of Fowler as a humanitarian figurehead at the time of the creation of the Government Stock Investment Company is provided by F. W. Chesson, *The Dutch Republics of South Africa: Three Letters to R.N. Fowler Esq. M.P. and Charles Buxton Esq. M.P.* (London: William Tweedie, 1871).

20. *Kerferd v. Dreyfus et al.*, February 6, 1875, C16/1019.

21. Motion for a Select Committee, HC Deb February 23, 1875, vol. 222 cc772–87. To the extent that the seeds of the Bolivian crop of litigation had been already planted, Bolivia might have been logically excluded. But what to say of the Honduras lawsuits, which were started in tribunal a few days before the Motion for a Select Committee was adopted?

22. *London Daily News*, April 23, 1875.

23. *Select Committee*, p. 210, Q 4780. This may be put in relation to the tension that Andrew St. George said existed at one point between Rose and Lewis (St. George, *Norton Rose*, 139) and to the fact that Rose had a connection with the Bolivian loan through his "friend" Erlanger. In late 1874 there was a big row about the partial suspension of amortization of Bolivian debt, which was scheduled to take place on January 1, 1875. When Bolivian authorities announced this, they related the decision to the fact that money kept in London for that purpose had dried up. Suspension of amortization technically amounted to a default, and the motivation signaled the imminence of an even more serious breach of contract in 1875. Compare with Neville B. Craig, *Recollections*, 65, who states that default would have occurred in June 1875.

24. Hansard, Loans to Foreign States Committee, Breach of Privilege, HC Deb April 16, 1875, vol. 223 cc1114–52; a summary of the facts pertaining to the breach of privilege debate is found in *Special Report from the Select Committee on Loans to Foreign States*, April 19, 1875, Parliamentary Papers 1875 (152).

25. "Things We Don't Like," *Fun*, May 1, 1875, 179.

26. "'Privilege' and Business," *Economist*, April 17, 1875, 453.

27. See, for example, *Economist*, April 17, April 24, May 8.

28. Markham was also involved as coeditor of the new anthropological manual: *Notes and Queries*. In addition, he was connected to the Anthropological Institute through the institute's Arctic committee: Clements R. Markham, ed., *Arctic Geography and Ethnology: A Selection of Papers on Arctic Geography and Ethnology* (London: John Murray, 1875), 278.

29. *Economist*, November 14, 1868, 1300–1301; BT 31/246/807.

30. The whole affair had the distinct outlook of a British vaudeville. Turquand had been the leader of the transformation of the Institute of Chartered Accountants into a royal institute in 1880, and the M.P. who had supported the bill in Parliament was Sir John Lub-

bock. The first vice president of the new Royal Institute of Chartered Accountants in 1880 had been Sir Robert Palmer Harding, the father of Anna Marion, whom George Earl Church married in 1898.

31. London Metropolitan Archive, CFB Archive, Transfer books, "Requests and applications to transfer certificates of permanent membership"; CLC/B/060/MS345; MS 34594 / 1 to 3.

32. H. Spencer, *Railway Morals*, 2.

33. "Avebury's Circle: The Science of John Lubbock FRS (1834–1913)," conference held at the Royal Society, March 22, 2013.

34. Earl of Cromer, "Government of Subject Races," 44.

# SOURCES

PRINCIPAL PRINTED PRIMARY SOURCES

*Parliamentary Papers*

P.P. 1839 (169) XIX. *First Report of the Commissioners Appointed to Inquire as to the Best Means of Establishing a Constabulary Force in the Counties of England and Wales.*

P.P. 1849 (1049) LVI. *Circular Addressed by Viscount Palmerston to Her Majesty's Representatives in Foreign States, Respecting the Debts Due by Foreign States to British Subjects.*

P.P. 1863 (3179) LXXIII. *Dispatches from Commodore Wilmot, Respecting His Visit to the King of Dahomey, in December, 1862, and January, 1863. Presented to the House of Commons by Command of Her Majesty, in Pursuance of Their Address.*

P.P. 1865 (3536) LVII. *Abyssinia. Papers Relating to the Imprisonment of British Subjects in Abyssinia.*

P.P. 1866 (3748) LXXV. *Abyssinia. Further Correspondence Respecting the British Captives in Abyssinia.*

P.P. 1867 (3774) XXVIII. *Report of the Commissioners Appointed to Inquire into the Existence of Corrupt Practices at the Last Election, and on Former Similar Occasions, for the Borough of Totnes; Together with the Minutes of Evidence and Appendix.*

P.P. 1875 (367) XI.1. *Report from the Select Committee on Loans to Foreign States, Together with the Proceedings of the Committee, Minutes of Evidence, Appendix and Index.*

ONLINE PERIODICALS

British Newspaper Archive http://www.britishnewspaperarchive.co.uk/

The Times Digital Archive 1785–1985 http://gdc.gale.com/products/the-times-digital -archive-1785–1985/

The Economist Historical Archive, 1843–2008 http://gdc.gale.com/products/the-economist -historical-archive-1843–2007/

*Publications of Anthropological and Ethnological Societies,
and of the Anthropological Institute*

Anthropological Society of London

*Journal*

*Journal of the Anthropological Society of London.* 8 vols. 1863–70, Trübner until 1868, Asher & Co. after 1869 (ending in June 1870) [The *Journal of the Anthropological Society* was numbered in Roman numerals; it contained summaries of meetings, minutes of anniversary meetings, presidential addresses, and more or less condensed abstracts of papers presented and summaries of discussion (occasionally full papers were printed). The *Journal of the Anthropological Society* and the *Anthropological Review/Anthropological Journal* are generally bound with one another but separate series do exist.]

*Review*

a. *Anthropological Review.* 8 vols. 1863–70 (ending in April 1870), Trübner until 1868, Asher & Co. after 1869 (ending in June 1870)

b. *Journal of Anthropology.* Nos. 1–3, July 1870–January 1871, Longman, Green & Co. [The *Anthropological Review* contained articles, a large fraction of which (unsigned) were supplied by the editor James Hunt with the intention to discuss "developments" in anthropology from the vantage point of the Cannibal Club. The *Journal of Anthropology* was the short-lived successor of the *Anthropological Review*, from July 1870 to January 1871. Both were numbered in Arabic.]

*Memoirs*

*Memoirs Read before the Anthropological Society of London.* 3 vols. 1863–64 (1865); 1865–66 (1866); 1867–69 (1870), Trübner. [This semi-periodical (it kept lagging behind discussions taking place at the Anthropological Society) contained the text of presentations made at the society. Presentations were often made without a paper, with the text contributed later.]

*Magazine*

*Popular Magazine of Anthropology.* Vols. 1–4 (1866), Trübner [This "laymen" *Magazine of Anthropology* was started in early 1866 by Hunt in an attempt to reach an even larger public, but discontinued afterwards; the *Magazine* also had a special issue, consisting in Pim's lecture "Negro and Jamaica"; Previous researchers (Van Keuren, "Human Science in Victorian Britain," p. 54) have implied that it succeeded the *Reader* which Anthropological Council member Bendyshe had purchased in 1865 but which would have gone bankrupt in 1866.]

Ethnological Society of London

*Transactions and Journal*

*Transactions of the Ethnological Society.* Vol. 1 (1861) to 7 (1869), John Murray
*Journal of the Ethnological Society of London.* Vol. 1 (1869)–Vol. 2 (1870), Trübner [The
   society's periodical was known as the *Journal of the Ethnological Society of London*
   until 1861 when it became the *Transactions of the Ethnological Society of London*. Its periodicity was somewhat haphazard in the first part of the 1860s, and it
   gave little "housekeeping information" (unlike the *Journal of the Anthropological
   Society of London*). In 1869 it went back to its previous title until the merger of the
   two societies. Upon moving with Trübner, the Ethnological Society's *Journal* began
   distinguishing between Roman numeral matter, containing minutes of meetings, and
   Arabic numerals for the "regular" papers.]

Anthropological Institute of London

*Journal*

*Journal of the Anthropological Institute of Great Britain and Ireland.* Vol. I (1872)–
   Vol. XXII (1893), Trübner & Co [This periodical took the succession of both Anthropological and Ethnological publications. In terms of presentation and structure, the
   continuity with the typesetting and organization of the *Journal of the Anthropological Society* and the *Anthropological Review* is striking. "Institutional continuity"
   is ensured by the first volume (1872) of the *Journal of the Anthropological Institute
   of Great Britain and Ireland*, which does contain a roman numerals "Appendix"
   showing the "Proceedings of the Anthropological and Ethnological Societies Prior
   to the Date of Amalgamation." Pre-amalgamation Anthropological Society proceedings correspond to material formerly reported in the *Journal of the Anthropological
   Society of London* for the period December 20, 1870 to February 14, 1871; it is found
   on pages i–xxxvi. Pre-amalgamation Ethnological Society proceedings take over the
   material reported in the *Journal of the Ethnological Society of London* for the period
   November 8, 1870 until January 24, 1871; they are found at pages xxxvi–clvii.]

## LEGAL CASES AND COURT REPORTS

The first number given—e.g., "[1866] B. 308"—is the case number;
the second—e.g., "(1880) 5 A.C. 176"—is the court number.

*Bee v. Ottoman Railway Company* [1866] B. 308 / n.a.
*Cooper v. Lloyd* [1874] C. 9 / n.a.
*Dent v. The Ottoman Railway Company* [1869] D. 151 / n.a.
*Fraser v. Lawrence* [1870] F. 20 / n.a.
*Kerferd v. Dreyfus* [1875] K. 8 / n.a.
*Kerferd v. Dreyfus* [1875] K. 19 / n.a.

Locock v. Smyth [1864] L. 102 / n.a.
National Bolivian Navigation Company v. The Public Works Company [1873] N. 42 / n.a.
National Bolivian Navigation Company v. Wilson n.a. / (1880) 5 A.C. 176
Oppenheim v. General Credit and Finance Company of London [1865] O. 14 / n.a.
Pickering v. Stephenson [1871] P. 163 / (1872) L.R. Eq. 322
Pike v. Nicholas [1868] P. 87 / (1869–70) L.R. 5 Ch. App. 251
Public Works Construction Company Ltd v. The National Bolivian Navigation Company
    [1873] P. 126 / n.a.
Reed Bros. & Co. v. Madeira and Mamore Railway Co. Ltd. et al. [1879] J46/584 / n.a.
Republic of Bolivia v. National Bolivian Navigation Company [1874] B. 142 / n.a.
Rubery v. Grant [1871] R. 139 / (1871–72) L.R. 13 Eq. 443
Society of Accountants in Edinburgh v. Corporation of Accountants n.a. / (1893) 20 R. 750
Vives v. Ottoman Railway Company [1869] U./V. 34 / n.a.
Wilson v. Church n.a. / (1878) 9 Ch. D. 552
Wilson v. Church n.a. / (1879) 13 Ch. D. 1
Wilson v. Church (No. 2) n.a. / (1879) 12 Ch. D. 454

ARCHIVAL SOURCES

Royal Anthropological Institute (London)

a. Anthropological Society of London

A-3, Council (Summary minutes) January 18, 1863–January 30, 1871
A-4, Ordinary meetings 1863–1871
A-5, Letters to the Anthropological Society of London. 1865–66; and [January 1, 1867],
    A–D, G–S (incomplete)
A-6, Anthropological Society of London Finance: Membership subscription ledgers,
    1863–1872
A-7, Anthropological Society of London: Finance: Receipted accounts, January
    1863–December 1866
A-8, Miscellaneous addresses, lists, circulars, etc. published by the Anthropological Soci-
    ety of London. 1863–68

b. Ethnological Society of London

A-1, Council Minutes (1844–1869)
A-2, List of members elected. 1844, 1846, 1868–71
A-31, Ethnological, Anthropological and Royal Anthropological Institute fellowships
    (1869–1955)

Disraeli Papers, Bodleian Library, Oxford

Stewart, R. W., Hand List of the Hughenden Papers, A list of the papers relating to Benja-
    min Disraeli (Earl's of Beaconsfield) and his family preserved at Hughenden Manor,

*near High Wycombe, Buckinghamshire*, London: The National Trust (typed 1961, revised 1968)
Ms Dep. Hughenden, 53/1 Captain Bedford Pim (1875–79)


*London Stock Exchange, Guildhall Library*

Minutes of the Committee for General Purposes (1865–1873); MS 14600/27 to 35
Stock Exchange Applications Honduras Loans, 1867 (24A/1443; MS18001/24A/1566); 1870 and 1872 (25A1749)


*Foreign Office (Kew Gardens)*

Ottoman Railway Company from Smyrna to Aidin, *Report of the Committee of Investigation to be Presented to Shareholders; at their adjourned eighteenth half-yearly general meeting, on Tuesday, May 21, 1867, at two o'clock*, London: Cousins. FO 78/2275
Honduras Inter-Oceanic Railway Company, FO 925/1208 [Map of Honduras and San Salvador, Central America, showing the line of the proposed Honduras, Inter-Oceanic Railway, territories of Indian Tribes and mines etc. with notes, by E. G. Squier, 1854. About 19 miles to an inch. Honduras Inter-Oceanic Railway Co.]
Ecuador Land Company Ltd. FO 25/34 [Prospectus, 1859]
Index of the Diplomatic Correspondence related to Central America: Costa Rica, Guatemala, Honduras, Nicaragua and Salvador, FO 566/30
Nicaragua Treaty: Mosquito Indians and claims of British subjects. Place and date of signature: January 28, 1860, FO 93/67/3, FO 94/537
Mosquito and Nicaragua: Corres. Affairs in Mosquito Part. 1 (January 1, 1860–December 31, 1873), FO 881/4013
Law Officers Reports, Mosquito, 1849–1867, FO 83/2311


*Record of the Meteorological Office (Kew Gardens)*

Ecuador: from William Wilson, BJ 7/919 [Letter from William Wilson, assistant secretary to Ecuador Land Company Ltd, Finsbury Square, regarding an expedition to survey territory for settlement by European immigrants and offering to supply meteorological observations.]


*Companies Registration Office Records (Kew Gardens)*
(by chronological order of incorporation, date of incorporation in parentheses)

Honduras Inter-Oceanic Railway Company Ltd. (1857), BT 41/301/1750 and BT 31/246/807
Ecuador Land Company Ltd. (1859), BT 41/215/1225
Foreign Lands and Mineral Purchase Company Ltd. (1864), BT 31/1025/1691C

Chontales Gold and Silver Mining Company Ltd. (1865), BT 31/1165/2494C
Nicaragua Railway Company Ltd. (1866), BT 31/1304/3339
Central American Association Ltd. (1866), BT 31/14380/2935 and BT 34/2441/2935
Public Works Construction Company Ltd. (1871), BT 31/1597/5329
Madeira and Mamore Railway Company Ltd. (1871), BT 31/1595/5315 and BT 34/81/5315
Honduras Inter-Oceanic Railway Company Ltd. (1873), BT 41/301/1750
Corporation of Foreign Bondholders (1873), BT 31/1884/7528

### Admiralty (Kew Gardens)

Orders in Council and Council Offices—Admiralty, and Ministry of Defence, Navy Department: Correspondence and Papers (various documents concerning *Gorgon* and Pim)

Orders in Council and Council Offices ADM 1/6183
From Captains P–Z, ADM 1/6155
Captains, ADM 1/6102

### Admiralty: Digests and Indexes

Index N–Q, 1861, ADM 12/695
Index N–Q, 1868, ADM 12/807
Index N–Q, 1869, ADM 12/825
Index N–Q, 1870, ADM 12/843

### Ship Log, *Gorgon*

April 27, 1859–November 29, 1859 (ADM 53/7795)
December 1, 1859–August 4, 1860 (ADM 53/7796)
August 5, 1860–March 23, 1861 (ADM 53/7797)
March 23, 1861–September 16, 1861 (ADM 53/7798)

### Ship Log, *Fury*

December 10, 1860–June 19, 1861 (ADM 53/6784) [inspected from April 1, 1861]

### Officers' Service Records (Series III)

Pim—Bedford—Clapperton—Trevelyan (Pim's file) ADM 196/16/296

### North America and West Indies Station: Correspondence, Reports and Memoranda

Central America and Greytown (Nicaragua) ADM 128/55

*Colonial Office (Correspondence, Original—Secretary of State) (Kew Gardens)*

Mrs. Hedgcock (question over land called Black River in Mosquito) CO 123/108

*Public Record Office, Domestic Records of the P.R.O. (Kew Gardens)*

John Russell Papers

Correspondence between John Russell and Richard Lyons (British minister in Washington, Mosquito, Mexico, political situation in pre-Civil War America), PRO 30/22/96 (July 28, 1859–April 30, 1860)

*Chancery (Kew Gardens)*

Court of Chancery: Clerks of Records and Writs Office: Pleadings 1861–1875

Kerferd v. Dreyfus ([1875] K8 and [1875] K19) (plaintiff George Briscoe Kerferd, trustee for the Honduras loan, defendants include bankers Adolphe Dreyfus and Martin Wolfgand Scheyer, as well as Edward Haslewood, Bedford Clapperton Trevelyan Pim, and the Republic of Honduras), C16/1019/K8 and C16/1019/K19

Oppenheim v. General Credit and Finance Company ([1865] O. 14), C16/291/O14.

*London Metropolitan Archives (Corporation of Foreign Bondholders)*

Certificates of Membership

CLC/B/060/MS34592/001 and 002 (1873)

Requests and Applications to Transfer Certificates of Permanent Membership

CLC/B/060/MS34594/001 (1873–1884)
CLC/B/060/MS34594/001 (1884–1894)
CLC/B/060/MS34594/001 (1894–1897)

*Tulane University, Latin American Library Manuscripts (New Orleans)*

Mosquito Territory, Nicaragua Documents, 1859–1867, Collection 50, Folders 1 to 6 (king of Miskito quit rent contracts)

*Church Collection, Hay Library*

Anonymous, *Report of Proceedings Connected with the Foreign Debt of the Republic of Ecuador, at a Meeting Held at the London Tavern, in the 28th of June, 1853*, London: Clay, 1853, xfHT C81

Committee of the Holders of Spanish Bonds, *Spanish Bonds: A Full Report of the Public Meeting Held on August 19, 1851, Further to Consider Bravo Murillo's Confiscatory Scheme for the Arrangement of the Debt, &c.; Letters of J. Capel, Sir H. Bayly and Mr. Tasker; Memorial of Mr. Wadsworth, with Lord Palmerston's Reply; Criticisms of the English Press etc.* London: John King, s.d. [handwritten: 1853].
Scattered Letters

*University College Special Collections, Routledge Archive (Nicholas Trübner & Co)*

Publication Books, 1863 to 1873 (100–102)
Publication Account Books, 1863 to 1873 (105–106)
Printed Catalogues (113)

Anderson, Sigurd. "Lawyers in the Civil War." *American Bar Association Journal* 48 (1962): 457–60.

Anonymous. "Berthold Seeman." *Journal of Botany, British and Foreign*, n.s., 1 (1872): 1–7.

———. "Colonel George Earl Church." *Geographical Journal* 35 (1910): 203–4.

———. *Foreign Loans and National Debts: A Review of the Report of the Select Committee on Foreign Loans*. London: Trübner, 1876.

———. *The Railway Directory for 1844, Containing the Names of Directors and Principal Officers of the Railways in Great Britain and Ireland*. London: Railway Times, 1845.

———. *William Digby Seymour: Proceedings before the Middle Temple Benchers, and the Northern Circuit Committee*. London: W. H. Cox, 1862.

Antrosio, J. "Marshall Sahlins, National Academy of Sciences, Napoleon Chagnon." *Anthropology Report*, February 25, 2013, http://anthropologyreport.com.

Archer, W. G. Preface to *The Kama Sutra of Vatsyayana*, translated by Sir Richard Burton and F. F. Arbuthnot, edited by W. G. Archer. London: Allen & Unwin, 1980.

Arendt, Hannah, *The Origins of Totalitarianism*, New York: Harcourt Brace, 1951.

Armstrong, John. "The Rise and Fall of the Company Promoter and the Financing of British Industry." In *Capitalism in a Mature Economy: Financial Institutions, Capital Exports and British Industry, 1870–1939*, edited by J. J. Van Helten and Y. Cassis, 115–38. London: Edward Elgar, 1990.

Arnold, A. J., and S. McCartney. *George Hudson: The Rise and Fall of the Railway King*. London: Hambledon & London, 2004.

Arondekar, Anjali. *For the Record: On Sexuality and the Colonial Archive in India*. Durham, NC: Duke University Press, 2009.

Asad, Talal, ed. *Anthropology and the Colonial Encounter*. London: Ithaca, 1973.

Atkinson, W. J. "Scientific Societies and the Admission of Women." *Nature* 79 (1909): 488.

Aubert, Vilhelm. "White-Collar Crime and Social Structure." *American Journal of Sociology* 58 (1952): 263–71.

Augar, Philip. *Death of Gentlemanly Capitalism*. London: Penguin, 2000.

Austin, J. L. *How to Do Things with Words: The William James Lectures Delivered at Harvard University in 1955*. Oxford: Clarendon Press, 196.

Baber, Zaheer. *The Science of Empire: Scientific Knowledge, Civilization and Colonial Rule in India*. New York: State University of New York Press, 1996.

Bacon, Francis, *Novum Organum: With Other Parts of the Great Instauration*, translated and edited by Peter Urbach and John Gibson. Chicago: Open Court, 1994.

Bagehot, Walter. "Why the Stock Exchange Is Likely to Have More and Greater Frauds in It Than Any Other Market." *Economist*, March 27, 1875.

Baigent, Elizabeth. "Markham, Sir Clements Robert (1830–1916)." In *Oxford Dictionary of National Biography*. Oxford: Oxford University Press, 2006.

Barnhart, T. A. *Ephraim George Squier and the Development of American Anthropology*. Lincoln: University of Nebraska Press, 2005.

Barr, W. "Franklin in Siberia? Lieutenant Bedford Pim's Proposal to Search the Arctic Coast of Siberia, 1851–52." *Arctic* 45 (1992): 36–46.

Barton, R. "Huxley, Lubbock, and Half a Dozen Others: Professionals and Gentlemen in the Formation of the X Club, 1851–1864." *Isis* 89 (1998): 410–44.

Batchelor, H. C. *Amazon Route to Bolivia: Its Physical and Commercial Impracticability Demonstrated by Extracts from a Work Just Published by Mr. Mathews Entitled "Up to the Amazon and Madeira Rivers, through Bolivia and Peru."* London: Little & Son, 1879.

Bates, Darrell. *The Abyssinia Difficulty: The Emperor Theodorus and the Magdala Campaign, 1867–68*. Oxford: Oxford University Press, 1979.

Baxter, Robert Dudley. *National Debts*. London: Rober John Bush, 1871.

Bayly, C. A. *Empire and Information: Intelligence Gathering and Social Communication in India, 1780–1870*. Cambridge, UK: Cambridge University Press, 1996.

Beasley, Edward. *Empire as the Triumph of Theory: Imperialism, Information, and the Colonial Society of 1868*. London: Routledge, 2005.

———. *Mid-Victorian Imperialists: British Gentlemen and the Empire of the Mind*. London: Routledge, 2005.

———. *The Victorian Reinvention of Race: New Racisms and the Problem of Grouping in the Human Sciences*. London: Routledge, 2010.

Beke, Charles Tilstone. *The British Captives in Abyssinia*. London: Longmans, 1867.

Bell, M., and C. McEwan. "The Admission of Women Fellows to the Royal Geographical Society, 1892–1914: The Controversy and the Outcome." *Geographical Journal* 162 (1996): 295–312.

Bemis, G. *Report of the Case of John W. Webster, Indicted for the Murder of George Parkman, before the Supreme Judicial Court of Massachusetts*. Boston: Little & Brown, 1850.

Berghahn, Volker R. "Philanthropy and Diplomacy in the 'American Century.'" *Diplomatic History* 23 (1999): 393–419.

Bidwell, Charles Toll. *The Isthmus of Panama*. London: Chapman & Hall, 1865.

Blake, Laurel, and Marysa Demoor. *Dictionary of Nineteenth-Century Journalism in Great Britain and Ireland*. Gent: Academia Press; London: British Library, 2009.

Blake, Robert. *Disraeli*. New York: St. Martin's, 1967.

Blakemore, H. *From the Pacific to La Paz: The Antofagasta (Chili) and Bolivia Railway Company, 1888–1988*. London: Lester Crook Academic, 1990.

Boase, G. C. "Seymour, William Digby (1822–1895)," in *Dictionary of National Biography*, edited by Sidney Lee, 335–36. London: Smith, Elder, 1896.

Bollaert, William. *Antiquarian, Ethnological and Other Researches in New Granada, Equador, Peru and Chile*. London: Trübner, 1860.

——. "Contribution to an Introduction to the Anthropology of the New World." *Memoirs Read before the Anthropological Society of London* 2, (1865/66): 92–152.

——. "The Jivaro Indians in Ecuador." *Transactions of the Ethnological Society of London* 2 (1863): 112–18.

——. "Maya Hieroglyphic Alphabet of Yucatan." *Memoirs Read before the Anthropological Society of London* 2 (1865/66): 46–54.

——. "Observations on the Past and Present Populations of the New World." *Journal of the Anthropological Society of London* 1 (1863): iii–x.

——. "On the Alleged Introduction of Syphilis from the New World: Also Some Notes from the Local and Imported Diseases into America." *Journal of the Anthropological Society of London* 2 (1864): cclvi–cclxx.

——. "On the Ancient Indian Tombs of Chiriqui in Veraguas (South-West of Panama), on the Isthmus of Darien." *Transactions of the Ethnological Society of London* 2 (1863): 147–66.

——. *W. M. Bollaert's Researches from 1823 to 1865: Principally on South American Subjects. A Bibliography*. London: publisher not identified, 1865.

Bonner, Robert E. "Slavery, Confederate Diplomacy and the Racialist Mission of Henry Hotze," 51 *Civil War History*, (2005): 288–316.

Boskin, Joseph. *Sambo: The Rise and Demise of an American Jester*. New York: Oxford University Press, 1988.

Bowler, Peter J. "Anthropology and Evolution." *Isis* 79 (1988): 104–7.

Bowman, I. "Geographical Aspects of the New Madeira–Mamore Railroad." *Bulletin of the American Geographical Society* 45 (1913): 275–81.

Braggion, F., and L. Moore. "The Economic Benefits of Political Connections in Late Victorian Britain." *Journal of Economic History* 73 (2013): 142–76.

Bramsen, Bo, Kathleen Wain, and Kathleen Brown. *The Hambros, 1779–1979*. London: Joseph, 1979.

Bravo, Michael. "Ethnological Encounters." In *Cultures of Natural History*, edited by N. Jardine, J. A. Secord, and E. Spary. Cambridge, UK: Cambridge University Press, 1996.

British Association for the Advancement of Science. *Notes and Queries on Anthropology for the Use of Travellers and Residents in Uncivilized Lands*. Charing Cross, UK: Stanford, 1874.

Broca, Paul, *Sur le volume et la forme du cerveau, suivant les individus et suivant les races*. Paris: Hennuyer, 1861.

Brodie, Benjamin Collins. "Address to the Ethnological Society of London, Delivered at the Anniversary Meeting on the 27th May 1853." *Journal of the Ethnological Society of London* 4 (1856): 98–103.

Brodie, Fawn M. *The Devil Drives: A Life of Sir Richard Burton*. New York: Norton, 1967.

Burrow, J. W. "Evolution and Anthropology in the 1860s: The Anthropological Society of London, 1863–71." *Victorian Studies* 7 (1963): 137–54.

Burton, Isabel. *The Life of Captain Sir Richard F. Burton.* 2 vols. London: Chapman & Hall, 1893.

Burton, Richard F. *Abeokuta and the Cameroon's Mountains.* 2 vols. London: Tinsley Bros., 1863.

———. *Explorations of the Highlands of Brazil, with a Full Account of the Gold and Diamond Mines.* 2 vols. London: Tinsley Bros., 1869.

———. *Letters from the Battlefield of Paraguay.* London: Tinsley Bros., 1870.

———. *A Mission to Gelele, King of Dahome, with Notices of the So-called "Amazons," the Grand Customs, the Yearly Customs, the Human Sacrifices, the Present State of the Slave Trade, and the Negro's Place in Nature.* 2 vols. London: Tinsley Bros., 1864.

———. *Scinde, or the Unhappy Valley.* 2 vols. London: Richard Bentley, 1851.

Cain, P. J. "Empire and the Languages of Character and Virtue in Later Victorian and Edwardian Britain." *Modern Intellectual History* 4 (2007): 249–73.

Cain, P. J., and Anthony G. Hopkins. *British Imperialism: 1688–2000,* 2nd ed. [1st ed., 1993]. Edinburgh, UK: Pearson, 2002.

Candolle, Alphonse de. *Histoire des sciences et des savants.* Geneva: Georg, 1869.

Carlyle, Thomas [calling himself Anonymous]. "Occasional Discourse on the Negro Question." *Fraser's Magazine for Town and Country* 40 (1849): 670–79.

Caudill, E. "The Bishop-Eaters: The Publicity Campaign for Darwin and on the Origin of Species." *Journal of the History of Ideas* 55 (1994): 441–60.

Chakrabarti, P. "The Travels of Sloane: A Colonial History of Gentlemanly Science." In *From Books to Bezoars: Sir Hans Sloane and His Collections,* edited by Alison Walker, Arthur MacGregor, and Michael Hunter. London: British Library Publishing, 2012.

Chambers, David, and Rui Pedro Esteves. "The First Global Emerging Markets Investor: Foreign & Colonial Investment Trust 1880–1913." *Explorations in Economic History* 52 (2014): 1–21.

Chapman, W. R. "Ethnology in the Museum: AHLF Pitt-Rivers (1827–1900) and the Institutional Foundation of British Anthropology." Diss., Oxford University, 1981.

Charnock, R. S., and C. Carter Blake. "Notes on the Wolwa and Mosquito Vocabularies." *Transactions of the Philological Society* 15 (1874): 350–53.

Chesson, F. W. *The Dutch Republics of South Africa: Three Letters to R. N. Fowler Esq. M.P. and Charles Buxton Esq. M.P.* London: William Tweedie, 1871.

Chevalier, Michel, and Thomas Ayon. *Convention pour l'exécution d'un canal maritime interocéanique sur le territoire de la République de Nicaragua.* Paris: P. Dupont, 1869.

Church, George Earl. "Desiderata in Exploration, II: South America." London: Royal Geographical Society, 1907.

———. *Explorations Made in the Valley of the River Madeira, from 1749 to 1868.* London: National Bolivian Navigation Co., 1875.

———. *Papers and Documents Relating to the Bolivian Loan, the National Bolivian Navigation Company, and the Madeira and Mamoré Railway Company, Ltd.* London: Dunlop, 1873.

———. *The Route to Bolivia via the River Amazon: A Report to the Governments of Bolivia and Brazil*. London: Waterlow, 1877.

Church, George Earl, and Clements R. Markham. *Aborigines of South America*. London: Chapman and Hall, 1912.

Clarke, H. "Anthropological Society of London," *Athenaeum* 2129 (1868): 210.

———. *Colonization, Defence, and Railways in Our Indian Empire*. London: John Weale, 1857.

———. *Contributions to Railway Statistics in 1846, 1847, & 1848*. London: John Weale, 1849.

———. *The Imperial Ottoman Smyrna & Aidin Railway: Its Position and Prospects*. London: Koehler Bros., 1861.

———. "On Anthropological Investigations in Smyrna." *Journal of the Anthropological Society of London* 4 (1866): xcviii–cii.

———. "On the Epoch of Hittite, Khita, Hamath, Canaanite etc." *Transactions of the Royal Historical Society* 6 (1877): 1–85.

———. "On the Financial Resources of Our Colonies." *Proceedings of the Royal Colonial Institute* 3 (1872): 130–47.

———. "On the Geographical Distribution of Intellectual Qualities in England." *Journal of the Statistical Society of London* 34 (1871): 357–73.

———. "On the Mathematical Law of the Cycle." *Herapath's Railway Magazine and the Annals of Science* 5 (1838): 378–80.

———. *Theory of Investment in Railway Securities*. London: John Weale, 1846.

Clavel, Damian. "Gregor MacGregor, the Poyais Bonds and the Latin American Sovereign Debt Market, 1822–1823: Financial Fraud or Failure?" MSc diss., University of Edinburgh, 2013.

Cohen, B. *Compendium of Finance containing an account of the origin, progress, and present state, of the public debts, revenue, expenditure, national banks and currencies of France, Russia, Prussia . . . and shewing the nature of the different public securities, with the manner of making investments therein*. London: Phillips, 1822.

Cohn, B. S. *Colonialism and Its Forms of Knowledge: The British in India*. Princeton, NJ: Princeton University Press, 1996.

Collinson, John. "Explorations in Central America Accompanied by Survey and Levels from Lake Nicaragua to the Atlantic Ocean." *Proceedings of the Royal Geographical Society of London* 12 (1868): 25–46.

———. "The Indians of the Mosquito Territory." *Memoirs Read before the Anthropological Society of London* 3 (1870): 148–56.

———. "On the Indians of the Mosquito Territory." *Journal of the Anthropological Society of London* 6 (1868): xiii–xvii.

Collinson, John, and B. Pim. *Descriptive Account of Captain Bedford Pim's Project for an International Atlantic and Pacific Junction Railway across Nicaragua*. London: J. E. Taylor, 1866.

Colonization and Commercial Co. of Bolivia. *Bolivian Colonization: Being a Prospectus of the Colonization and Commercial Co. of Bolivia. Incorporated in San Francisco, California, January 25th, A.D. 1870*. San Francisco: Alta California, 1870.

Committee of the Holders of Spanish Bonds, *Spanish Bonds: A Full Report of the Public Meeting Held on August 19, 1851, Further to Consider Bravo Murillo's Confiscatory Scheme for the Arrangement of the Debt, &c.; Letters of J. Capel, Sir H. Bayly and Mr. Tasker; Memorial of Mr. Wadsworth, with Lord Palmerston's Reply; Criticisms of the English Press etc.* London: John King, [1853?].

Conklin, A. L. *In the Museum of Man: Race, Anthropology, and Empire in France, 1850–1950.* Ithaca, NY: Cornell University Press, 2013.

Conrad, Joseph, *Lord Jim, a Tale.* London: Blackwood, 1900.

———. *Nostromo, a Tale of the Seaboard.* New York: Harper & Bros., 1904.

Coombes, Rebecca. "Fraternal Communications: The Rise of the English Masonic Periodical." 2012, http://www.masonicperiodicals.org/.

Coote, C. Holmes. *A Report on Some of the More Important Points Connected with the Treatment of Syphilis.* London: John Churchill, 1857.

———. "Pim, Bedford Clapperton Trevelyan (1826–1886)." In *Dictionary of National Biography*, vol. 45, edited by Sidney Lee, 306–7. London: Smith, Elder, 1896.

Costeloe, Michael P. *Bonds and Bondholders: British Investors and Mexico's Foreign Debt, 1824–1888.* Westport, CT: Praeger, 2003.

Cowell, F. R. *The Athenaeum Club and Social Life in London.* London: Heinemann, 1975.

Craig, Neville B. *Recollections of an Ill-Fated Expedition to the Headwaters of the Madeira River in Brazil.* Philadelphia: J. B. Lippincott, 1907.

Cromer (Earl of) [Evelyn Baring]. "On the Government of Subject Races." *Edinburgh Review* (1908). [Reprinted in Earl of Cromer, *Political and Literary Essays, 1908–1913.* London: Macmillan, 1913, 3–53.]

Darwin, C. *The Descent of Man, and Selection in Relation to Sex.* London: John Murray, 1872.

Daunton, M. "'Gentlemanly Capitalism' and British Industry, 1820–1914." In *State and Market in Victorian Britain: War, Welfare and Capitalism.* Woodbridge, UK: Boydell, 2008.

Davis, Joseph Barnard. "Anthropology and Ethnology." *Anthropological Review* 6 (1868): 394–99.

Dawson, Frank Griffith. *The First Latin American Debt Crisis: The City of London and the 1822–25 Loan Bubble.* New Haven, CT: Yale University Press, 1990.

———. "MacGregor, Gregor (1786–1845), Soldier and Adventurer." In *Oxford Dictionary of National Biography.* Oxford: Oxford University Press, 2004.

de Gobineau, A. *The Moral and Intellectual Diversity of Races, with Particular Reference to their Respective Influence in the Civil and Political History of Mankind,* Philadelphia: J. B. Lippincott and Co., 1856.

De Kalb, C. "A Bibliography of the Mosquito Coast of Nicaragua." *Journal of the American Geographical Society of New York* 26 (1894): 241–48.

Dennis, Philip A., and Michael D. Olien. "Kingship among the Miskito." *American Ethnologist* 11 (1984): 718–37.

Desmond, Adrian, and James Moore. *Darwin's Sacred Cause: Race, Slavery and the Quest for Human Origins.* London: Houghton Mifflin, 2009.

Devereux, William Cope. *A Cruise in the "Gorgon"; or, Eighteen Months on H.M.S. "Gorgon" Engaged in Suppressing the Slave Trade on the East Coast of Africa. Including a Trip up the Zambezi with Dr. Livingstone.* London: Bell & Daldy, 1869.

Disraeli, Benjamin. *An Inquiry into the Plans, Progress, and Policy of the American Mining Companies.* London: John Murray, 1825.

Draper, Nicholas. *The Price of Emancipation: Slave-Ownership, Compensation and British Society at the End of Slavery.* Cambridge, UK: Cambridge University Press, 2010.

Drayton, Richard. "Knowledge and Empire." In *The Oxford History of the British Empire,* vol. 2, *The Eighteenth Century,* edited by P. J. Marshall, 72–82. Oxford: Oxford University Press, 1998.

———. "Maritime Networks and the Making of Knowledge." In *Empire, The Sea and Global History: Britain's Maritime World, c. 1760–c. 1840,* edited by D. Cannadine, 231–52. New York: Palgrave Macmillan, 2007.

Drummond Wolff, Henry. *Rambling Recollections.* 2 vols. London: Macmillan, 1908.

Duguid, Charles. *The Story of the Stock Exchange: Its History and Position.* London: Grant Richards, 1901.

Eastwick, Edward B. *Venezuela; or, Sketches of Life in a South American Republic, with the History of the Loan of 1864.* London: Chapman & Hall, 1868.

Edwards, J. R. *A History of Financial Accounting.* London: Routledge, 1989.

Ellingson, Terry Jay. *The Myth of the Noble Savage.* Berkeley: University of California Press, 2001.

Elliot, Arthur D. *Life of George Joachim Goschen, First Viscount Goschen, 1831–1907.* 2 vols. London: Longmans, 1911.

Endersby, J. *Imperial Nature: Joseph Hooker and the Practices of Victorian Science.* Chicago: University of Chicago Press, 2008.

Evans, David Morier. *City Men and City Manners.* London: Groombridge, 1852.

———. *Facts, Failures and Frauds: Revelations, Financial, Mercantile, Criminal.* London: Groombridge, 1859.

Ewald, Charles. *The Right Hon. Benjamin Disraeli, Earl of Beaconsfield KG and His Times.* 2 vols. Edinburgh, UK: William McKenzie, 1881.

Eyre, E. J. *Journals of Expeditions of Discovery into Central Australia and Overland from Adelaide to King George Sound.* 2 vols. London: Boone, 1845.

Feuchtwanger, E. J. "Spofforth, Markham (1825–1907)." In *Oxford Dictionary of National Biography.* Oxford: Oxford University Press, 2004.

Finkelstein, David. *The House of Blackwood: Author–Publisher Relations in the Victorian Era.* University Park: Pennsylvania State University Press, 2012.

Flandreau, Marc. "Audits, Accounts, and Assessments: Coping with Sovereign Risk under the International Gold Standard, 1871–1913." In *International Financial History in the Twentieth Century: System and Anarchy,* edited by M. Flandreau, C.-L. Holtfrerich, and H. James. New York: Cambridge University Press, 2003.

———. "Sovereign States, Bondholders Committees, and the London Stock Exchange in the Nineteenth Century (1827–68): New Facts and Old Fictions." *Oxford Review of Economic Policy* 29 (2013): 668–96.

————. "The Vanishing Banker." *Financial History Review* 19 (2012): 1–19.

Flandreau, Marc, and Juan Flores. "Bonds and Brands: Origins of Foreign Debt Markets." *Journal of Economic History* 69 (2009): 646–84.

————. "The Peaceful Conspiracy: Bond Markets and International Relations During the Pax Britannica." *International Organization* 66 (2012): 211–41.

Flandreau, M., J. Sławatyniec, and Alberto Gamboa. "The Dart and Elliott Associates of Victorian Britain: Foreign Debt Vulture Investors in the London Stock Exchange, 1827–1866." Working paper, Graduate Institute of International Studies, forthcoming.

Flandreau, M., and Frédéric Zumer. *The Making of Global Finance.* Paris: Organisation for Economic Co-operation and Development, 2004.

Flynn, John Steven. *Sir Robert N. Fowler, Bart., M.P.: A Memoir.* London: Hodder & Stoughton, 1893.

Fodor, G. "The Boom That Never Was? Latin American Loans in London, 1822–1825." Discussion Paper no. 5. Universita' degli studi di Trento, 2002.

Fortes, Meyer, and Edward Evan Evans-Pritchard. *African Political Systems.* London: Oxford University Press, 1940.

Fortune, T. *Fortune's Epitome of the Stocks and Public Funds: Containing Every Necessary Information for Perfectly Understanding the Nature of Those Securities, and the Mode of Doing Business Therein: With a Full Account of All the Foreign Funds and Loans.* London: Sherwood, Gilbert & Piper, 1833.

Fyfe, A., and B. Lightman, eds. *Science in the Marketplace: Nineteenth-Century Sites and Experiences.* Chicago: University of Chicago Press, 2007.

Galton, Francis, *Hereditary Genius.* London: Macmillan, 1869

————. "On Composite Portraits." *Journal of the Anthropological Institute* 8 (1879): 132–44.

Gallagher, John, and Ronald Robinson, "The Imperialism of Free Trade." *Economic History Review* 6 (1953): 1–15.

Galtung, Johan. "Scientific Colonialism." *Transition* 30 (1967): 11–15.

Gondermann, Thomas. "Die Etablierung der Evolutionslehre in der Viktorianischen Anthropologie: Die Wissenschaftspolitik des X-Clubs, 1860–1872." *NTM Zeitschrift für Geschichte der Wissenschaften, Technik und Medizin* 16 (2008): 309–31.

Graham, Walter James. "*The Athenaeum.*" In *English Literary Periodicals*, 317–21. New York: Nelson, 1930.

Green, Abigail. *Moses Montefiore: Jewish Liberator, Imperial Hero.* Cambridge, MA: Belknap, 2010.

Griffiths, Dennis. *Encyclopedia of the British Press, 1422–1992.* London: Macmillan, 1992.

————. *Plant Here the "Standard."* London: Palgrave Macmillan, 1995.

Gripari, Pierre. *Tales of the Rue Broca.* Indianapolis, IN: Bobbs-Merrill, 1969.

Gross, John. *The Rise and Fall of the Man of Letters: A Study of the Idiosyncratic and the Humane in Modern Literature.* New York: Macmillan, 1970.

Grove, Richard H. *Green Imperialism: Colonial Expansion, Tropical Island Edens and the Origins of Environmentalism, 1600–1860.* Cambridge, UK: Cambridge University Press, 1995.

Gugliotta, A. C. "Tavares Bastos (1839–1875) e a Sociedade Internacional de Imigraçao: Um Espaço a Favor da Modernidade." Manuscript presented at the "Usos de Passado" conference, available at http://www.rj.anpuh.org/resources/rj/Anais/2006/conferencias /Alexandre%20Carlos%20Gugliotta.pdf.

Hall, Catherine. "The Economy of Intellectual Prestige: Thomas Carlyle, John Stuart Mill, and the Case of Governor Eyre." *Cultural Critique* 12 (1989): 167–96.

———. "Imperial Man: Edward Eyre in Australasia and the West Indies, 1833–1866." In *The Expansion of England: Race, Ethnicity and Cultural History*, edited by Bill Schwarz, 129–68. London: Routledge, 1996.

Hanham, H. J. *Elections and Party Management: Politics in the time of Disraeli and Gladstone*, London: Longmans, 1959.

Hanke, Lewis. *A Note on the Life and Publications of Colonel George Earl Church.* Providence, RI: Friends of the Library of Brown University, 1965.

Harcourt, Freda. "Disraeli's Imperialism, 1866–1868: A Question of Timing." *Historical Journal* 23 (1980): 87–109.

Harland-Jacobs, Jessica, *Builders of Empire: Freemasons and British Imperialism, 1717–1927.* Chapel Hill: University of North Carolina Press, 2007.

Harries, Patrick. *Butterflies and Barbarians. Swiss Missionaries and Systems of Knowledge in South-East Africa.* Athens: Ohio University Press, 2007.

———. "From the Alps to Africa: Swiss Missionaries and Anthropology." In *Ordering Africa: Anthropology, European Imperialism, and the Politics of Knowledge*, edited by Helen Tilley and Robert J. Gordon, 201–24. Manchester, UK: Manchester University Press, 2007.

Haslewood, Edward. *New Colonies on the Uplands of the Amazon.* London: King, 1863.

Heartfield, James. *The Aborigines' Protection Society: Humanitarian Imperialism in Australia, New Zealand, Fiji, Canada, South Africa, and the Congo, 1837–1909.* New York: Columbia University Press, 2011.

Hecht, S. B. *The Scramble for the Amazon and the "Lost Paradise" of Euclides da Cunha.* Chicago: University of Chicago Press, 2013.

Hendricks, H. *A Plain Narrative of Facts.* London: Stephen Couchman, 1824.

Henty, G. A. *The March to Magdala.* London: Tinsley Bros., 1868.

Herschel, J. F. W. *Manual of Scientific Inquiry Prepared for the Use of Her Majesty's Navy and Adapted for Travellers in General.* London: John Murray, 1849.

Heuman, Gad. *The "Killing Time": The Morant Bay Rebellion in Jamaica.* Knoxville: University of Tennessee Press, 1994.

Hewitt, Martin. *The Dawn of the Cheap Press in Victorian Britain: The End of the Taxes on Knowledge, 1849–1869.* London: Bloomsbury, 2014.

Hilton, Boyd. *The Age of Atonement: The Influence of Evangelicalism on Social and Economic Thought, ca. 1795–1865.* Oxford: Oxford University Press, 1988.

Ho, Karen. *Liquidated: An Ethnography of Wall Street.* Durham, NC: Duke University Press, 2009.

———. "Situating Global Capitalism: A View from Wall Street Investment Banks." *Cultural Anthropology* 20 (2005): 68–96.

Hodgson, B, H. *Essays on the Languages, Literature, and Religion of Nepal and Tibet: Together with Further Papers on the Geography, Ethnology, and Commerce of Those Countries.* London: Trübner, 1874.

Holt, T. C. *The Problem of Freedom: Race, Labor, and Politics in Jamaica and Britain, 1832–1938.* Baltimore, MD: Johns Hopkins University Press, 1992.

Holtfrerich, C. L. *Frankfurt as an International Financial Center: From Medieval Trade Fair to European Trading Center.* Munich: Beck, 1999.

Hooker, J. D. "On the Struggle for Existence amongst Plants." *Popular Science Review* 6 (1867): 131–39.

Hooker, J. R. "The Foreign Office and the 'Abyssinian Captives.'" *Journal of African History* 2 (1961): 245–58.

Hoover, Nora Kelly. "Victorian War Correspondents G. A. Henty and H. M. Stanley: The 'Abyssinian' Campaign 1867–1868." Unpub. Ph.D. diss., Florida State University, 2005.

Horn, Ignace Edouard. *Annuaire international du crédit public.* Paris: Guillaumin, 1861.

Howitt, Caroline. A. "Stevenson and Economic Scandal." *Journal of Stevenson Studies* 7 (2010): 143–56.

Howsam, Leslie. *Kegan Paul: A Victorian Imprint; Publishers, Books and Cultural History.* London: Kegan Paul, 1998.

Hozier, Henry. *The British Expedition to Abyssinia, Compiled from Authentic Documents.* London: Macmillan, 1869.

Hughes, J. R. T. *Fluctuations in Trade, Industry and Finance: A Study of British Economic Development, 1850–1860.* Oxford: Clarendon, 1960.

Hunt, Frederick Knight. *The Fourth Estate: Contributions towards a History of Newspapers, and of the Liberty of the Press.* 2 vols. London: Bogue, 1850.

Hunt, James. *The Negro's Place in Nature.* London: Trübner, 1863.

Ionesco, Eugène. *Théâtre, I.* Paris: Gallimard, 1954.

James, William. *The Meaning of Truth (a Sequel to Pragmatism).* London: Longman, Green, 1909.

Jameson, J. F. "The London Expenditures of the Confederate Secret Service." *American Historical Review* 35 (1930): 811–24.

Jenks, L. H. *Migration of British Capital to 1875.* New York: Knopf, 1927.

Jerrold, Blanchard. *The Best of all Good Company. [A day with] Dickens, Scott, Lytton, Disraeli, Thackeray, Douglas Jerrold. With portraits, etc. First series.* London: Houlston, 1871.

———. "On the Manufacture of Public Opinion." *Nineteenth Century* 13 (1883): 1080–92.

Jevons, W. S. *Investigation in Currency and Finance.* London: Macmillan, 1884.

Kapil, Raj. *Circulation and the Construction of Knowledge in South Asia and Europe, 1650–1900.* Basingstoke, UK: Palgrave Macmillan, 2007.

Kebbel, T. E., ed. *Selected Speeches of the Earl of Beaconsfield.* 2 vols. London: Longmans Green, 1882.

Keller, J., and F. Keller. *Exploración del Rio Madera en la Parte Comprendida Entre la Cachuela de San Antonio y la Embocadura del Mamoré.* La Paz: Imprenta de la Union Americana, 1870.

———. "Relation of the Exploration of the River Madeira." In *Explorations Made in the Valley of the River Madeira, from 1749 to 1868*, edited by George Earl Church, 1–76. London: National Bolivian Navigation Company, 1875.

Kellett, E. E. "The Press." In *Early Victorian England*, vol. 2, edited by G. M. Young. Oxford: Oxford University Press, 1934.

Kelly, John D., B. Jauregui, S. T. Mitchell, and J. Walton, eds. *Anthropology and Global Counterinsurgency*. Chicago: University of Chicago Press, 2010.

Kennedy, Dane, *The Magic Mountains: Hill Stations and the British Raj*. Berkeley: University of California Press, 1996.

Kennedy, K. *The Highly Civilized Man: Richard Burton and the Victorian World*. Cambridge, MA: Harvard University Press, 2007.

Kiernan, V. G. "Foreign Interests in the War of the Pacific." *Hispanic American Historical Review* 35 (1955): 14–36.

Kirchberger, Ulrike. *Aspekte Deutsch–Britischer Expansion: Die Überseeinteressen der Deutschen Migranten in Großbritannien in der Mitte des 19. Jahrhunderts*. Stuttgart: Steiner Verlag, 1999.

Krader, Lawrence. "Karl Marx as Ethnologist." *Transactions of the New York Academy of Sciences*, ser. 2, 35 (1973): 304–13.

Kuklick, Henrika. "The British Tradition." In *A New History of Anthropology*, edited by H. Kuklick, 52–78. London: Blackwell, 2009.

———. *The Savage Within: The Social History of Anthropology, 1885–1945*. Cambridge, UK: Cambridge University Press, 1991.

Kuper, Adam. *Anthropology and Anthropologists, the Modern British School*. 3rd ed. London: Routledge, 1996.

Kynaston, David. *The City of London: A World of Its Own, 1815–1890*. London: Pimlico, 1995.

Laing-Meason, Malcom Ronald. *The Bubbles of Finance: Joint Stock Companies, Modern Commerce, Money Lending and Life-Insuring, by a City Man*. London: Sampson Low, 1865.

———. *The Profits of Panics: Showing How Financial Storms Arise, Who Make Money by Them, Who Are the Losers, and Other Revelations of a City Man*. London: Sampson Low, 1866.

Lamoreaux, N. *Insider Lending: Banks, Personal Connections, and Economic Development in Industrial New England*. Cambridge, UK: Cambridge University Press, 1996.

Landes, D. *Bankers and Pashas: International Finance and Economic Imperialism in Egypt*. Cambridge, MA: Harvard University Press, 1958.

Leppert, Richard, and Bruce Lincoln. "Introduction to a Special Issue—Discursive Strategies and the Economy of Prestige—of *Cultural Critique*." *Cultural Critique* 12 (1989): 5–23.

L'Estoile, Benoît de, Federico Neiburg, and Lygia Sigaud. *Empires, Nations, and Natives: Anthropology and State-Making*. Durham, NC: Duke University Press, 2005.

Levi, Leone. "Of the Progress of Learned Societies Illustrative of the Advancement of Science in the United Kingdom during the Past Thirty Years." In *Report of the Thirty-Eighth Meeting of the British Association for the Advancement of Science*, 169–73. London: John Murray, 1868.

———. "The Scientific Societies in Relation to the Advancement of Science in the United Kingdom." In *Report of the Forty-Ninth Meeting of the British Association for the Advancement of Science*, 458–68. London: John Murray, 1879.

Lewis, Martin W., and Kären Wigen. *The Myth of Continents: A Critique of Metageography*. Berkeley: University of California Press, 1997.

Livingstone, D. N. "Post-Darwinian British and American Geography." In *Geography and Empire*, edited by A. Godlewska and N. Smith, 134–35. London: Blackwell, 1994.

Lobban, Michael. "Commercial Morality and the Common Law: or, Paying the Price of Fraud in the Later Nineteenth Century?" In *Legitimacy and Illegitimacy in Nineteenth Century Law, Literature and History*, edited by M. Finn, M. Lobban, and J. Taylor, 119–47. London: Palgrave, 2010.

Lobingier, Charles S. *Supreme Council 33rd Degree Part 1 or Mother Council of the World of the Ancient and Accepted Scottish Rite of Freemasonry, Southern Jurisdiction, United States of America*. Louisville, KY: Standard Printing, 1931.

Lorimer, D. *Colour, Class and the Victorians: English Attitudes to the Negro in the Mid-Nineteenth Century*. Leicester, UK: Leicester University Press, 1978.

Mackenzie, Kenneth R. H. *Royal Masonic Cyclopaedia*. 2 vols. London: Kenneth R. H. Mackenzie, 1877.

Maginn, W. *Whitehall, or the Days of George IV*. London: Marsh, 1827.

Malchow, H. L. "God and the City: Robert Fowler (1828–91)." In *Gentlemen Capitalists: The Social and Political World of the Victorian Businessman*. Palo Alto, CA: Stanford University Press, 1992.

Malinowski, Bronislaw. "Practical Anthropology." *Africa: Journal of the International African Institute* 2 (1929): 22–38.

———. *Sexual Life of Savages in North-Western Melanesia*. London: Routledge & Sons, 1929.

Marchand, L. A. *The Athenaeum: A Mirror of Victorian Culture*. Chapel Hill: University of North Carolina Press, 1941.

Markham, Sir Albert H. *The Life of Sir Clements R. Markham, FRS, KCB*. London: John Murray, 1917.

Markham, Clements R., ed. *Arctic Geography and Ethnology: A Selection of Papers on Arctic Geography and Ethnology*. London: John Murray, 1875.

———. *The Fifty Years' Work of the Royal Geographical Society*. London: John Murray, 1881.

———. "On Crystal Quartz Cutting Instruments of the Ancient Inhabitants of Chanduy (near Guayaquil in South America)." *Journal of the Anthropological Society of London* 2 (1864): lvii–xli.

Marx, Karl. *Capital: A Critique of Political Economy*. 3 vols. Chicago: C. H. Kerr, 1909.

Mason, Joan. "The Admission of the First Women to the Royal Society of London." *Notes and Records of the Royal Society of London* 46 (1992): 279–300.

Mathew, W. M. "The First Anglo-Peruvian Debt and Its Settlement, 1822–49." *Journal of Latin American Studies* 2 (1970): 81–98.

Mathews, Edward D. *Up the Amazon and Madeira Rivers, through Bolivia and Peru*. London: Sampson Low, 1879.

Matthews, Elizabeth. *From the Canadian Arctic to the President's Desk:* HMS Resolute *and How She Prevented a War*. London: Alto, 2007.

Maury, M. F. *Letter on the Physical Geography of the Nicaraguan Railway Route, from M. F. Maury to Bedford Pim.* London: J. E. Taylor, 1866.

McLeod, R., and P. Collins. *The Parliament of Science: The British Association for the Advancement of Science, 1831–1981.* Northwood, UK: Science Reviews, 1981.

McMillan, N. "Robert Bruce Napoleon Walker, F.R.G.S., F.A.S., F.G.S., C.M.Z.S. (1832–1901), West African Trader, Explorer and Collector of Zoological Specimens." *Archives of Natural History* 23 (1996): 124–41.

Meszaros, P. F. "The Corporation of Foreign Bondholders and British Diplomacy in Egypt 1876 to 1882: The Efforts of an Interest Group in Policy-Making." Ph.D. diss., Loyola University Chicago, 1973.

Michie, R. C. *The London Stock Exchange: A History.* Oxford: Oxford University Press, 1999.

———. "The Social Web of Investment in the 19th Century." *Revue Internationale d'Histoire de la Banque* 18 (1979): 158–75.

Midgley, C. *Women against Slavery: The British Campaigns, 1780–1870.* London: Routledge, 1995.

Mitchell, Timothy. *Rule of Experts: Egypt, Techno-Politics, Modernity.* Berkeley: University of California Press, 2002.

Monypenny, W. F., and G. E. Buckle. *The Life of Benjamin Disraeli, Earl of Beaconsfield.* 6 vols. New York: Macmillan, 1916–20.

Moore, J. B. *History and Digest of the International Arbitrations to Which the United States Has Been a Party.* 6 vols. Washington, DC: Government Printing Office, 1898.

Morris, Robert, *Freemasonry in the Holy Land, or Handmarks of Hiram's Builders.* Chicago: Knight and Leonard, 1876.

Morris, R. J. "Clubs, Societies and Associations." In *The Cambridge Social History of Britain, 1750–1950,* vol. 3, *Social Agencies and Institutions,* edited by F. M. L. Thompson, 395–444. Cambridge, UK: Cambridge University Press, 1992.

———. "Voluntary Societies and British Urban Elites, 1780–1850: An Analysis." *Historical Journal* 26 (1983): 95–118.

Murchison, J. H. *British Mines Considered as Means of Investment.* London: Mann, 1854.

Murray, J., and P. Hubert. *Anthropology and the Government of Subject Races.* Port Moresby, Papua New Guinea: Edward George Baker, 1921.

Myatt, F. *The March to Magdala: The Abyssinian War of 1868.* London: Leo Cooper, 1970.

Napoleon III (Louis Napoléon Bonaparte). *Le Canal de Nicaragua.* In *Oeuvres de Napoléon III.* Paris: Plon, 1856.

Naylor, R. A. *Penny Ante Imperialism: The Mosquito Shore and the Bay of Honduras, 1600–1914: A Case Study in British Informal Empire.* London: Fairleigh Dickinson University Press, 1989.

Newman, Aubrey. "Masonic Journals in Mid-Victorian-Britain." Masonic Press Project, King's College, 2012, http://www.masonicperiodicals.org/.

Newman, James L. *Paths without Glory: Richard Francis Burton in Africa.* Dulles, VA: Potomac Books, 2009.

Odlyzko, A. "The Collapse of the Railway Mania, the Development of Capital Markets and the Forgotten Role of Robert Lucas Nash." *Accounting History Review* 21 (2011): 309–45.

Olien, Michael. D. "E. G. Squier and the Miskito: Anthropological Scholarship and Political Propaganda." *Ethnohistory* 32 (1985): 111–33.

———. "Were the Miskito Indians Black? Ethnicity, Politics and Plagiarism in the Mid-Nineteenth Century." *New West Indian Guide* 62 (1988): 27–50.

Oliver, Samuel Pasfield. *Description of Two Routes through Nicaragua*. Gosport, UK: J. P. Legg, 1867.

———. *Off Duty. Rambles of a Gunner through Nicaragua, January to June 1867*. London: Taylor & Francis, 1879.

Olivier de Sardan, Jean-Pierre. *Anthropology and Development: Understanding Contemporary Social Change*. New York: Zed Books, 2005.

Olson, J. S. *The Indians of the Central and South America: An Ethno-Historical Dictionary*. Westport, CT: Greenwood, 1991.

Patterson, Thomas C. *A Social History of Anthropology in the United States*. Oxford: Berg, 2001.

Patton, Mark. *Science, Politics and Business in the Work of Sir John Lubbock: A Man of Universal Mind*. London: Ashgate, 2007.

Pels, Peter. "Professions of Duplexity: A Prehistory of Ethical Codes in Anthropology." *Current Anthropology* 40 (1999): 101–36.

Pels, Peter, and Oscar Salemink. *Colonial Subjects: Essays on the Practical History of Anthropology*. Ann Arbor: University of Michigan Press, 2000.

Perkin, Harold. *Origins of Modern British Society, 1780–1880*. London: Routledge, 1969.

Pi de Cosprons, Honoré [calling himself Duc du Roussillon]. *Mémoire sur l'origine Scytho-Cimmérienne de la langue Romane*. London: Author, 1863.

———. *Origines, migrations, philologie, et monuments antiques*, 2 vols. Paris: Lacroix, 1867.

Pietz, William. "The Problem of the Fetish, I." *Res* 9 (1985): 5–17.

———. "The Problem of the Fetish, II: The Origin of the Fetish." *Res* 13 (1987): 23–45.

———. "The Problem of the Fetish, III: Bosman's Guinea and the Enlightenment Theory of Fetishism." *Res* 16 (1988): 105–23.

Pim, Bedford. "Extract from a Paper by Commander Bedford Pim, R. N., on a New Transit-Route through Central America." *Proceedings of the Royal Geographical Society of London* 6 (1861): 113.

———. *Gate of the Pacific*. London: Lovel, Reeve, 1863.

———. *The Negro and Jamaica*. London: Trübner, 1866.

———. "Proposed Interoceanic and International Transit Routes through Central America." In *Report of the Thirty-Third Meeting of the British Association for the Advancement of Science, Held at Newcastle upon Tyne in August and September 1863*, 143–45. London: John Murray, 1864.

———. "Remarks on the Isthmus of Suez, with Special Reference to the Proposed Canal." *Proceedings of the Royal Geographical Society of London* 3 (1858/59): 177–206.

Pim, Bedford, and Berthold Seemann. *Dottings on the Roadside, in Panama, Nicaragua, and Mosquito*. London: Chapman & Hall, 1869.

Pitt-Rivers, Augustus H. [Lane Fox]. *An Address Delivered at the Opening of the Dorset County Museum*. Dorchester, UK: James Foster, 1884.

———. *The Evolution of Culture, and Other Essays.* Oxford: Clarendon, 1906.

———. *Primitive Warfare: Illustrated by Specimens from the Museum of the Institution; A Lecture at the Royal United Service Institution.* London: United Service Institution, 1867.

Platt, D. C. M. *The Cinderella Service: British Consuls since 1825.* London: Longman, 1971.

———. *Finance Trade and Politics in British Foreign Policy, 1815–1914.* Oxford: Clarendon Press, 1968.

Pouchet, G. *The Plurality of the Human Race.* London: Longman, Green, Longman & Roberts, 1864.

Proudon, P.-J. *Manuel du Spéculateur à la Bourse,* 5th ed. Paris: Garnier, 1857.

Quevedo, Q. "The Madeira and Its Head-Waters." In *Explorations Made in the Valley of the River Madeira, from 1749 to 1868,* edited by George Earl Church, 167–88. London: National Bolivian Navigation Company, 1875.

Qureshi, Sadiah. *Peoples on Parade: Exhibitions, Empire, and Anthropology in Nineteenth-Century Britain.* Chicago: University of Chicago Press, 2011.

Rainger, Ronald. "Race, Politics, and Science: The Anthropological Society of London in the 1860s." *Victorian Studies* 22 (1978): 51–70.

Ray, Alfred. 1873, *Bolivian Loan and the Contract of the Public Works Construction Company: For the Building of the Madeira and Mamoré Railway.* London: Dunlop, 1873.

Reade, W. Winwood "Efforts of Missionaries among Savages." *Journal of the Anthropological Society of London* 3 (1865): clxiii–clxxxiii.

———. *Savage Africa: Being the Narrative of a Tour in Equatorial, Southwestern and Northwestern Africa.* New York: Harper Bros., 1864.

Reinhart, C. M., and K. S. Rogoff. "Serial Default and the 'Paradox' of Rich-to-Poor Capital Flows." *American Economic Review* 94 (2004): 53–58.

———. *This Time Is Different: Eight Centuries of Financial Folly.* Princeton, NJ: Princeton University Press, 2009.

Reining, Conrad C. "A Lost Period of Applied Anthropology." *American Anthropologist* 64 (1962): 593–600.

Richards, Evelleen. "Huxley and Woman's Place in Science: The 'Woman Question' and the Control of Victorian Anthropology." In *History, Humanity and Evolution: Essays for John C. Greene,* edited by James R. Moore, 253–84. Cambridge, UK: Cambridge University Press, 1989.

———. "The 'Moral Anatomy' of Robert Knox: The Interplay between Biological and Social Thought in Victorian Scientific Naturalism." *Journal of the History of Biology* 22 (1989): 373–436.

Richardson, Christopher. *Mr. John Diston Powles: or, the Antecedents, as a Promoter and Director of Foreign Mining Companies, of an Administrative Reformer.* London: Author, 1855.

Robb, G. *White-Collar Crime in Modern England: Financial Fraud and Business Morality, 1845–1929.* Cambridge, UK: Cambridge University Press, 1992.

Rocher, Rosane, and Ludo Rocher. *The Making of Western Indology: Henry Thomas Colebrooke and the East India Company.* London: Routledge, 2012.

Rodgers, Nini. "The Abyssinian Expedition of 1867–1868: Disraeli's Imperialism or James Murray's War." *Historical Journal* 27 (1984): 129–49.

Rosenthal, Eric, *On 'Change through the Years: A History of Share Dealing in South Africa*. Johannesburg: Flesch Financial Publications, 1968.

Rubenson, S. *King of Kings: Tewodros of Ethiopia*. Addis Ababa: Haile Sellassie University, Nairobi: Oxford University Press, 1966.

Rubinstein, W. D. "Jewish Top Wealth-Holders in Britain, 1809–1909." *Jewish Historical Studies* 37 (2002): 133–61.

———. *Men of Property: The Very Wealthy in Britain since the Industrial Revolution*. Rutgers, NJ: Rutgers University Press, 1981.

Said, Edward. *Joseph Conrad and the Fiction of Autobiography*. Cambridge, MA: Harvard University Press, 1966.

———. *Orientalism*. New York: Pantheon Books, 1978.

St. George, Andrew, *A History of Norton Rose*, Cambridge: Norton Rose, 1995.

Salit, Charles R. "Anglo-American Rivalry in Mexico, 1823–1830." *Revista de Historia de América* 16 (1943): 65–84.

Salvucci, R. "Capitalism and Slavery: Not What Eric Williams Had in Mind." H-LatAm, 2012, http://www.h-net.org/reviews/showrev.php?id=31959.

———. *Politics, Markets and Mexico's "London Debt," 1823–1887*. Cambridge, UK: Cambridge University Press, 2009.

Sampson, Marmaduke Blake. *Central America and the Transit between the Oceans*. New York: S. W. Benedict, 1850.

———. *Criminal Jurisprudence Considered in Relation to the Physiology of the Brain*. London: Samuel Highley, 1851.

Sandage, Scott. *Born Losers: A History of Failure in America*. Cambridge, MA: Harvard University Press, 2009.

Sandburg, Carl. *Storm over the Land. A Profile of the Civil War Taken Mainly from Abraham Lincoln: The War Years*. New York: Harcourt, Brace & World, 1942.

Sayers, R. S. *Gilletts in the London Money Market, 1867–1967*. London: Clarendon, 1968.

Schaffer, Simon, Lissa Roberts, Kapil Raj, and James Delbourgo. *The Brokered World: Go-Betweens and Global Intelligence, 1770–1820*. Sagamore Beach, MA: Science History Publications, 2009.

Schumpeter, J. *History of Economic Analysis*. Oxford: Oxford University Press, 1954.

Schwarz, William. *The Expansion of England: Race, Ethnicity and Cultural History*. London: Routledge, 1996.

Scott, J. D. "Hambro and Cavour." *History Today* 19 (1969): 10.

Scratchley, Arthur. *On Average Investment Trusts*. London: Shaw & Sons, 1875.

Scroggs, William O. *Filibusters and Financiers: The Story of William Walker and His Associates*. New York: Macmillan, 1916.

Scully, William. *Brazil: Its Provinces and Chief Cities: The Manners and Customs of the People etc*. London: John Murray, 1866.

Sebrell, Thomas E., *Persuading John Bull: Union and Confederate Propaganda in Britain, 1860–65*, Lanham, MD: Lexington Books, 2014.

Seed, Patricia. *American Pentimento: The Invention of Indians and the Pursuit of Riches*. Minneapolis: University of Minnesota Press, 2001.

———. *Ceremonies of Possession in Europe's Conquest of the New World, 1492–1640.* New York: Cambridge University Press, 1995.

Seemann, Berthold. *History of the Isthmus of Panama.* Panama City: Star & Herald, 1867.

———. *Narrative of the Voyage of* HMS Herald. 2 vols. London: Reeve, 1853.

Semmel, Bernard. *The Governor Eyre Controversy.* Jamaica: MacGibbon & Kee, 1962.

Sera-Shriar, Efram. *The Making of British Anthropology, 1813–1871.* London: Pickering & Chatto, 2013.

———. "Observing Human Differences: James Hunt Thomas Huxley and Competing Disciplinary Strategies in the 1860s." *Annals of Science* 70 (2012): 1–31.

Shapin, S. *A Social History of Truth: Civility and Science in Seventeenth-Century England.* Chicago: University of Chicago Press, 1994.

Sillitoe, P. "The Role of Section H at the British Association for the Advancement of Science in the History of Anthropology." *Durham Anthropology Journal* 13 (2005): 1–17.

Silva Ferro, Ramón de. *Historical Account of the Mischances in Regard to the Construction of a Railway across the Republic of Honduras; the Failure of the Loans Solicited . . . and the Principal Difficulties Which Have Occurred.* London: C. F. Hodgson & Son, 1875.

Sinclair, D. *The Land That Never Was: Sir Gregor MacGregor and the Most Audacious Fraud in History.* Cambridge, MA: Da Capo, 2004.

Smart, P. E. "John Lubbock." In *Dictionary of Business Biography,* vol. 3, edited by D. J. Jeremy, 873–76. London: Butterworths, 2004.

Smith, G. Barnett. "Beke, Charles Tilstone, 1800–1874." In *Dictionary of National Biography,* vol. 4, edited by Leslie Stephen, 138–40. London: Smith Elder, 1885.

Soltau-Pim, Sophia. *Job and Fugitive Pieces.* London: Gee, 1885.

Spencer, Herbert. "Morals of Trade." *Westminster Review* (April 1859). [Reprinted in H. Spencer, *Essays: Moral, Political and Aesthetic.* New York: Appleton, 1881, 107–48.]

———. *Railway Morals and Railway Policy.* London: Longman, Brown, Green & Longmans, 1855.

Spevack, M. "Benjamin Disraeli and the Athenaeum Club." *Notes and Queries* (March 2005): 68–70.

Spinner, Thomas J. *George Joachim Goschen: The Transformation of a Victorian Liberal.* Cambridge, UK: Cambridge University Press, 1973.

Squier, Ephraim George. *Waikna, or, Adventures on the Mosquito Shore.* New York: Harper & Bros., 1855.

Stansifer, Charles. L. "The Central American Career of E. George Squier." Diss., Tulane University, 1959.

———. "E. George Squier and the Honduras Interoceanic Railroad Project." *Hispanic American Historical Review* 46 (1966): 1–27.

Stapleton, Darwin H. "Joseph Willits and the Rockefeller's European Programme in the Social Sciences." *Minerva* 41 (2003): 101–14.

Stauder, Jack. "The 'Relevance' of Anthropology to Imperialism." In *The "Racial" Economy of Science: Toward a Democratic Future,* edited by S. Harding, 408–33. Bloomington: Indiana University Press, 1993.

Steinmetz, George. *The Devil's Handwriting: Pre-Coloniality and the German Colonial State in Qingdao, Samoa, and Southwest Africa*. Chicago: University of Chicago Press, 2007.

———, ed. *Sociology and Empire: The Imperial Entanglements of a Discipline*. Durham, NC: Duke University Press, 2013.

Stevenson, R. L. "Body Snatcher." *Pall Mall Gazette Christmas Extra*, 1884.

———. "Our City Men: No. 1—A Salt-Water Financier." *London* 1 (1877): 9–10.

———. "Paris Bourse." *London* 4 (1877): 88.

Stewart, Robert. *The Foundation of the Conservative Party, 1830–1867*, London: Longman, 1978.

Stocking, George W., ed. *Colonial Situations: Essays on the Contextualization of Ethnographic Knowledge, History of Anthropology 7*. Madison: University of Wisconsin Press, 1993.

———. *Race, Culture, and Evolution: Essays in the History of Anthropology*. Chicago: University of Chicago Press, 1968.

———. *Victorian Anthropology*. New York: Free, 1987.

———. "What's in a Name? The Origins of the Royal Anthropological Institute (1837–71)." *Man*, n.s., 6, (1971): 369–90.

Strangeways, T. *Sketch of the Mosquito Shore, including the Territory of Poyais, etc.* Edinburgh, UK: Leith, 1822.

Sullivan, Alvin, ed. *British Literary Magazines: The Romantic Age, 1789–1836*. Westport, CT: Greenwood, 1983.

Suzuki, T. *Japanese Government Loan Issues on the London Capital Market, 1870–1913*. London: Athlone, 1994.

Svoboda, Alexander. *The Seven Churches of Asia: With Twenty Full-Page Photographs, Taken on the Spot, Historical Notes, and Itinerary*. London: Sampson Low, Son and Marston, 1869.

Taylor, James. *Boardroom Scandal: The Criminalization of Company Fraud in Nineteenth Century Britain*. Oxford: Oxford University Press, 2013.

Thorner, Daniel. *Investment in Empire: British Railway and Steam Shipping Enterprise in India, 1825–1849*. Philadelphia: University of Pennsylvania Press, 1950.

Thorpe, T. E. "Scientific Societies and the Admission of Women Fellows." *Nature* 80 (1909): 67–68.

Tilley, Helen. *Africa as a Living Laboratory: Empire, Development and the Problem of Scientific Knowledge*. Chicago: University of Chicago Press, 2011.

Tilley, Helen, and Robert J. Gordon. *Ordering Africa: Anthropology, European Imperialism, and the Politics of Knowledge*. Manchester, UK: Manchester University Press, 2007.

Timbs, J. *Clubs and Club Life in London: With Anecdotes of Its Famous Coffee Houses, Hostelries, and Taverns, from the Seventeenth Century to the Present Time*. London: John Camden Hotten, 1872.

Tournès, Ludovic. "La Fondation Rockefeller et la naissance de l'universalisme philanthropique américain." *Critique Internationale* 35 (2007): 173–97.

Trollope, Anthony. *Letters of Anthony Trollope*, vol. 2, *1871–1882*. Stanford, CA: Stanford University Press, 1983.

———. *The Way We Live Now*. London: Chapman & Hall, 1875.

Turnbull, Paul. "British Anthropological Thought in Colonial Practice: The Appropriation of Indigenous Australian Bodies, 1860–1880." In *Foreign Bodies: Oceania and the Science of Race, 1750–1940*, edited by Bronwen Douglas and Chris Ballard, 205–28. Canberra: Australian National University E-Press, 2008.

Urry, James. *Before Social Anthropology: Essays on the History of British Anthropology*. New York: Routledge, 1993.

Valenzuela, Javier. "The Construction of the Smyrna–Aidan railway in Southwestern Anatolia 1856–1866: A Discussion." MA diss., University of Texas at El Paso, 1975.

Van Keuren, D. K. "Human Science in Victorian Britain: Anthropology in Institutional and Disciplinary Formation, 1863–1908." Diss., University of Pennsylvania, 1982.

———. "Museums and Ideology: Augustus Pitt-Rivers, Anthropological Museums, and Social Change in Later Victorian Britain." *Victorian Studies* 28 (1984): 171–89.

Vogt, K. C. *Lectures on Man: His Place in Creation, and in the History of the Earth*. London: Longman, Green, 1854.

Wake, Charles Staniland. *Chapters on Man: With the Outlines of a Science of Comparative Psychology*. London: Trübner, 1868.

———. "Tribal Affinities among the Aborigines of Australia." *Journal of the Anthropological Society of London* 8 (1870/71): xiii–xxxii.

Walford, Edward. *Portraits of Men of Eminence: Captain Bedford Pim, R.N., M.P.* London: Provost, 1874.

———. "St Martin-in-the-Fields." In Edward Walford, *Old and New London: A Narrative of Its History, Its People, and Its Places*, vol. 3, *Westminster and the Western Suburbs*, 149–60. London: Cassell, Petter & Galpin, 1878.

Waller, J. C. "Gentlemanly Men of Science: Sir Francis Galton and the Professionalization of the British Life-Sciences." *Journal of the History of Biology* 34 (2001): 83–114.

Wallerstein, Immanuel. *Unthinking Social Science: The Limits of Nineteenth-Century Paradigms*. 2nd ed. Philadelphia: Temple University Press, 2001.

Walpole, Sarah. "Nuts and Bolts: A Survey of the Pre-1871 Anthropological Institutes." Manuscript presented at the Workshop on the History of the Royal Anthropological Institute, December 9–10, 2015.

Ward, Thomas Humphry. *History of the Athenaeum, 1824–1925*. London: Athenaeum Club, 1926.

Watson, T. "Jamaica, Genealogy, George Eliot: Inheriting the Empire after Morant Bay." University of Northern Carolina, 1997, http://english.chass.ncsu.edu/jouvert/v1i1/WATSON.HTM.

Watts, Cedric. "Four Rather Obscure Allusions in Nostromo." *Conradiana* 28 (1996): 77–80.

Waugh, F. G. *Members of the Athenaeum Club from Its Foundation, 1824 to 1887*. London: Athenaeum, 1894.

Weber, Max. "Stock and Commodity Exchanges" (translation of "Die Börse," 1894). *Theory and Society* 29 (2000): 305–38.

Wheatly, H., and P. Cunningham. *London Past and Present: Its History, Associations, and Traditions*. Cambridge, UK: Cambridge University Press, 2011.

White, Richard. *The Middle Ground: Indians, Empires, and Republics in the Great Lakes Region, 1650–1815.* New York: Cambridge University Press, 1991.

———. *Railroaded: The Transcontinentals and the Making of Modern America.* New York: W. W. Norton, 2011.

Wiener, Martin Joel. *Reconstructing the Criminal: Culture, Law and Policy in England, 1830–1914.* Cambridge, UK: Cambridge University Press, 1990.

Williams, Eric. *Capitalism and Slavery.* Chapel Hill: University of North Carolina Press, 1944.

Williams, M. W. *Anglo-American Isthmian Diplomacy, 1815–1915.* Washington, DC: American Historical Association, 1916.

Wilson, Leonard G. "A scientific libel: John Lubbock's attack upon Sir Charles Lyell." *Archives of Natural History* 29 (2002): 73–87.

Wilson, Sarah. "Law, Morality and Regulation: Victorian Experiences of Financial Crime." *British Journal of Criminology* 46 (2006): 1073–90.

———. *The Origins of Modern Financial Crime: Historical Foundations and Current Problems in Britain.* London: Routledge, 2014.

Winkler-Dworak, M. "The Low Mortality of a Learned Society." Working paper, Österreichische Akademie, 2006.

Wisnicki, Adrian S. "Charting the Frontier: Indigenous Geography, Arab-Nyamwezi Caravans, and the East African Expedition of 1856–59." *Victorian Studies* 51 (2009): 103–37.

Wood, Henry Trueman. *A History of the Royal Society of the Arts.* London: John Murray, 1913.

Wright, T. *The Life of Sir Richard Burton.* 2 vols. London: Everett, 1906.

Yelvington, Kevin A. "A Historian among the Anthropologists." *American Anthropologist* 105 (2003): 367–71.

# PERIODICALS

(excluding anthropological and ethnological reviews)

*Anglo-Brazilian Times*
*Anglo-Egyptian and Levant Herald*
*Athenaeum*
*Birmingham Daily Post*
*Brisbane Courier*
*British Medical Journal*
*Caledonian Mercury*
*Chester Chronicle*
*Civil Engineer and Architect's Journal*
*Economist*
*Edinburgh Gazette*
*Engineer*
*Evening Post*
*Examiner*

*Freeman's Journal*
*Freemasons' Magazine and Masonic Mirror*
*Fun*
*Grey River Argus*
*Herapath's Railway Magazine*
*Leicester Journal*
*Liverpool Daily Post*
*London*
*London Daily News*
*London Standard*
*Morning Chronicle*
*Morning Post*
*Navy*
*New York Herald*
*New York Times*
*Newcastle Journal*
*Norfolk Chronicle*
*Norfolk News*
*Pall Mall Gazette*
*Punch*
*Railway Register and Record of Public Enterprise for Railway*
*Reader*
*Sheffield Daily Telegraph*
*Sheffield Independent*
*Times* (London)
*Vanity Fair*
*Weekly Stock and Share List* (also known as *Cracroft's List*)
*Westminster Review*

# INDEX

Page numbers in italic refer to figures/illustrations; tables are noted with t.